PERGAMON GENERAL PSYCHOLOGY SER

Editors : Arnold P. Goldstein, *Syracuse Universit*
Leonard Krasner, *SUNY, Stony Brook*

A New Morality from Science: Beyondi:

PGPS-32

A New Morality from Science:
Beyondism

RAYMOND B. CATTELL
Research Professor and Director of the
Laboratory of Personality and Group Analysis
University of Illinois, and of
the Institute for Research on
Morality, Colorado

PERGAMON PRESS INC.

New York · Toronto · Oxford · Sydney · Braunschweig

PERGAMON PRESS INC.
Maxwell House, Fairview Park, Elmsford, N.Y. 10523

PERGAMON OF CANADA LTD.
207 Queen's Quay West, Toronto 117, Ontario

PERGAMON PRESS LTD.
Headington Hill Hall, Oxford

PERGAMON PRESS (AUST.) PTY. LTD.
Rushcutters Bay, Sydney, N.S.W.

VIEWEG & SOHN GmbH
Burgplatz 1, Braunschweig

Printed in the United States of America
08 016956 2 (H)
08 017192 3 (S)

Contents

v

Preface

Doubtless the thoughts of many scientific men are converging today on the possibility that ethical values might, in some way, be erected on the firm foundation of science. My own belief in this possibility was tentatively expressed some forty years ago in my first book, *Psychology and Social Progress*, in what I then called—and continue to call here—cooperative competition. That book was written primarily to convince the general public (there being then six men in Britain whose full-time profession it was to research in psychology) that advance in psychology as a science was vital to mankind. It argued that political rules-of-thumb were no longer a sufficient basis for social construction in modern societies.

However, I was under no illusion that fuller knowledge alone would suffice. Indeed, as men set aside their perennial repetitions of blind "solutions" and become radically more creative, they may waste their time on strife even more than before, for progress is something about which good men disagree sharply and bad men are indifferent. Pasteur had urged his politically contentious students, "Vivez dans la calme des laboratoires"; but the disciplined fair-mindedness and dedication to truth which brought serenity there—the belief that all would end well in the affairs of science as such—could guarantee nothing if we could not apply them to the pursuit of moral certainties too.

If the reader of a preface is entitled to some glimpse of the author and the machinery of construction of the book, then I have to tell him that my belief then expressed that a solution to ethics lies in science has never deserted me. But pursuit of the possibilities has been grievously interrupted. Between my 1933 book, which might be called a devoted work of late adolescence, and the present book stretches a life of scholarship and research, issuing in thirty books and some three hundred and forty articles in technical branches of experimental, clinical, personality, social, and methodological-statistical areas of psychology. Beneath these intellectually detached productions ran the subterranean heat of the original conviction which brought me from chemistry to psychology, breaking out only in three brief eruptions, a discussion series paperback on psychology and morals (1938), an ethics chapter in a symposium on science and social reform (1944), and an article (1950b) explicitly, but baldly, defining the concept of Beyondism as now developed here. My excuse for so long neglecting what I felt to be so important — and also for what may be a haste and lack of finish in this final production — is the daily uninterruptible work of the laboratory. When hunting with a keen pack of fellow researchers, the chase cannot be stopped. Besides, one has misgivings about one's right to neglect contributions which, however small, constitute concrete bits of new knowledge in order to go off on some speculative venture that may prove just one more of philosophy's wild goose chases.

An accumulation of three bleak realizations eventually forced me to take painful leave of my good research companions. First, it became increasingly clear over the years that no one in the social sciences was actually getting down to the ethics problem in the fundamental way that it seemed to me it required. My "asides" on the matter mentioned above, after some moments of puzzled discussion by a few colleagues, had been put aside as if they fitted into no customary mode of thinking. Second, from observing that few men and fewer women could make any sense out of it, I perceived that a much more systematic introduction and far more illustration to make the setting and the application more real in everyday life were necessary. And granting that three score years and ten is a proper innings, I knew that a task of this large magnitude could not be postponed. Third, and more happily, I experienced those quickenings in this area of thought which tell an author, as surely as a pregnant woman, that a live entity is ready to be born.

My first reason — that social scientists have neglected to face the job — may be questioned, with surprise, by some social scientists. They will claim that sciences — and especially the social sciences — have never been more concerned with values and a sense of social responsibility than over the last fifty years — and, especially, over the last decade. They will point out that the meetings of the American Association for the Advancement of Science have been distraught for the last four or five years with debates about the social and moral duties of science. To this I answer, "Yes, but in precisely the wrong sense". These social scientists complacently believe that they have the moral values already in their hands—Christianity, Democracy, Humanism, or whatever—and that the only problem is one of informing science with our current brand of "revealed" morality. To a scientist whose thinking has not, by habit, become compartmentalized, however, the religious and intuitive systems which gave sanction to these values themselves belong to a pre-scientific, dogmatic era. Shocking as it may seem, the traditional, revered values may themselves be wrong. Indeed, we may be engaged in the very dangerous process of pouring the new wine of science into the old bottles of "revealed" theology. The movement has to be in the opposite direction.

To one who has deliberately, temporarily kept himself politely silent, but basically quite agnostic with regard to many fervent popular assumptions, it has seemed that religious and scientific truth must be ultimately reducible to one truth, and that is likely to be by scientific discovery. It is not, therefore, a question of bringing morality into science, as these social scientists have supposed, but of developing morality *out* of science. The idea evokes less indignation now than it did in my 1933 book; but nine out of ten people still find it hard to follow the argument, and prefer to avoid the difficulty. I still cannot realistically expect that it will evoke the necessary patient thought and serious research except among a small minority. All original thought and experience is lonely—as some of the most original scientists and artists have ruefully, but reliably, confessed. And as for the *reception* of ideas even within my own limited and specialized research contributions, it has been perfectly clear that the more trivial and banal among them have been better received than those showing more original thought or offering a more fundamental and subtle solution to an old problem.

But this is not the only reason why I have allowed forty years to go by before returning to a fuller scale of presentation, for it is my experience in scientific work that fragile ideas brought too early into the market place of general discussion and debate are often coarsened and lost rather than developed. In the minds of those who hear them—and, alas, frequently also in the minds of those who attempt to propound them—what is really new gets stamped into the gross common coinage of existing verbal conceptions—as Francis Bacon complained—with the standard misconceptions of the period forever imprinted upon them. Physical isolation is not essential for intellectual incubation, but it helps; and I am indebted to a granite eyrie in the misty Dartmoor of my youth, as well as to my ridge in the Rocky Mountains, where much of this was written, for the necessary solitude.

Communication, whenever one man is asked by another to see matters of everyday life from a strange new angle, faces two obstacles — a cognitive difficulty and an emotional insult — for the latter is unavoidable where values are concerned; and its inevitable presence compounds the difficulties in the former. To be sure, we aim to lessen the "insult," in this case by asking the reader if he will momentarily hold his emotions in Euclidean detachment from the reasoned conclusions — as in some domain of make-believe — on such disturbing issues as war, sex, social rights, and the like. But the bulk of mankind "leads" with its heart and is, perhaps rightly, suspicious in everyday life of the heartless person who does not. So the reader is specifically being asked for a moment to step out of those useful "prejudices" of daily life and discipline himself to entertain some "as if" reasoning. Even so, the fact is that both the reader and the writer will have their emotionalities; and the writer confesses to some scarcely containable annoyance with those sociologists who have so long desecrated the temple of science by ignoring the evidence of human genetics. He is only a little less impatient with those humanists who judge the conclusions of traditional religion as "superstition," while promulgating moral laws of their own, equally subjective, and, indeed, half the time borrowed from "revealed" religion.

This much autobiographical confession may be of help to the reader; but for the rest, this Preface will simply steer the reader regarding the structure of the book itself. The first part — Chapters 1 through 5 — consists of a statement of the basic principles, but gives also some perspective on their historical roots and contrasts. The second part proceeds to their applications in the modern world — destructively, in terms of the existing systems with which they clash; constructively, in terms of the new institutions which they call for in the society of the future. The first part is presented with a definite logical sequence and dependency, as abstract principles, like geometry, can be. The second has to shape itself with respect to a wide range of particulars, and where the necessary research is rarely available to give the reliability required. The sequence of I and II would, admittedly, appeal to Euclid more than to a sophisticated educator like Herbart, for it leaves till last the matters of immediate interest to the ordinary reader instead of beginning with the familiar and that which is of daily relevance. But with new ideas, this logical sequence is the only one truly intelligible; and I can only promise the reader who finds this a bit demanding that he will come to the gingerbread ornament of current cultural "gossip" in due course. Actually, it is for Part II, despite perhaps its greater readability, that I feel more apologetic, for the wealth of detailed social scientific research that is needed to sustain particular conclusions there simply does not exist; and if it did, no book of this size could hope to document it.

Indeed, the writer is painfully aware of these and other shortcomings in a book that, in spite of being more difficult to write than thirty others he has worked upon, is also foredoomed to fall short more than these others of what it should be. For, over and above the shortage of scientific material, there is the greater problem of writing in general terms what should be written in technical terms and, indeed, in mathematical formulae. I am confident that one day it *will* be possible to make such a presentation and to write in elegant equations what has here often had to be put into contorted (and, as some will complain) repellent jargon. To popularize a mature and relatively finished science, like making a simplified sketch of a complex building, is difficult but possible. To write truthfully about an immature science is like attempting a condensed sketch of a building half erected and hidden by the construction scaffolding. The need for condensation of integration from such diverse and variously immature sciences as sociology, economics, history, psychometrics, clinical psychology, group dynamics, and behavior genetics has not helped elegance or literary grace.

Yet, when all is said, to bring into the field of discussion by intelligent and educated general readers these rough-hewn major ideas is more important than to aim at any perfection of a book as a book. The reader is asked to be a sympathetic midwife at the birth of ideas that are momentous for our time, and which he will encounter increasingly from other sources in the near future. In due course, each facet of this new science[1] of morality is likely to be developed in less cramped and more elegant form. Perhaps one may take consolation from the wisdom of Bacon that "as the births of living creatures at first are ill-shapen, so are all innovations."

Although, as confessed above, I have deliberately shielded the incubation of the central concepts here from distortions through premature contacts with fashionable trends, yet it has been a pleasure to realize, especially in the last decade, that several original writers—they stand in my bibliography—have put out to sea in the general direction of my present explorations. The quality of thought in these writings—particularly in the use of genetics, in sophistication of evolutionary inference, in striving toward mathematical models, and, above all, in fearless integrity of thought—is most heartening[2]. At the same time, it will be evident from the bibliography that I have also gone much further back into the past for good thought in this area than do my brother social psychologists when they commonly make up their references. The spirit of science is older than organized research, and great minds are too few in any one century to throw thirty centuries away.

Finally, for the reader's guidance, let me point out that, although concentration into bare essentials has often been taken as a necessary goal in the main text, I permitted myself that redundancy in echoing the text in the notes which good education and communication theory urge. The reader set for a fast pace should omit these; the reader who can browse a little and likes to get the flavor of repetition in new perspectives will, I hope, enjoy them. In any case, for the systematic student, I have set out, point by point, a summary in the last section of every chapter; and especially for the more abstract issues, I believe these condensations will contribute to clear perspective.

RAYMOND B. CATTELL

University of Illinois
March 1972

NOTES TO PREFACE

[1] The statement that we are watching the birth of a new science will naturally provoke the scientific reader to ask what its boundaries, its methods, and its professionally trained servants are likely to be. Hitherto, the area of social observations, inferences, and generalization here covered has been considered much too wide for the experimental social psychologist to whom I would give a central role. It has, at least traditionally, been the area of historians, sociologists, economists, and, also, of many writers "of no fixed professional abode." It is probably wise to demand that, if a scientist is to invade this area, it should be an experimental researcher, to get the full discipline of a scientific tradition; but, obviously, he must also be a psychologist experienced in social phenomena as viewed in history, sociology, and cultural anthropology, if his work is to have a comprehensive contact with the social data and issues involved. "Experimental," of course, should not be taken in the narrow "brass instrument" sense of "manipulation," for in multivariate experiment, as I have argued elsewhere (1966), there are elegant statistical methods which permit the social psychologist to approach history, political science, economics, and population genetics in an experimental spirit and in search of predictive laws. Unfortunately, though it is thus becoming increasingly evident that social psychology is logically the core science ultimately containing the explanatory principles needed in economics, sociology, and other relatively descriptive or specialized social sciences, social psychologists have a long way to go before they can make good this promise. It is with deep embarrassment that I, as a social psycho-

logist, have had to fall back in tackling this broad field on findings of so incomplete a nature and theories so close to mere surmise that I may be accused of having a split personality, with respect to the standards I express, for example, in my *Handbook of Multivariate Experimental Psychology*. However, I can only say that, in terms of factual support and soundness of basis in method, the theories I entertain are certainly no less reputable than those of, say, Marx, Russell, or Toynbee, with which they have to contend as present rivals in this area; and they are certainly closer in spirit to the tenor of scientific research. Fortunately, Beyondism, as here developed, does not attempt a greater precision than the scientific approximations used to support it warrant, for much of Part II is frankly given as conjective. Although its many developments, as in Part II, will need an immensely more organized realm of exact research to sustain them, the really indispensable central propositions in Part I are too broadly based across the domain of science to need detailed change. This very concern with the fundamentals of the structure has meant that its walls are bare. One would give much to see what even fifty years may do to the enrichment of its furnishings. One would like to know, for example, what a Haldane or Fisher of 2050 A.D. will have to tell about the mutual induction between culture patterns and genetic configurations; or to hear what a genius of the dynamic calculus of motivation has unravelled and confirmed by then about the relation of sexual moral patterns to cultural creativity.

[2] Despite these recent encouraging sounds of great and lively company, it remains true, as I said at the beginning, that the steep and thorny path of progress in this area is one which relatively few will follow. Consequently, I am more than usually indebted to those who have given help in shaping perspective and checking the clarity of communication. Notably, I wish to say how grateful I am to Professor J. L. Horn of the University of Denver for some profound psychological observations, to Professor J. R. Royce of the Center for Advanced Study in Theoretical Psychology for philosophical evaluations, to R. J. Throckmorton for the shrewd comments of an educational psychologist, to Professor Marilee Clore for stimulating criticisms from the standpoint of an historian, to Dr. H. Weckowitz for reactions of a political scientist, to Dr. Ivan Scheier for the wisdom of a practicing psychologist (given with the forthrightness of a former fellow author), and to Dr. Robert Graham (whose book, *The Future of Man*, appeared this year) for the practical wisdom of an executive who is also a scientist. Although at times they have disagreed with me and with each other, I am sure they have substantially reduced what Macaulay (1897) aptly described as those "mistakes [that] must reasonably be committed by early speculators in every science." I am greatly indebted to them for thus bringing the wisdom of a committee of social scientists to bear where exactitude is still not possible for the calculations of an individual.

Part I Basic Principles of an Evolutionary Ethics

CHAPTER 1

Three Gateways to the Understanding of Life

1.1 UNDERSTANDING LIFE: DISCOVERING MORAL GOALS

Culturally, we live in momentous times – times in which values are in a ferment. Our generation is cursed with the anguish of moral conflict and blest with an unprecedented opportunity for major reconstruction. How shall we train and mobilize our minds and souls for this confrontation?

The concern of every sane and thoughtful man with what life is about boils down to, "What am I?", "Where am I?", and "What ought I to do?". The last question, which most distinguishes man from the lower animals, introduces moral values, which are the center of all values. For the highest type of *homo sapiens*, no question is so important as that of the moral purpose of life; and the deepest happiness is achieved only through some understanding of it. The aim of this book is to find out what our general scientific knowledge and the psychology of human nature have to say, in the freedom of the modern atmosphere, about the roots of morality.

To understand morality, we have to understand life itself as far as we can; and men have traditionally gone in at one of three gateways in seeking that understanding: religion, the arts, and science. After sympathetically examining the inspiration of religion; the intuitive, emotional message of the arts and literature; and the methods of truth-testing which have grown up in science, some of us at least, may be convinced that this last – the most austere and sometimes emotion-starved path – is actually the best. Nevertheless, before any such decision is made, it behooves us to look at the different modes of knowing and at our human equipment for knowing. Without losing ourselves in tomes of philosophy, we may yet

3

hope in this introductory chapter, to reach the main sense of the vast body of cogitation on this question. Thence we can legitimately move to our main theme, which is the derivation of ethics from science.

Far more people have taken their morality from religion than from any other source; and our first step should, therefore, be a naturalistic, and, hopefully, unprejudiced examination of what religion has meant, historically, psychologically, and as a logical basis for moral values.

1.2 A RIDDLE COUCHED IN THREE QUESTIONS

Among human cultural activities, religion has been one of the hardiest and most pervasive in effects on everyday life, as that remarkable old catalogue *The Golden Bough* (Frazer, 1890) may remind us. Religion fathered the first profession; it fills vast sections of the world's libraries; and the spires, domes, and minarets that express its call on human devotion pierce the sky on countless horizons. It survived the disproof of the claim for God's children that they stand at the center of the universe. It rose again after the guillotine slash of rationalist logic at the French Revolution. It continues despite the inexorable advance of nineteenth and twentieth century science, which marches with the youthful militancy of new knowledge into the long sacred views about the nature of our world, the origins of life, and the manner of God's creativity.

The young are the hope of every new cultural development, yet religion and morality — except when dressed as a Crusade — have never been an enthusiasm of the young. At adolescence, the intelligent young become much concerned with morality and justice, but scarcely with the dogma and moral scruples of intuitive religion. Indeed, today, among the mainly scientifically educated adult generation of Western culture and the atheistic or secular Russian and Asian cultures, the religious spirit is barely tolerated, as a puzzling, and, at worst, misleading anachronism. Nevertheless, one must admit that whatever the social role of religion should or should not be, there must either be some tremendous and mysterious vitality in its ways of thinking and feeling, or some addictive weakness in human nature. This is surely true of the whole range from religious emotionalism to the philosophical religions, and, over time, from the pre-Socratic period, through Stoicism, Epicureanism, down to present Existential adjustments.

What is the common element in this persistent claim upon human thought? Surely, when dogma, ritual, priesthood, and ornate accretions of superstition are set aside, the common appeal of religions is not only that they seem to answer to the same tormenting questions as do the sciences or the arts, but that they do in a way which gives richer emotional satisfaction. Throughout history, wherever the daily stress and struggle for survival eased – as when primitive man, the hunt finished and the food eaten, watched the lights go on in the quiet evening sky – vast and vague surmises would arise. The answers of religion, though not as astonishing as those later to be offered by science, satisfyingly filled the intellectual twilight. And even today, when science throws a brilliant arc light into our lives, it is still only through a crack in the door, and beyond this narrow beam we are still haunted by the wildest speculations.

Regardless of our opinions of the relative values of the answers from religions and other sources, we should recognize – though those born in the generations of intellectual warfare between science and religion may find it hard to do so – that the questions which religion and science have asked are virtually identical. First: "Where am I? What is the nature of this universe in which this small, pulsating bit of protoplasm finds itself?" Secondly: "What am I? What are the properties – the limitations, the needs, the full possibilities – of this bit of living matter I call myself?" Thirdly: "What shall I do?"

The first two questions concern the stage and the actor. But what is the play? What is the *purpose* of the individual's appearance? Here the individual seeks an answer to, "What *ought* I to do?"; and religion gives him an answer in terms of a greater purpose and plan. The fact that science disagrees in several ways with the answer given by religion still does not detract from the debt we owe religion for having helped to raise the question. And the fact remains that the emotionally more primitive approach of religion has attracted the bulk of mankind to that gateway. Only in the last century of more universal and intellectually disciplined education has an increasing section of the population been able to tolerate or embrace with some enthusiasm the scientific world view. Our task is to examine the validities of these answers by science and by religion, as well as of that given in the more direct emotional answer of art.

In doing so, let us recognize that we shall encounter some obstacles from the fact that hitherto the majority of mankind has not been in the habit of attempting to reason individually and independently of authority on these questions. In earlier times – perhaps with a realistic regard for the average citizen's lack of training – authorities have preferred to do the reasoning. Today, there is enough spent on education and enough leisure for large numbers seriously to devote themselves to fine reasoning on issues beyond the banal problems of everyday. Indeed, a truly participating, democratic culture can only be maintained on the basis of such moral sophistication.

Yet even in our fortunate age, the big questions, "Where am I?", "What am I?", and "Why am I?" tend to be set aside in the busy period of practical responsibilities between the brief fresh freedom of adolescence and the equally brief serenity of stocktaking in age. In the first place, the averagely intelligent majority find their early sensible and serious concern soon blunted and stultified by failure to get comprehensible answers. Later, distracted by the drain of economic needs, professional ambition, and family cares, they are compelled to settle for ready-made, approximate solutions. It is surely a sane solution to adjust to the approximate answer and the traditional authority — except for those whose vocation it is — as philosophers, priests, and scientists — to pursue the questions over all their lives. At least it was in other ages, though with the advance of the social sciences and the provision of leisure, there may now be both material and time for every citizen to become a serious student of these problems. Otherwise, for most people, questions of basic values rise into poignant illumination only when crises thrust upon them some sharp point of moral decision, some endurance of a crushing disappointment, a deep love affair, or the heard but unbelieved summons to one's own imminent death. These, whom Thoreau believed to be "the mass of men" leading "lives of quiet desperation" can today, if they wish, find their way more surely to a serenity of reasoned insight. However, perhaps what is said in this book has its best chance of being useful to the intelligent who are also young in mind. These are readers ready to follow an argument wherever it may lead, yet disciplined enough to be critical in reasoning and mature enough to consider momentarily unpleasant conclusions.

The motive force to participate in this odyssey of social thought will, however, in most readers, spring from a realization that with the decline of the moral authority of revealed religion we are in imminent danger of entering a general moral morass. The mere advance of scientific knowledge about the workings of society cannot save us from that. No increase in the general level of education — still less any rise in the noise level of mass communication — can be a substitute for the patient and creative pursuit of necessary, new, ethical values.

1.3   CONCERNING THE COMPETENCE OF SCIENCE TO ANSWER

To the scientist, it would seem a natural conclusion that scientific research is, in principle, capable of approaching answers to all three of the above questions. Is not the word science, by its derivation, concerned with *knowing*; and knowing recognizes no artificial boundaries between the physical, biological, and psychological domains.

To the basic questions, "Who and what am I?", science returns the partial answer which the modesty and caution of its methods dictate, namely, that I am a member of a species *homo sapiens*, with an interesting position — anatomically, physiologically, and mentally — in the taxonomic tree of life. I am built of proteins — polypeptide chains — and minerals, according to a blueprint written in the genetic code of my chromosomes. By a process not unlike that in a chain of fire-crackers (but self-restoring), a vast number of electro-chemical signals circulate in the neurons of my brain; and, in some, as yet mysterious, way this generates awareness of the world around me and myself. Similarly, and with similar rather large unknowns admitted into its equations, science can tell me *where* I am; on a planet with the rare temperature suitable for life, circling a rather nondescript, middle-sized star, rather far out on the swirling arm of a galactic nebula, in a boundless space, illuminated by countless galaxies extending indefinitely — as far as the eye with its present technical aids can see.

Answers of this nature, enriched every year with new facets and heightened in precision, have been presented to the first two questions by mutually critically alert scientists; and except for doubts by epistemologists, whose business it is to ask how we know that we know (and whose viewpoint we shall duly take into account), there has been no real doubt about their acceptance. They are, moreover, given as factual systems that are admittedly incomplete, couched in theories that are recognized as likely to change in structure; and with the understanding that science proceeds by successive approximations.

But when we come to the third question, "In this defined setting, what ought this defined person to do?", the whole firmament of scientific and social discussion may well seem to go into convulsions. Throughout the nineteenth century (and much of the twentieth — as witness the Scopes trial), educated people agreed that science should deal with "*What* is life?" but not with "Why?". The revealed religions have exclaimed aghast that "science has nothing to do with defining moral values!" Strangely enough, surveys indicated that most scientists agreed with them. Drawbridge's

(1932) study, if repeated, would almost certainly show scientists now more willing to speculate—but only as we take the younger men in this generation.

Quite apart from the still persisting strong popular resistance to science trespassing on religion, and the doubts of conscientious scientists and epistemologists concerning the capacity of science to handle religious phenomena, the psychologist himself might doubt that science can ever perform the *emotional functions of religion*, as traditionally set. Does not its very ideal of a coldly cognitive, detached and analytical approach negate the very possibility of profound emotional experience? To this, as a psychologist one must eventually respond with an emphatic "No!" A very real perception of the admirable qualities of the loved one does not destroy love. The aeronautical scientific analyses which enable us to fly, or the bacterial analyses which preserve for us more robust health, do not detract from the emotional experience of soaring in flight or the poetic appreciation of good health on a day in spring. We learn, in time, to attach our emotional responsiveness to the things which in fact have strong effects upon our lives, regardless of their initial appeal or lack of appeal to our primal instincts. People can get excited about a recondite enzyme analysis that promises an end to cancer, or very angry about an increase in an air pollution index from 0.125 to 0.130. Science can create its own world of emotion. Nevertheless, the combined result of these difficulties and pre-existing emotional loyalties to religion was to beget, mostly between 1850 and 1950, a bitter conflict between science and religion. Psychologists and rationalists have found much food for reflection on the weakness of human nature and the socio-economic power of religious organizations, as they survey the diverse causes of emotional hostilities and socio-religious pressures which for a hundred years have harassed attempts such as the present to bring science to bear on moral issues. Since these obstacles can by no means be said to have disappeared, the tactician may wish to study them in more detail elsewhere, (Cattell, 1938; Draper, 1898; Freud, 1928; Hirsch, 1931; Huxley, 1957; Simpson, 1926; White, 1896).

Leaving these battles to history as irrelevant at the present stage of educated thought, we must beg leave here to walk through the thin theological picket line, engaging in no polemics but proceeding directly on our course of basic reasoning. For the central theme of this book, as indicated on the opening page, is that a third avenue to moral values exists in science itself. It aims to demonstrate that by paying full regard to logical consistency, and making constant appeal to scientific experiment and empirical observation, valid moral rules and a profoundly satisfying emotional relation to man and the universe can be reached on a scientific basis. That is to say, we propose to give a thorough trial to the proposition that science can answer also the third question above.

Nevertheless, so complex a question, so demanding of original thought, as that of the derivation of moral goals and ethical rules from science, must be approached cautiously and systematically. The next step-by-step development of the argument will begin in Chapter 3 below and proceed through 4 and 5. The present chapter and Chapter 2 must be considered a skirmish with the problems likely to be involved, and a brief survey of the areas of knowledge needed. The present chapter, as stated symbolically in its title, proposes to make a relatively naturalistic evaluation of the activities of religion, science, and the arts as alternative gateways to truth. In so doing, it will glance at the properties of our instruments of understanding: the human mind, logic, and scientific method. Thus, the question becomes, "What is the methodological competence respectively of science and its rivals in our existing cultural institutions to answer the questions at stake?" Let us begin with religion.

In the first place, we must recognize that religion is many other things to man besides what science claims to be — namely, a method simply for pursuing truth. Religion has also always been more even than ethics and morality, and certainly more than the cognitive system of beliefs about God and the Universe, in which it claims, with varying degrees of logic or appeal to historical origins, that its central moral values are embedded. For example, it provides an immense aesthetic experience, as in the bright Buddhist temples or the majestic chants that rise in our great cathedrals. By contrast, science is, so far, an aesthetic experience only for the few. A religion is also a socio-political community, in which education and welfare are fostered, and in which family recreational and other social life can be experienced. (Again, the social community of science, at present, means something only to the few.) For centuries, men have received the sacrament of birth, married, and died in the ritual of their religions. Furthermore, many simple communities have satisfied in one institution virtually all needs for knowledge of the universe, ethical direction, social support, and aesthetic experience.

When an institution which, for most of two or three thousand years in Western culture, has so comprehensively and organically satisfied social, emotional, aesthetic, and moral needs begins to go to pieces, it is not surprising that the community experiences so widespread a malaise as we now see. (That "going to pieces" is documented, psychologically, in my earlier book (1938), and, philosophically, in the recent "God is Dead" theme of theological writings.) No one is likely to claim that science, *as it is now known* and with present boundaries to its operations, can provide a home for more than a fraction of the emotional needs met by religion, and which now rise protestingly like a flock of birds disturbed on an old roost. Indeed, to most of those not actively engaged in science, its activities may seem but the glitter of a few instruments, and its voice but the dry rustle of papers and the metallic clicking of a computer. Artists and literary men in particular have quite failed to realize the richness and depth of emotional life which the dedicated scientists, rapidly growing in number, gain from their world view and their involvement in research.

This blindness of the contemporary arts to the emotional message of science is a specific social and cultural problem to which we must return. Meanwhile one must freely admit that whereas it is open to the experience of most educated men and women to perceive the deeper emotional meaning which science gives to our attitudes to our universe it is comparatively difficult to catch what it is saying about moral feeling for our fellow men. Regarding the latter, for example, there is a popular view that it has a sinister message or none at all. The skeptical, cerebral and carping procedures of debate among scientists in the cold correctness of science suggests to many indeed, that it is inherently inimical to any emotional warmth, akin to that in religious and artistic experience (however much the inner fires of science may show in the sparks flying from scientific debate!). The superficiality of this view we shall hope to explain, but certainly the public seems to have concluded that science lacks the immediate emotional experience they require. William James, in his *Varieties of Religious Experience*, contrasts the "oceanic" feeling of religious acceptance with the intellectual hair-splitting of theory – implying that science could, at best, be the theology and exegesis of religion. Darwin wistfully speaks of the loss of aesthetic experience that accompanied his increasing scientific concentration on analysis of the natural world – though nature had been his first love. And anyone familiar with scientific conferences will know with what sanitary care all appeals to emotion, aesthetic feeling (and, alas, sometimes even intuition!), are meticulously excluded from the laboratory.

The reaction of the robust natural scientist to the above will be that exclusion of emotion from immediate, technical reasoning is one thing, but the whole impetus of research activity under emotional drive and aesthetic pleasure quite another. Further, he will vehemently insist that he experiences in the fellowship and values of his scientific community all the emotional and aesthetic satisfaction that the typical religious community offers. Finally, he will maintain that reason and emotional life do not have to stand at odds, but that depth of intellectual understanding and richness of emotional experience go hand in hand. While, as a fellow scientist, I realize, at first hand, that this is true for those who have the appropriate background, I must still, in the name of realism concerning our existing mixed-up society, insist that: (a) throughout most of the history of science from Aristotelian times, science has not played this role; (b) that even today it has demonstrated its capacity to do so only for a minority; (c) that it has not yet fully thought out and brought out convincingly its capacity for emotional, religious, aesthetic leadership. Current assumption in literature persists in the theme that science is a merely technical world. This can be documented from C. P. Snow's (1959) admission of "two cultures" to the latest popular literature or journalism forever pursuing, remote from science, some emotional "salvation" in this or that artistic or theatrical fad.

What is usually not contested is that science is eminently fitted for and successful in the handling of the *cognitive foundations of belief*, as shown by the quality of its answers to the first two of the basic questions. But here the scientist is surely right in saying that since emotion not based on correct cognitive perception is the definition of madness, he has the real key also to the third question — that of emotional purpose. In short, sane emotional life is, as in Plato's metaphor of the charioteer, a condition in which the steeds of emotion are directed by the cognitive reality-perceptions of the charioteer.

Everywhere in nature the higher organism releases its emotion only on the basis of vital cognitive discernments, either in love or war. The discernments may be partly unconscious, as in falling in love; but neither in our loves nor our hostilities can we survive if the rein is given to emotion. Why should we expect rules to be different in regard to religion and morality? In these also the emotional assent ought ideally to *follow* rational cognitive examination. But this is not what happened in the history of religions. To the person already emotionally attached to this or that traditional religious belief, a demand for a critical examination or the argument that comparable emotional satisfaction can be found in newer constructions may come hard and be resented. Every human being has a legacy of emotional vitality to spend. As those who bestow it in drunkenness, tawdry ideals, or drugs, learn to their cost the vitality of emotion disappears when squandered, and wise men know that it lasts longest when invested in accordance with external realities. If science is providing our most accurate picture of the realities of our universe, then religion, as an emotional experience, has to ally itself with science.

1.4 HUMANITY AND THE EVER-OPEN GATEWAY OF RELIGION

The above is an interim statement of how one might aspire to an integration of science and religion. Throughout this book we shall pursue this in more detail, on the one hand, with the development of a new moral value system out of science; and, on the other, with understanding what the emotional reception of that system is likely to be in our existing society. To appreciate what these latter problems would be in presenting a new synthesis, let us examine a little longer what religion, at present, means for the emotional life of society. With this we will proceed also to examine in more detail the relative claims of religion, literature, and art to provide understanding of the meaning of the human drama.

Such fuller examination is justified also on the grounds that we should seek fairly to understand any established system that we would presume to replace by a different approach. For example, a new social system which dismisses religion as "the opium of the masses" is falling into a crude oversimplification in the judgment of any objectively enquiring social scientist. Such crudity would be a fatal flaw in the planning of a great society. What must not be overlooked is that despite a minority of mankind finding purpose and solace in literature and the arts, and a still smaller minority in science, the broad gateway to emotional adjustment through which most of humanity has thronged over the centuries has been that of religion. This tenacity of religion, it is true, has seemed no virtue to rationalists, from Voltaire to Marx and Russell, who have compared it to an octopus insinuating its tentacles into the business of all kinds of peer institutions, economic and political.

This pervasiveness of religious systems—for better or worse—is comprehensible to the psychologist and has been noted by anthropologists and sociologists and social psychologists from Frazer (1890) and Durkheim (1915) to Freud (1913) and Baetke (1962). A very relevant instance in regard to the problems of our own age is the way in which political democracy (and even socialism in its birth in the Christian socialism of Kingsley and others) has become intertwined with religious values until any attempt to develop it into more rationally enlightened forms is strangled by essentially religious conservatism. There is no question that political democracy was born historically out of *practical* community experience, notably in the independent births in the Greek city-states and the Icelandic Althing. Though initially independent of religion, it has today two new forms, one in which it has been "captured" by the Christian-Judaic religious values and another (still independent of religion) by Marxian (and Confucian) economic values. In many other fields of human invention it could be shown that it is a natural tendency for clear cut innovations to become suffused with the emotionality and cognitive vagueness inherent in religion. Other examples are hygienic laws, e.g., prohibition of smoking and drinking, becoming involved in the religious values of Mohammedanism, and patriotic duties becoming sustained by Shintoism in Japan.

There is every hope nowadays of understanding the dynamics of this history, in as much as cultural anthropology, sociology, and psychology have turned the searchlight of at least a *descriptive* scientific examination upon the age-old procession of priest and congregation, religious ritual sacrifice, saint, mystic, messiah, and prophet (Archer, 1958; Baetke, 1962; Cattell, 1933a, 1938; Frazer, 1890; Freud, 1913, 1928; Graves, 1946; Joad, 1930; McDougall, 1934; Mead, 1952; Mowrer, 1967; Schröder, 1960; Tillich, 1955; Unamuno, 1926; Westermarck, 1932). They have given us a good descriptive taxonomy, a set of sound developmental generalizations, and some admittedly still speculative interpretive psychology. They have traced religion back to misty beginnings in animism and magic. Although the definition of religion — as communication with some supernatural power — has to be broad to include the diversity of social activities that may, in some generic sense, be called religious, a central "type" of behavior and belief emerges. In fact the main functions of this core institution, over and above cosmic explanation, are twofold: to satisfy the frustrated emotionality of the individual, and to aid primitive societies in gaining some degree of social integration.

All primary emotional needs — from sex to hunger, and self-assertion to security — are catered for in some form by religion. Psychologically, these needs can be seen to use religion as a safety valve when culture has brought maladjustment into their biological functioning. That is to say, these drives tend to become woven into the texture of religion to the extent that they are deprived or frustrated by the natural or cultural environment. It has often been said that religion is sustained by hope and fear. Hope is a feature of all approaches to understanding life and rationalists have not particularly objected to its role. But the eighteenth century Encyclopedists, from Montesquieu to Diderot, Morelly and Voltaire, detested the central role of fear and awe. This component in religion upsets the sociological rationalist today, and its earlier prevalence becomes understandable only when we consider the boundless insecurity of life among primitive men. For more years than not, man has lived precariously between the incomprehensible thunders, earthquakes, and floods of nature, and the battle, murder, and insane cruelties of his fellow men. In fear he "projected" (as the clinical psychologist would explain it) gods and demons, fetishes, and the ghosts of the departed (but still threatening) ancestors of the tribe. These had to be propitiated by sacrifices and rituals and appealed to by prayers.

However, if we are to give full and fair psychological perspective, we must recognize that religion harnessed much else besides fear and dependency. There was also joy and veneration for the all-giving sun and whatever gods might be. And it ministered to the last pitiful clinging of primitive man to his loved ones as they passed over the verge of life into what he dreamed and desperately hoped might be a land of immortality.

The more bold among the priests and shamans sought with strange rituals, magic stones, and cabalistic signs to gain some control over this strange world which they and their fellows had created by animistic projection. Prayer passed sometimes from abject supplication to attempts at cajoling and even outwitting the spirit world. This effort at more willful control became magic, which at some point in the various institutional re-shufflings generally tended to split off from religion. Probably its self-assertiveness, and its Faustian self-interest proved indigestible to the basic attitudes of a religion of fear, conservatism, and social solidarity. For thousands of years, magic (black or white) wandered in doubtful repute until, with growing power of reasoning and some initial gains in real control of the world, magic joined with craft skills and folklore recipes and became the crooked parent of science. What religion and magic-as-science have always had in common is the possession of an entertaining set of explanatory myths and a map of the universe, purporting to tell us the underlying nature of the world of appearances. But the two maps became increasingly different. Religion began to regulate social life, in the beginnings of a moral law, founded, at first, on "irrational" taboos, totem laws, and ancestor worship. For this it developed priests and "churches," offered purification from guilt, and showed well-chosen but thorny paths to heaven. Its map became more emotional than cosmic. Thus, religions have had, first, a function which may be broadly called "emotional consolation," and only secondarily a task of explaining the cosmos and creation; and it is in this role that we still have to deal with them today.

The growth from these dim and chaotic beginnings of the religious and moral systems that sustain us at the present moment is one of the most remarkable, though still poorly psychologically understood, aspects of history. Some of these developments, like Hinduism and Chinese folk religion, seem to have grown into elaborate rituals, firm moral laws, and even sacred writings without the intervention of any single, conspicuous "master founder." They bear the stamp everywhere of "trial and error" evolution, and of many cooks having contributed to the broth. Because they had their beginning in folk customs, in disconnected proverbs and aphorisms, and in practical, small-community ethics, rather than in the master system of a philosopher, they have also shown little systematic and logical "renovation" by reformers and codifiers.

Our greater historical awareness of the more dramatic and creative birth of Buddhism, Christianity, Mohammedanism, and the Mosaic code should not blind us to the fact that these are only more conspicuous variants of the inspired mutations which sparkled with lesser intensity over the whole development of religion. As the brighter stars in the sky tend to be nearer to us, so those historically closer revelations which permit the retention of records blaze with the warmth of concrete personalities, and with messages which seem correspondingly more significant. And it is in their recent, historical instances (we can only guess what happened earlier), the prophets claim explicitly that they received revelations directly from divine sources. It may be descriptively useful later in our discussions consistently to refer to systems of belief resting on the above foundations as *revealed religions.* Such religions have their main values established by intuitions, on the part of a few leaders, who experience such intense conviction that they believe it to be of divine origin. Characteristically these beliefs are further shaped by what may be called a social verification. That is to say, by cultural variation and natural selection—including the survival or non-survival of the groups which adopt the beliefs—the revealed religions that survive are stamped with the hallmark of a pragmatic truth.

Certain general trends are evident in the characteristic histories of most religions. One is a transition from countless gods and demons—created by the first fine, careless rapture of prolific artistic animism, as in the pantheons of Egypt, Persia, Greece, Rome, and the Nordic tribes—toward an ultimate monotheistic simplification. A curious phenomenon is the coagulation of the good spirits, on the one hand, and the evil spirits, on the other—prior to the final step of monotheism—into God and Devil, or the Zoroastrian Mazda and Ahriman, or Bogu and Besu, and so forth. They coincide with emphasis on heaven and hell, and an appropriate Dantean adjustment of rewards and punishments to the individual's ethical responsiveness during his earthly lifetime. But finally, in the most recent stage of civilizations, the Devil is mysteriously dead, the fires of Hell go out, and the Monotheistic Deity loses his human features, to shine remotely as a Divine Being or as a philosopher's abstraction in the form of a First Cause or an Immanent Principle.

Among these trends one sees also an increasingly explicit expression of *moral goals*, and acknowledgement of their integral connection with the religious beliefs; an increasing organization of a *"church,"* with a roster of priests, holy days, educational appendages, monastic orders, etc.; and an elaboration of *rituals, prayers, sacrificial offerings, definitions of sin, and ways of redemption.* Some observers have claimed to see in what is taken to be the present obsolescent phase of religion certain atavistic reversions into what they describe as "heretical splinter groups," "ghetto ministers" with bizarre social interpretations (and repetitions of such primitive lapses as occurred at Münster in the Peasants' Revolt in Germany) or into mysticism and "spiritualism," or above all, into sheer confusion and doubt about moral values. Probably, fairly reliable patterns of difference could be traced regarding such elaborations, perversions, and excrescences between periods of religious vigor and religious decay.

Socially, one of the most important trends we see in the history of religion, notably since the Renaissance, has been toward increasingly clear, and legally expressed separation of the citizen's obligations to church and state. Moral awareness has drawn sharper lines between what must be rendered to God and what to Caesar. Russia, less trusting after the historical outcome of religious toleration, by the civilized Roman state, has drawn this church-state line as a veritable "cordon sanitaire," confining religious cells within the larger body of the state. Indeed, even apart from these secular religions, there has existed since the Renaissance only an uneasy truce between religious and national cultural loyalties. Indeed, over much of the world, it would defy the insight of the political scientist, of the attitude measurement of the psychologist to say whether universalistic revealed religion or the values of idealized national cultures might yet win supremacy for the ultimate loyalty of the educated man.

The tendency of intellectuals – as instanced by Wells' description of Roman Catholicism as an "historical ruin" – is to assume that if there is any conflict of patriotic loyalty with universalism, the latter is going to take the form of a quite new secular, rationalist universalism, But this may overlook the realities of primitive emotional mechanisms and needs in the masses. The fact is, in any case, that the national, democratic values of citizenship have in any case been permeated by the universalistic religions. The question is only whether this penetration at the practical level into unconscious civic judgments, the rituals, holidays, special congregational social activities, and voting patterns really implies any commensurate attachment to the spiritual and intellectual statements of such religions.

Only in those uncommon instances where civil, political authority and religion coincide, as in the theocracies of the Puritans, the Calvinists, the Mormons, Israel (and, briefly, in Savonarola's Florence), has the penetration of social life by religious institutions tended to produce an appreciable domination also of the average citizen's intellectual views of the world. But otherwise, the penetration of religion into daily customs has often been a successful but empty imperialism of trivia, in which for that matter, other cultural movements have been apt to end. A true unification of religious ethical values with an intellectual view of our universe has never been a broad social and civic product. It has appeared only in elite groups as a small and isolated growth, as in Plato, the scholastic philosophers, Spinoza, and the scholarly theologians of recent centuries in Western culture.

In attempting to reach an intellectual and psychological understanding of religion, additional meaning is added, as usual, from comparisons — noting the departures from whatever is central and common. It has been said by anthropologists seeking a common-core operational formula for religion that all religions have at least a priest; a book; a temple, church, or meeting place; and a set of social rituals habitually repeated[1]. However, there are many deviations, and few communicants happy in any one religion seem to realize how profoundly other religions differ. Even when one can run over a list of functional elements — prayers, belief in after life, procedures of expiating guilt, church organizations, priests and prophets, cosmogonies, gods — which are alike, there is generally something important that is very different. It is variously mentioned as the spiritual aim, the Weltanschauung, the orientation to life, and the emotional perception.

The understanding of these differences of emotional meaning — which come out not only in such sharp contrasts as that of the Buddhist Nirvana and the Viking's Valhalla, but also in the philosophical forms of religion, e.g., the goals of Stoicism and Epicureanism — lies, one suspects, more in innate temperament than anything else. After a lifetime of quantitative study of the dimensions of temperament, the present writer must, in this matter, leave the reader to evaluate this emphasis on temperament as a conclusion of experience, not easily documented, at least in this limited space. One brief argument for this dependence on temperament would be that the geographical boundaries of religions tend to follow racial rather than political or historical paths[2]. (It should be added, however, that the variation of temperament we are discussing is much larger than racial variation. In any case, since even liberal thinkers are only just beginning to escape from prejudices over race, it is too slender a piece of evidence to survive tendentiousness.) The equally empirical conclusion of William James — that even the differences of the philosophical systems of great intellects are most readily explained by their temperaments — was equally annoying to rationalist strongholds.

Thus, both religious systems as a fabric of dogma and ritual and accompanying moral beliefs contain psychological, temperamental elements which are quite difficult to catch in the typical anthropological or sociological catalogue. Nor can we hope — if our temperaments are different — fully to "empathize" into the pulse beats, and the gut chemistry, of a chemically different being. Some may scoff at this in regard to fellow man — but it becomes obvious if we stretch the gap to an orangutan, a stick insect, or an oyster. Nevertheless, where the poet fails, the experimental psychologist may yet succeed — not in empathically apprehending, but, at any rate, in formulating the difference in a way to permit predictions[3].

However, our purpose here is not that of pursuing a taxonomy of religious mood qualities; but mainly to point out: (a) that the existence of these emotional qualities in "religious truth" is one important feature which denies any simple equation with scientific truth; but (b) that, nevertheless, the derivation of moral values from science, which we shall pursue here, recognizes the possibility, indeed, the scientific necessity, of deriving these values in part from a determination by temperament differences.

The first necessary difference in science and religion in regard to "truth" is obvious in that the scientist seeks deliberately to separate the emotional reaction from the cognitive reaction. In every science scientists try so to describe the cognitive picture that it can be intelligible (as a basis for operations) to scientists in other specialities and other cultures — to the deaf and the color blind and those of totally different temperaments, and ultimately to a being from outer space.

It is a consistent corollary of this difference of emotional meaning found by the comparative psychology of religion that truth-testing in religion requires the notion of "emotional truth." Religion and science may be asking much the same questions, but they are definitely not giving the same kinds of answer. Actually, as becomes evident when we turn to art, acceptance of emotional truth is not peculiar to religion, and needs to be studied as a "method" of understanding in its own right, as in the following section. If dependence on "emotional truth" is, indeed, a feature of religion — and also a fallacy — then it behooves the modern world to beware of those activist religious liberals or sentimentalists who assert, first, that science and religion are seeking the same truth; and, secondly, that moral relativism is acceptable, in that all religions present answers that are somehow "different versions of the same truth." Encountering the enormous differences of values in the communicant for Buddhism, for Christianity, for Zoroastrianism, or in any other religion, they say of each: "It is true for him." The word "true" is obscure here. As far as the "cognitive map" of the universe is concerned, these "truths" disagree with each other and with the map of science. And, as far as the required moral direction of action is concerned — the "emotional truth" — our later arguments will be that this also is too inaccurate — too haphazardly shaped by irrelevant aspects of local economies, history, etc. — to constitute "truth". One must distinguish, on the one hand, between relativism, which is haphazard and denies objective truth, and on the other the existence of special moral values that are correct in relation to the racial biology and historical position of the group concerned. The latter is like the course plots of two ship captains who want to get to the same port but adopt different compass directions because they are at different starting points. The former represents an unwillingness to accept the task of getting to the same port at all; and the "tolerant" accommodating acceptance of "moral relativism" in this sense has no relation to true "positional relevance to real values" as we may call the latter.

There lived in Bloomsbury, in the present writer's student days, an old lady who assiduously attended philosophical society meetings and who would exhort us periodically to adopt her belief that "All religions are equally true, especially Theosophy." But moral relativism in practice can go beyond even this, to the view that the same person may hold different "emotionally acceptable" truths for different occasions! Thus, *apropos* of the strange mingling of Shintoism and Buddhism in the Japanese culture, Schoeps (1967) tells us "Many Japanese are Shintoists on happy occasions; on sad occasions, they prefer to be Buddhists." Religious preferences also appear in the life cycle of the individual; and these are not entirely to be explained merely by the greater knowledge of the older person, but have something to do with temperament expressed in the metabolic rate!

1.5   AN EXAMINATION OF OUR EQUIPMENT FOR KNOWING: RATIONAL, EMPIRICAL, AND EMOTIONAL TESTS OF TRUTH

A new angle must be explored in our purpose of scrutinizing the nature and validity of the three major institutions — religion, science, and the liberal arts — by which men have sought to put together their world views. It is necessary now to look at the very *instruments* of knowing — our senses, our reasoning, our emotions — and to travel over that domain which philosophers call epistemology, before the judgment bench of which science also has to be brought to trial. One must ask about the legitimacy and potency of intuition, logic, empirical evidence, and "revealed" truth. This chapter initially set out to examine the religious, aesthetic, and scientific paths to truth, as ongoing social institutions, whereas now (before turning to the arts and science itself) we are pausing to examine *instruments and methods* that might be used in varying degrees by each of these institutions. Such an analysis leads to abstract and difficult branches of philosophy and psychology. But we must, at least, state a position before proceeding.

The difference between science, on the one hand, and religion and the arts, on the other, is obvious in general terms to everyone. Science applies empirical search, as in experiment and discovery, along with explicit logical reasoning, whereas leaders of religion have applied intuition, and have made claims also to a direct, divine revelation. Incidentally, if we define intuition as reaching a conclusion without explicit *awareness of* (not without *resort to*), all the logical steps and factual supports taken into the final judgment, then both the bulk of everyday reasoning and some of the finest first steps in science itself must also be recognized as the product of intuition. Here the only difference of science and religion is that in the former the intuition is subsequently checked by explicit logic and experiment. Parenthetically, there are steps in religious creativity, notably among the medieval Scholastics, which have used explicit logic as definitely as in science; and there have been what amount virtually to experimental tests — as with the priests of Baal — but they have been founded upon *a priori* premises which a scientist would regard as too elaborate for reduction to scientific postulates.

A source of confusion about the role and meaning of intuition is that it is almost always — except in the realm of scientific theory and detection — secondarily contaminated by an appeal to "emotional truth." That is to say, the path to the conclusion is affected both by the desire to reach a conclusion that is sound in the sense of corresponding to external realities, and also in the sense of being emotionally satisfying. The resulting compromise, if it is not explicitly examined, is apt to sacrifice the first goal to the second. For example, Copernicus's appeals to scientific truth were powerless for fifty years against an intuitive conclusion by his opponents based on the greater emotional appeal of the earth being the center of the universe.

Reasoning about reasoning is full of semantic pitfalls. When we are told, in that appealing French aphorism, that "There are truths of the heart which the head cannot recognize," it looks as if we are speaking of emotional truths but actually it may be dealing with a powerful intuitive experiential process which is as free of secondary "emotional truth" beguilements as is any scientific intuition. The "truth of the heart" may simply be an unconscious but realistic and logical analysis of a long collation of life experiences. The explicit logical process cannot reach these same conclusions because it has no way of recognizing coding and putting into syllogistic form the complex experiences which actually occurred to the given individual. And logic, without a computer, may be incapable of explicitly handling — in any practicable time — such a multiplicity of facts. This I believe, is the defensible sense in which so fine a thinker as Dean Inge (1926) continues to assert that religion embraces truths unknown to science. Indeed, the chief fault of logic in practice (especially in that journalistic and conversational use of "logic" which I have below called "tea table intellectualism") is that it sets up too few and too simple and stereotyped concepts and premises for logic to operate upon — relative to the real facts in a complex reality. There is usually nothing wrong with the logic *per se*.

Thus, in contrasting science with religion and the arts, one may, by a rough use of the word intuition, say that science proceeds by logic and experiment, while religion, literature, and the arts rest on intuition. But this misses the more vital part of the distinction that we are seeking, for there are *two* kinds of intuition, one shared by all these approaches, and one claimed by religion and aesthetics, but denied by science. Let us call these *cognitive intuition*, as so far described above, and *emotional intuition*, a new concept to be described in the following section, on art and literature.

Whenever a scientist asserts that, through adherence to logic and experiment (and an initial approach through cognitive intuition), science aspires to an objective truth, to be contrasted with the results of other alleged paths to truth, some sophisticated philosopher is sure to make the counter assertion that even scientific truth is inadequate and relative. Respectable schools of philosophy claim other avenues to truth. Since the whole argument in this book for a morality based on science stands or falls by the capacity of science to approach a true picture of our world, such attacks must be dealt with.

Accordingly, let us turn to epistemology, the branch of philosophy which deals with the nature of our knowing (and the related ontology, which deals with the nature of being). Here the first Rubicon to be crossed is that of *Solipsism*, which developed from Berkeley. If waking experience and dream cannot be distinguished, the whole universe could be a dream by one man (an extremely ingenious man). There is no logical way of crossing the obstacle of Solipsism; and it must surely be set aside, ultimately, by probability, and the privileged position reasoning has always given to the *simplest* explanation. After absolute Solipsism, we meet the objection which Hume developed that we can never "know" the external world directly: that our sensations are different from what causes them. This does not bother the scientist, who is happy to manipulate the external world by a model of well-fitting referents or symbols. Our experiences of, say, red and green, are different from the vibrations which cause them; but from indirect evidence we know there is a constancy of reference. What may temporarily worry the scientist — or, at least, slow him up — is that the range of our senses obviously does not correspond to the range of possible incoming information. The dog has no color perceptions differentiating the red and green wave lengths; and until a generation ago, we had no perception of radio waves. But in time there is no reason why those sources of information should not be translated into our sensory range. Thirdly, we have to consider possible limitations in our symbol system for representing what we meet. If language were our only system (and, unfortunately for us, at the hands of those who demand only "freedom of speech," it *is* often the only syntax and basis of logic), we should be in trouble. But science has developed many flexible symbol and syntax systems, largely as new branches of mathematics.

Actually, two objections can be raised to the adequacy of science, first, that its "senses" are not complete enough; and, second, that what we know as logic, on which the "syntax" of theories rests, is not good enough. The answer to the first has been given, that our avenues of sensing *are* initially inadequate, but that indirect ways, "translating" into the avenues we have, can apparently be found, provided all incoming information interacts in one "universe." The second doubt is both alarming and fascinating. The restriction of the domain — the earth's surface—in which we, as biological organisms, have grown up and learnt our adaptations may well mean that not only our sensorium but also our logical expectations are impoverished. The philosophers here seem to have been less aware of the danger than the scientists, and have confidently supposed that the logic developed in our small corner of the universe must "fit" everywhere. Perhaps scientists have become more thoughtful since such experiences as Planck's discontinuity, Einstein's bent space, and the recent disconcerting plurality of incomprehensible physical particles. Perhaps it is possible, in new domains for which our nervous system was not evolved, for an object to be in two places at once, for influences to be exerted without intervening media, for two and two to exceed four, or for time to stand still. The answer to this problem is surely the same as that concerning the sensorium: that we must be prepared to build up the new logical rules as our experience broadens. Meanwhile, faith in the integrated homogenous quality of the universe may perhaps justify our expecting that even when experience is restricted to one small corner most of the workings of the universe will be demonstrated, though caution now dictates that our "laws" be commonly considered only probabilities.

The only essential and stable meaning we can give to the statement "This is true" is that there is a parallelism—an "isomorphism" between some symbolical statement—"the theoretical model"—and external facts, processes, and predictions. Conflicts over the capacity of science to deliver truth concern mainly three sources of error, all of which can be diminished. The first is an undue dependence on the soundness of existing forms of *a priori* logic, as just discussed. Actually, scientists are aware that they carry as a legacy from their historical origin as participants in "rational, philosophic" activities, (e.g., the philosophy of Aristotle and, still more, Plato, which obstructed scientific empiricism in the early Renaissance) *an undue emphasis on the sacred completeness of logic.* Even modern philosophers, including Russell, have a penchant for basic postulates and logical steps which are self-evident, while others seem to think that logic and even knowledge can be developed out of language. Our position here (consistent, incidentally, with Chomsky's (1957); that language is partly innate in form) is that genetic selection has shaped our nervous system so that certain repeating relations in our environment are felt to be "self-evident." By the same argument we must be prepared to agree that the restrictedness of our past environment means that our logic is incompletely evolved for the larger universe. Science has to be empirical not only about "facts" but also about "reasoning." When Adams and Leverrier predicted that a planet would appear at a certain place and verified it, they applied a typical empirical check on scientific calculation and logic. When the transit of Mercury later failed to fit precisely the same intermediate reasoning, and required Einstein's re-formulation of what had been a kind of logic, an empirical check was being made on the mode of reasoning itself. As we shall see, it is in the field of social science that the psychologist has the greatest right to object to faith in rational, *a priori*, (especially purely verbal) "logic" unchecked by exact measurement.

The second weakness of science is that it is *initially* as subject to the emotional biases of the participants as any other method, as witness the medical profession's treatment of Harvey and Pasteur, the denial by the scholars of the Catholic Church of the arguments of Copernicus and Galileo, or Soviet Russia's denial of the genetics of individual and racial differences (Lysenkoism).

The third weakness of scientific truth is that (as the pessimist sees it) it is forever changing, or (as the constructive scientist sees it) it is always in process of growth by successive approximations. The first and second weaknesses above can, by discipline, be reduced. Doubts about scientific truth because of this third characteristic only require the doubter to become educated about the process of creative, exploratory thinking, as analyzed psychologically by Johnson (1968), Taylor and Barron (1963), Vidal (1971), and many others. It is formalized in what the present writer (1966) has analyzed as the inductive-hypothetico-deductive (IHS) spiral. Our social applications of science simply have to take account of this.

1.6   THE GATEWAY OF THE ARTS AND LITERATURE

Science, it has been argued, depends on two tests of truth, that concerning predicted facts and that concerning internal logic or syntax, though both are in the end empirically tested. These *can* take the form of cognitive intuition which is, however, different from emotional intuition. It is different, first, in that it can be brought to explicitness, if required, and, therefore, fully communicated, whereas a gain of emotional truth through emotional intuition cannot be reliably transmitted. The differences of the arts and the sciences cover much more than the dependence of the former on emotional truth, especially when one comes to consider, as we shall in Chapters 8 (Mass Media) and 9 (Art and Morals), the social, institutional role of the arts. But here we are concerned purely with the arts as a way of gaining truth about our world.

The appeal of the arts – of music, poetry, painting, sculpture, literature, and the performing arts (drama) – as a gateway to understanding life has always been very great. For this reason religion – which believed it had the organized answer – did well through history to harness the arts to its purposes. In questioning, as we shall here, any claim of the arts to any fundamental way of discovering truths different from that of science, we are not questioning their tremendous educational value for emotional adjustment, nor are we questioning the possibility of giving a real psychological meaning to emotional truth reached by art and emotional intuition. Our argument will be, however, that in the end the latter draws its correctness from cognitive truth.

Claims for a direct emotional knowing or testing of truth go back into the mists of antiquity. But the sense in which it is claimed in early times is different from the technical sense to be studied here – and far more easily dismissed. The central feature of this early "emotional knowing" is a purely subjective conviction of truth – from music, intoxication, art, or religious mysticism – which fails, by any realistic test one can apply. Many people have come out of nitrous oxide anaesthesia (and more recently a hundred other drugs) with a sense of absolutely profound discovery, yet muttering only some banal cliché[4]. There is no convincing argument that famous instances cited among the more respected annals of religion and art have any more claim to validity than these instances from intoxication. Yet, hidden in such phrases as "Truth is Beauty: Beauty Truth," the mystics have again and again asserted this supra-sensory and direct emotional path to truth[5].

A modern psychologist-philosopher of eminence who has sought to shape and utilize the emotional truth concept is Royce (1965), but it turns out that the criterion in which he takes refuge (when the scientific criteria of fact and logical consistency are denied) is the goodness of fit of the emotional attitude to the emotional norms of society! But delusion and insanity are not just being emotionally different from other people, and being emotionally similar is no guarantee that one is not merely sharing some popular delusion.

A sense *can* be defined in which an emotional reaction is truthful, as being *adapted* to a situation. Indeed, a whole class of insanities – the affective psychoses – are recognized precisely by their absence of that appropriateness. The individual is manically elated over something that does not justify it, and profoundly depressed by what is sometimes realized by the patient himself to be only a trifle. The touchstone for "truth" here is whether the behavior is biologically adaptive. The unduly depressible individual will not eat or avail himself of what is necessary for life, and the unduly elated individual will wear himself out or break his neck by jumping a chasm too broad for leaping. The correctness or truth of the emotional attitude is thus in the last resort tested by the same empirical and rational tests as cognitive truth. Yet, an additional feature is here being taken into account, namely, the appropriateness of the emotional, energy-expenditure-determining reaction in relation to the cognitive realities of the external situation. (In relation to the fundamental postulate, of course, that it is desirable for the organism and species to survive!) What is "emotionally true," therefore, carries the extra meaning that it maximally aids survival. But this quality of the emotional response is something quite different from the emotional conviction of truth. It may be, however, that our sense of beauty is a signal, built into us by natural selection, that certain emotional balances are more adaptive than others. This is only a speculation to be psychologically investigated.

Meanwhile, much of the learning and evaluation of truth in the arts does not hinge upon their having any strange "emotional path to truth" of their own, but rests on the same basis as "truths of the heart" in simple cognitive intuition. Tolstoy's *War and Peace*, Shakespeare's *Tempest*, Goethe's *Sorrows of Werther*, Tennyson's *Locksley Hall*, Beethoven's *Eroica Symphony*, Leonardo's *Virgin of the Rocks*, or Michelangelo's *Last Judgment* convey to us, with superb stirring of the imagination, a sane emotional response to truths that are in fact reached by the same respect for experience and logic as is characteristic of the cognitive intuitions of science.

Later, we may have occasion to criticize second-rate art as pandering to unrealistic, autistic, wishful thinking; and favoring a contemporary humanistic philosophy in art which is dangerously flattering to man. But art and literature *in themselves* need have no such bias. The author of the Psalms, Dante, Donne, Baudelaire, Hölderlin, and Nietzsche have, in quite diverse ways, avoided anthropocentrism, though it may be a standing weakness of the arts to over-value man. The point to be made here, however, in regard to gaining an understanding of our universe and ourselves, is that there exists no magic gateway in the arts. Indeed, unlike science, which can organize its knowledge architectonically so as to move on from generation to generation, the artists and writers of past generations stand on the same ground; and if a particular artist towers over the heads of the others, it is not by virtue of something which all have built, but because his personal stature enables him to do so.

1.7 SUMMARY

(1) The aim of this first chapter has been to ask what man has done in the past to answer the questions that have haunted him since the dawn of thought: "Where am I?", "What am I?", and "What should I do?".

(2) Answering these questions has largely been, until a few centuries ago, the task of religion; and, to this day, some people still believe that the first two questions should be referred to revealed religion, while virtually all believe that religion should handle the last. Actually, as is increasingly realized by educated people, science has, by now, far more comprehensively and reliably answered the first two. It is the object of this book to show that it is also the only sound basis for obtaining an answer to the third.

(3) Actually, modern man finds before him three main gateways to systems of understanding himself and his world; religion, science, and the activities of art and literature. A brief examination is given to each, partly as a social institution, but mainly concerning its particular methodological qualifications for reaching verifiable knowledge.

(4) A condensed psychological and anthropological analysis has been made of the complex functions which great religions have typically performed throughout history. This is intended as a basis for later comparisons with the functions of other social institutions. The corresponding analysis of the *social* functions of literature and the arts in respect to morality has been almost entirely postponed to a later chapter (Chapter 8), since it needs to be considered in a strictly contemporary context. However, just as religion has a major role in socio-emotional life — quite apart from the claims to philosophical truth we are here primarily examining — so the arts have an adjustive "cathartic" and "condenser function" (Chapter 8) for human emotions denied expression elsewhere by cultural pressures.

(5) A more comprehensive analysis of what these three avenues for the pursuit of understanding may mean in their wider social and institutional roles remains to be made below. But in their intrinsic properties as instruments for seeking truth one must immediately conclude that religion and art, in epistemological evaluation, lack the validity of the instrument which we call scientific method. Science uses empirical exploration, along with checking of inferences, reached by an explicit logical syntax. This results in knowledge which is communicative and cumulative. All three share the use of *cognitive intuition*, in some of their

phases; but religion and the arts use, in addition, *emotional intuition* and the concept of emotional truth.

(6) In spite of the methods of science being already able, by the verdict of history to demonstrate far greater success, they need to be subjected to intensive epistemological examination. We conclude that both our sensorial equipment for perceiving the world and our logical habits for understanding it are imperfect products of evolution but capable of extension. The meaning of "the truth of a statement" is its "fit" to the outer world including its predictive value, tested by fact and logic. Provided we discount the classical, absolute worship of logic, and recognize that a grand logic has still to be evolved by empirical generalization extending beyond local logics, science is our most dependable avenue to truth.

(7) A meaning can be given to emotional truth as "the most appropriate emotional response for survival in a given species in relation to a given environmental situation." As such, it is really only the familiar cognitive truth with a behavioral corollary attached to it. But the expression is actually constantly in use as if it connoted some entirely new species of truth-testing. Thus it appears popularly as the degree of emotional conviction experienced, and philosophically as the fit of the emotional response to a cultural norm. In the first sense, it is not independent of scientific, cognitive truth; and in the two last, it involves a quite false use of "truth."

(8) Both because of this intrusion of emotional intuition into their values, and also through their lack of the systematic empirical and rational, logical tests typical of the inductive-hypothetico-deductive spiral of science, religion and the arts cannot be accepted as avenues to new truths. Their roles may be to provide emotional education in truths otherwise established.

However, both revealed religion and the arts and literature offer a supreme expression of the same use of cognitive intuition, as is found also in the first steps in science. But they lack the instruments and organizations necessary to carry its fruits beyond the outcome of merely individual achievement. One may guess that the great religions have reached appreciably valid conclusions, but they have undoubtedly done so by processes with which no self-respecting scientist would want his work to be associated. Since better truth-finding processes now exist— methods which advance knowledge from generation to generation— religious and ethical genius is better expressed through these new channels. The love which men of education and spiritual sensitivity have for great literature and religious creativity must not blind them to the fact that these are not gateways to transmissible, verifiable new truths.

1.8   NOTES FOR CHAPTER 1

[1] Even this core of attributes is not fully possessed by at least half the living and dead religions which cultural anthropologists have studied. For example, there is no priest in Hinduism and Shintoism; there is no book (even metaphorically) among the Australoids, and there is virtually no ritual in Quakerism or Unitarianism. The fact that Hinduism "broadened down from precedent to precedent," with no single great founder, with no prophets, in the Hebrew sense, and only occasional reformers, gives it a very different sanction and spirit indeed from Christianity. The fact that Judaism has always awaited a Messiah and the coming of the kingdom of the one true god gives it again, a totally different temper from the mannerly or diplomatic accommodation to an emperor, to an infidel or pagan neighbor, along with the accommodation to the primitive pursuit of superstition such as has been achieved in the urbane procession of Confucianism. And what could be more different than the Buddhist's world-weary desire to get out of the cycle of life; the ancient Greek's dauntless intention (with Zeus's aid) to snatch what he can from the web of Fate, and the Viking's boisterous confidence that he will drink with Thor and Odin in Valhalla?

[2] Some of the clearest evidence that certain religions are adopted more readily by certain temperaments — especially when such temperaments are homogenously massed together where there can be cumulative effects — is to be found in the new conceptual developments from cross-cultural psychological measures (McDougall, 1921; Cattell, Breul and Hartmann, 1952; Darlington, 1969). In terms of finding direct attitudinal measures of religious value to correlate substantially with personality measures the work of Morris (1956) is particularly striking. In the less direct terms of geographical mapping of racial distributions the largely cognitive, community-oriented, unemotional religion of the Mongolian peoples in Taoism and Confucianism, for example, is in obvious contrast to the fierce emotional abstractness of conceptions in the Semitic peoples. Again, it has been noted that in Europe, notably in France, the low countries, and Germany, Protestantism has fairly closely followed the distribution of Nordic settlement. Economics, conquest, climate, and historical accidents play important parts too, but it is difficult to escape the conclusion from an open-minded investigation that biological temperament also plays a part in deciding the emotional quality of the religion which most readily prevails. (See especially Lynn, 1971.)

[3] The methods are too technical to admit adequate brief discussion here. They hinge on the psychologist's ability to describe a stimulus, situation or course of action by vector quantities utilizing the factors obtained in temperament and motivation (Sells, 1962; Horn, 1966; Cattell, 1965). To do this between distinct populations requires certain assumptions about the equivalence of factors in different bio-social groups (Cattell, 1971). The whole approach is one of describing perceptual differences in terms of dimensions of the perceivers.

[4] An instance with a little more appeal than others has been offered by William James,

who awoke from sleep with the tremendous revelation ringing in his head:

> Hogamus, higamous;
> Men are polygamous.
> Higamous, hogamous;
> Women monogamous!

At best it can be said of the experience that it created a form somewhat different from the limerick!

[5] Examples of conviction from emotional intuition (which we should distinguish from emotional truth in the other, empirical sense above) are as numerous and incoherent as sand grains in the desert. I therefore take yesterday's magazine at random and read of a young actress (drama being a field where emotional truth is so frequently called truth) exclaiming (*Life*, April 23, 1971): "'When I left the West Coast I was a liberal. When I landed in New York I was a revolutionary.' Just what she means by that, Jane Fonda is constantly — and not always consistently — redefining. 'I didn't have time to sit down with books and get a historical analysis and put it all into perspective,' she says. 'It was an emotional, gut kind of thing.'"

CHAPTER 2

The Origins of Present Uncertainty and Confusion

2.1 MORAL CONFUSION AND THE RECESSION OF REVEALED RELIGIONS

The argument to this point has been essentially that, in spite of the powerful grip on the emotional thinking of man which religion and the arts have acquired, the only dependable path toward clarifying that which art and religion seek to understand is the path of scientific method. Our emotional life needs to be shaped afresh, in sanity, from this new source. But the proposition is strange to most ears, and at this juncture in history will kindle enthusiasm only in a minority.

Yet, as this chapter proposes to show, this very point in history is one at which a new Dark Age of confusion threatens to overwhelm society. Fateful moral decisions have to be made, from the outcome of which the whole bio-cultural experiment which is mankind may either go downward to the dust or climb to the brightness of new horizons. Ominous though the spectacle may be, it behooves us, before we propose the voyage which later chapters will undertake, to look at the mounting confusion in this contemporary point of departure. For only when the seriousness of the situation is appreciated will citizens be prepared for those sacrifices of the familiar comforts of religious and artistic consolation, and for the realistic and unusual reasoning that are now required.

It is apparently the rule in history — as a new genre of historical research is revealing — that people do not know what is happening to them at the time it is happening. Previously, historians have made the mistake of supposing that the writings which come down to us — actually the works of a few odd, wise men — represented what people generally were thinking at the time. But what Noah foresaw, or Francis Bacon wrote at the birth of Renaissance science, or H. G. Wells said in predicting the date of World War II, were evidently viewed as the merest "epiphenomena" by the masses at the time, who believed that matters of current importance lay in quite different directions. Our contention here is that we are living in a generation of confusion and dissolution of values, the full magnitude and consequence of which is not generally realized.

People are, of course, aware of confusion, and of the strong attempts to abrade values and exacting standards by which society has been maintained. But men of experience know that such attacks by the unadjusted are as old as the untamed ocean's attack upon the land, and they rightly suspect that almost every age has claimed the distinction of being an age of anxiety. Nevertheless, a psychologist with the perspective of a cultural anthropologist, and a sense of history, must conclude that the perplexities which thinking people express today, and the turmoil in the behavior of less thinking people, go unusually deep. It is our contention that beyond and beneath the present economic, political, and behavioral disorganization lies primarily the disintegration of the imperatives of dogmatic "revealed" religious values. To many this is a chance to revert to sensual and feral man: to a few it is the great opportunity to raise ethics on to a rational basis, consistent with the developments which science is bringing into the rest of our lives.

Intellectual confusion, ethical uncertainty, and social disturbance are, of course, three different things – but they have some common roots and common consequences. As far as social crisis is concerned, the "developed" countries are, at the moment, seemingly more comfortably remote from pestilence, world wars, and famine than at any time in history. But the rising tide of crime and addiction to drugs represents a new and strange threat, the limits of which, in our ignorance, we cannot predict. And as to moral uncertainty, intellectuals can rightly say it is a sign of educational maturity to be able deliberately to question tradition and to tolerate uncertainty long enough to investigate and re-evaluate. But men and countries have to act as well as think, putting their reputations and their lives "on the line" every day. And in matters of morals we quickly pass from polite uncertainty to positively lethal confusion unless the toleration of uncertainty is accompanied by a determination and a research method to end the uncertainty. Criminals are the chief beneficiaries from prolonged abrogation of moral standards. In the academic's aim to "investigate and re-evaluate" the emphasis must be on hard scientific investigation, not sheltered, dilettante, speculative doubts.

There is a type of writer, often calling himself and called by others a "liberal" (though the true liberal is more responsible) for whom a great variety of beliefs are equally sound, and for whom variety cannot be too great. He is seldom a scientist, for a scientist believes that one theory is more correct than another—or should soon be made so—while a thousand other alternatives are clearly wrong. Neither is the type of writer to whom the label liberal has mostly nowadays become attached—by a bastard descent from the true nineteenth century liberal—usually a careful or serious student of the life of societies. For there is every indication that there exists an *optimum variety* of values in a healthy society, which stops far short of infinity. When that optimum is not reached the members of that society, though perhaps happy—as Spartans, Christian monks and Communists can be happy in a global peace of mind—yet lack the seeds of growth. When that optimum is exceeded, mutual trust and willingness to sacrifice for society are apt to decline and important group undertakings "turn awry and lose the name of action" in the endless friction of incompatible values.

Around such terms as "relativism," "authority," "deviation," "authoritarian," "freedom," "responsibility," a host of misunderstandings cling today, some of them willful. Until social science steps in to define, quantify and calculate, the discussion of social questions in most current writings will continue to move in a semantic quagmire. For example, no difference is drawn, either by the person who loves or the person who detests deviation, between such totally distinct varieties of non-conformity as: (1) experimental deviation based on *originality*, (2) mechanical deviation based on a *desire to be different* (for whatever reasons), and (3) criminal, *anti-social deviation*, which resembles the two first only in being deviant. Similarly, the schools have apparently not taught the distinction between freedom and license, or the fact that, in the social world—as more obviously in the natural world where the laws of physics and chemistry prevail—the greatest freedom of self-expression is enjoyed by those who grasp and respect the nature of laws.

The untrained newspaper reader is served a semantic trick, again, by writers from Mussolini to Popper or Leary, who have mixed up authority and authoritarian. Authority may be highly desirable—in medicine, for example, where there are professional and quack treatments, in scholarship, where authorities recognize, for example, correct and incorrect translations, and so on. Indeed, the aim of knowledge is to clarify authority. On the other hand, authoritarian interference with harmless freedoms may stultify the growth of individual character or convert social progress from an evolutionary to a wasteful, hate-producing revolutionary mode.

These few words must suffice for the present to warn that we have no intention of accepting here the game of stereotypes—liberal, fascist, free, communist, etc.—which, used by knaves to make a trap for fools, at present characteristically drags discussion into foul emotional whirlpools.

As pointed out above although the sinister, significant rise in criminal, anti-social action throughout most "civilized" countries over the last twenty years; or the increasing alienation and identity crises of youth; or the emotionality cult against logical analysis (Hofstadter, 1963); or the rising suicide and drug addiction, have associations with (a) economic causes, notably an increasing real standard of living (leisure and lack of hunger promote experiment), yet the most highly correlated association is (b) the appearance of "moral relativism" or the decay of agreed ethical authority. The vacuum in real life pressures occasioned by the first, through sudden technological alleviation of the severity of life is not our concern here. If Bernard Darwin was right in his book *The Next Million Years* (1952), the period of lushness will be a quite temporary respite from reality pressures and an only momentary proliferation of temptation, whereas the pressure of the problem of rational moral direction has begun in earnest and will stay with us indefinitely.

Genuine moral relativism theoretically need not mean any loss of moral determination, but only an experiment with different kinds of values, each ardently and uniformly accepted in its own community. Such *group* experiment, with special conditions discussed later, is vital for progress. But without the deliberate planning of these special conditions there is little doubt that in most societies relativism—especially *individual* moral relativism—will coincide with deterioration of standards. For, (a) the splitting up into ethical splinter groups has historically necessarily tended to coincide with the breakdown of a monolithic and powerful "universal" religion, (b) the experiencing of too great a diversity of values—as child guidance clinic experience soon showed (Burt, 1925; Cattell, 1937b)—is apt to breed in the individual a casualness about *all* values, (c) in the universal tendency of the Freudian id to avoid moral responsibility and inhibition, the claim that moral systems are merely relative is one of the best defenses for the unscrupulous intellectual to use as an evasive rationalization. (The *experimental* relativity defined here, by contrast, does not say all are equally correct, but that *one* is correct, though we do not yet know which.) And, (d) although experiment and deferred justment are conditions of progress, if lack of "closure" persists too long one may suspect that insufficient effort and intelligence are being applied to find the true solution.

For, apart from the minor relativity due to people starting from different historical positions and temperaments, an ethical goal, as we have suggested, must by its nature be monolithic. Experimentation and variation are only different means to an end. And if moral scientists dally too long over giving authority to some solution, we may doubt, as we would with physicians or scientists who take too long to find the remedy for a disease, whether they are effective scientists. One wonders how anyone with the barest acquaintance with science can write (as Popper does) as if uncertainty and toleration *ad lib* were virtues in and of themselves. It is a mistake to interpret liberalism as indefinite toleration (e.g., of fascism), or to fail to see that in science toleration really means initial unbiased entertainment of any idea, but is properly followed by fierce intolerance of experimental error, or of sloppy thinking.

The theologians of the twentieth century have either fought a gallant rearguard action "in defense of God" or have seemed to capitulate to contemporary philosophy in a despairful admission that "God is dead." So long as moral relativism ranged only *within* a religion, as in the innumerable interpretations of the Christian bible (Mangasarian, 1960) which Catholicism regarded as mistaken, the loss of prestige of moral authority was small. Respect for ethical behavior will deteriorate on a grander scale in an age of complete intellectual individualism in moral interpretation. Organized traditional religions complain today that "men no longer acknowledge sin and no longer seek divine mercy." But they are quite unprepared to abdicate from the position of revelation and seek ethical values in research. A cynic or a radical might put this down simply to the fact that religions are highly economically endowed, as going concerns with vested interests. (Are they not the largest real estate owners in most countries?) But there is no reason to accuse revealed religion of lack of sincerity or conviction. Nevertheless, if the chief institutions charged with maintaining ethical values and morale in society *are* resting on an untenable intellectual position, in the eyes of most educated people, their obstinate obstruction can only add to the severity of the final disorder. Those of us who participated in the warfare between science and religion (the present writer's books in 1933 and 1938 are instances) hoped that in loosening the superstitious grip of "revealed" religion the way might, in this generation, be opened to a rational ethics. Today the present writer would put more emphasis on the *construction of a new source of positive values*, and less on the demolition of obstructions. Yet it remains true that a sufficient force of men to work on the new is possible only if the endowments of the old are re-distributed. Whether the spectacle attracts or repels, the fact is that there is a receding tide of faith, whose "melancholy roar" the perceptive ear of Matthew Arnold heard so early on "Dover Beach." This falling tide may leave some stranded or wrecked cultures, unless the pilots are indeed alerted to the danger.

2.2   ARE THE SOCIAL SCIENCES YET SCIENCES?

The promise of a new moral authority (a non-authoritarian and rational authority) in science which we pursue here must not be mistaken for any plan simply to expand the offerings of the social sciences as they are at present being given to us. A glance at an average sample of what has been put forward by activists, in the name of social sciences, is enough to revolt both a scientist and a moralist—not to mention the man in the street and his political representatives. But this need not continue; the biological sciences have followed the physical sciences to an effective maturity, and there is no reason why, with massive support, the social sciences should not do the same. Let us therefore ask how and why their light on important human affairs has been either wavering or positively misleading.

First, one must recognize that the social sciences came as a late child to the scientific family. Among their peer sciences psychology, sociology, and economics were recognized late and were late to receive support. Only recently have many universities, including Oxford and Cambridge, admitted such studies as psychology and sociology to the circle of academic sciences, and only in the last year has the National Science Foundation begun to direct research funds through a specifically assigned branch, as in the older sciences. Meanwhile, the practical leaders actually in charge of social affairs—administrators and elected political representatives—are disinclined to aid what they view, in many cases with justice, as the activities of impractical reformers or emotional revolutionaries impertinently disguised as scientists.

The gist of the criticisms from other scientists should also be stated: (a) that the social sciences have not shown themselves technically as capable as the physical sciences in discovering basic laws, and, therefore, in ministering to society's problems with an adequate technology. The physical and biological sciences have had cause, from the standpoint of precise methodology and objectivity, to be embarrassed by this third erratic sister. And (b) that they constantly confuse means with ends—science with values—and either naively or as political partisans, preempt to themselves as "reformers," the right (as one senator expressed it) "to tell other people how to live." They *may* have some rights of that kind, but only if they achieve the stature of real sciences (Malinowski, 1937)[1].

With substitution of rigorous research, objective measurement, models and calculation for the too long tolerated loose discussion of pretentious and bogus "theories" these immaturities of the social sciences will surely cure themselves. Unfortunately, the more serious disablement—that of indiscriminately mixing value judgments with strictly scientific inferences and predictions—is a more insidious disorder, likely to be eradicated more slowly. In fact, it is likely to require a deliberate extirpation, based on a re-education of students in the social sciences to a new discipline of thinking. As Illinois University Vice-President Lanier (1971) recently said, nowadays a social scientist who makes himself "part of the action" is often likely to make himself also "part of the problem" and we need "a code of ethics among social scientists comparable to that required of physicians and psychologists." Social scientists inevitably have to make recommendations in which socio-moral values are also involved, but only by learning explicitly to set their ethical assumptions in one part of a document and their scientific findings in another, before putting them together, will social scientists gain the respect of fellow scientists and win acceptance by a now justly circumspect social and political leadership.

In the physical sciences, by the nature of the data and laws, a deliberate or accidental contamination of conclusions by personal values posing as science is far less of a danger. But there have been instances—e.g., Oppenheimer, writers in the *Journal of the Atomic Physicists*, Soddy, Carrel, Haldane, Russell—where the prestige of science has been consciously or unconsciously used to support left or right political values, or in Russell's case, deviant standards in sexual morality. In these cases the writer is usually not deliberately using his scientific prestige in any more direct way than to show it as a badge of intelligence. The reasonably alert reader has no difficulty in sorting out the writer's personal opinion— offered as that of an experienced scientist, but admittedly loyal to some value system—from an *ex cathedra*, authoritative scientific statement by a specialist in the field. Several comments by the present writer, here similarly offer only the *experience* not the documented proofs of a senior psychologist. In the social sciences, unless new safeguards for separating personal from objective research conclusions are deliberately and explicitly introduced, contamination by propagandist values will become a perennial, insidious and deadly danger.

The danger is not only that politicians and private institutions with axes to grind will find tame or corruptible social scientists to support their positions. The greater danger which recent experiences both here and abroad, e.g., Lysenkoism in Russia, have revealed is that partisans primarily political in interest and intention either accidentally or deliberately infiltrate the ranks of science. In the case of the Lysenko episode, and comparable events in Nazi Germany, the disturbing realization to scientists was that the exile or death of those ejected from their academic positions followed what *seemed* initially to be severe technical criticism by fellow scientists, but was actually politically staged. Incidentally, an unanticipated result of the "democratic" process initiated in some universities of having promotions of professors depend partly on a grade given by students has been pressure from the usual "leftness" of the young against independent minded social science professors lacking the fashionable coloration. Thus instead of the professor teaching the austere objectivities of a science, the classroom becomes from the beginning an emotional invasion of science.

Important though this issue of maintaining the ethical standards of science within science may be, it should not be confused with the main theme of this book. Effectively projecting the accepted community values of fair play, honesty and justice into the social dealings of science is vitally important. But our larger and more radical task here is to show how ethics may be developed for society *out* of science, not to bring the existing revealed ethics of society *into* science. Its aim is to develop a basis for moral values as a special *branch of science*, rooted in the objectivity of science itself.

2.3 THE NATURE OF THE PRESENT CONTRABAND VALUES IN APPLIED SCIENCES

Our explorations of the methods of human knowing have led to the conclusion that *in principle* science can bring moral knowledge. If so, this branch of science will be the most valuable of all, defining the finest ways to spend our lives. Meanwhile, the interactions of science and religio-moral systems are actually far less happy than this ideal form, as just seen, and it is a necessary preparation for the newer outlook to look more closely at what has happened under the old, in order that the two approaches may be more clearly separated. Let us in fact take cognizance of the ways in which social scientists have been smuggling values from all manner of revealed religious and arbitrary political beliefs into their recommendations, as if these were a part of their science.

Since every social scientist has values, and only the best have the austerity to keep them separately stated from their scientific findings, an attack on this contraband, like the zeal of any revenue officer, is likely to be interpreted by each as an attack on his own views. It is perhaps necessary to be explicit that the reference here is to *any* arbitrary values. If we give a little weight (or should we say emphasis) to Marx's discovery on the economic determination of historical movements (with a little help from Freud on the origins of "reason"), we see that it is natural that academics, considering themselves less well paid than businessmen of comparable ability, should on an average, tend to write more frequently with implicit "left wing" than "right wing" values. When business issues a report on social problems it is likely to be equally "right" in its emphasis.

The example of "right" and "left" also brings out the effect of emotional oversimplification, for it is hard *by any objective analysis to find any necessary, inherent logical coherence in what is now called "left" or "right."* Research described later shows that it is in fact impossible to accept on factor analytic evidence that just *one* dimension – or even two or three – can account for the *real* diversity of choices in political action. The distortions through people identifying themselves with the right-left stereotype are simpler and easier to detect than those more serious permeations of "logical" conclusions that come from the complex philosophy of a revealed religion – say Buddhism or Catholic Christianity – or from the inbred "rationalist" tradition handed down from the French Encyclopedists into the modern Humanist position, or even from the vague and transient philosophy of Hippiedom. By "rationalist," in the specific sense just touched on, we shall mean the belief that the application of reason and logic, without empirical investigation, is a sufficient basis for revising social and ethical values. It shows itself in the bulk of social reform writings from the eighteenth through twentieth centuries, from Voltaire through Bradlaugh and Ingersoll to Popper and Russell. Methodologically it certainly appeals to the serious student, as being *one* step in advance of revealed religion, but as far as science is concerned, it lacks the second step and therefore offers only a very erratic basis for ensuring progress.

Humanism (not to be confused with Humanitarianism!) must be left for the moment largely to define itself by the dictionary though it will be studied much more closely in Chapter 7. It is a consensus of refined values in human affairs as expressed in classical, Renaissance, and modern

virtually turns the philosopher's enshrinement of reason upside down, considering words and reason as devices given to hide our motives both from others and ourselves.

By "rationalist" here we mean, however, an historical tradition (growing lustily in the late eighteenth century) intending to apply reason to social matters previously handled by superstition, dogma and rule of thumb.

Today this is not enough, in the first place because of the shabby psychological tricks that we know reason can play, e.g., in defenses emanating from the unconscious, and, secondly, because reason without empiricism is only a half of science, and a dangerous half. The rationalist tradition is today still strong, especially in the ways in which would-be-progressive journalists, dramatists and radical politicians approach social problems. It therefore needs to be clearly dissected from the new applications of social science *per se* that we envisage as a true source of progress.

For today the persisting fashionableness in the "intelligentsia" of the unmatured rationalist movement (which, after Greece, began in the Renaissance and flowered with the French "Enlightenment" and the American revolutions) presents the true progressive with a distracting nuisance. This is recognized by many sophisticated people who today fully endorse the thesis of Carl Becker's *Heavenly City of the 18th Century Philosophers* (1932) that the rationalists of the Enlightenment were, in their habits of thought and truth-testing, much nearer to the medieval philosophers than to the modern physical (and ultimately social) scientists. For at present the pseudo social scientists and certain activist social reformers often work unquestioning, with methods and values that are still uncritically based on the "liberal enlightenment."

Rationalism was admittedly the indispensable spearhead in the break through which delivered us from the rule of hide-bound tradition, superstition and political domination by ancient dogmatic religions. But in retrospect its contribution has been more evident and impressive in its exposures of inconsistencies and absurdities in revealed religious dogmas and old socio-political habits — as superbly done by Voltaire, Volney, Diderot, Hume, Locke, Comte, Mill, Russell, and many others — than in any real creation of new values and goals for mankind. Reformist, not to say revolutionary, enthusiasm is unfortunately stirred up more by such witty rapier thrusts against authority and convention than by the essentially scientific, constructive work of say, the Webbs or Spencer or Owen or Lenin. Unfortunately, in emphasizing a purely negative use of criticism it often attacks values and practices which are vital to survival, as, indeed ridicule and the political wit or cartoonist can always easily do. For rationalism—especially as verbal reasoning devoid of research—is inherently incapable of recognizing the empirical laws which sometimes make an incomprehensible practice vitally necessary to society. Joined with respect for the natural sciences—especially biology—rationalism could have succeeded. But in the seventeenth and eighteenth centuries no biological science effective for man existed. And the offspring of rationalism in the obsolete brand of liberalism persisting today continues to think that reason is literary and philosophic rationalism, not the painstaking discovery and complex creativity of science. The more empirically enquiring liberals of the nineteenth century at least originally took a far more scientific approach. Mill (1863), Comte (1905), Malthus, Veblen (1899), Marx (1890), Bentham (1834) and de Noüy (1947) had respect for social laws that did not necessarily work quite as reason would like. (Indeed, liberals generally at that time recognized that individual liberty and equality, for example, both extolled by the rationalism of the French Revolution, were, in fact, incompatible.) But many followers, nevertheless, still tried to thrust a personal philosophy upon the laws of nature, and it is this pure rationalizing tradition, rather than that which developed into social science, that we are here concerned to inspect for its dangers.

There are offshots of rationalism, especially in literary satire and the cartoon, so esteemed by the youthful gallery, in which the smartest reasoning is that which is unreasonable and escapes the discipline of cold fact. The simplest and cheapest form of originality is to turn God (or anything else) exactly upside down, as Oscar Wilde loved to do. If reason sometimes unearths paradoxes then—the rebel believes—the unearthing of paradoxes must be constructive progress. To anyone continuing this last degeneration of rationalism, it suffices, for example, if he has been brought up in a Western ethic of productive work to write in praise of idleness. Or if, in a religious background, where a firm conscience has conventionally been assumed a source of serenity, he attacks it as the origin of all neurosis, and so on. The most eminent in the history of rationalism, such as Rousseau, Voltaire or Shaw, are far from free of the sheer vanity and hubris which might be thought to belong only to a decadent phase. Rationalistic satire, as when Oscar Wilde tells us that consistency is the last refuge of the unimaginative, or that an ideal marriage is a contradiction since marriage is not an ideal institution, convinces, by its brilliance, the immature; but the entertaining flash of half-truth is accepted by the experienced person merely as showing the limitations of verbal symbolism. Even in the sober domain of science, pure wit and "rationality" have frequently attacked and ridiculed great discoveries, as when Liebig and Wöhler were ridiculed for claiming to create organic compounds that were "logically impossible", Harvey for claiming that blood makes a complete circuit, Newton for saying that white light is composite, or Einstein for his "absurdities" in the general relativity theory.

Logical reasoning carried out in the setting of hard philosophical thought by a Kant, a Hume or a Leibnitz is a very different and more creative tradition than what we see in these mere rebounds of the ball of wit from accepted religious morality or conventional social authority. But even the serious constructions of rationalist philosophers — with the single exception of some contributions to mathematics — stand today like empty houses which the march of empirical science has passed by. Indeed, the progress of social science in the next fifty years will undoubtedly leave all the existing inventions of political scientists and the Utopias of social theorists as useless museum pieces.

As we now follow rationalism in its excursions into our present domain of concern — that of moral truths — we find it disabled by the same defects as have marked its action in other areas. First, as indicated above, there is a naive continuing cultural movement which regards reason without science as enough to handle all socio-political value problems and, secondly, and more seriously in this domain, there is an unawareness of what psychology has brought out regarding the real role of reasoning in human behavior. Rationalism, beginning in pride of human intellect and encouraged by the remarkable constructions of philosophy and mathematics became, in our cultural history, a system given to reasoning by deductive methods, without the sense of any need for inductive checks. Brilliant inference from first principles, rather than the patient watchfulness and learning of empiricism, has been admired and preferred in philosophy for two thousand years — until it was chastened and gave way, as late as the nineteenth century, to the discipline of scientific investigation. But it gave way only in the physical sciences, where the power of the more mature method was obvious. A mathematical logician like Russell (in *Marriage and Morals*) or a rhetorical logician like Shaw (in *The Doctor's Dilemma*), and thousands of lesser writers, can still talk nonsense in the name of rationalism.

The second weakness of reason — or, at least, a discrepancy between the powers assigned to it by philosophers as a supreme instrument and the findings of psychologists about its natural functions — needs a little more description. Throughout the realm of animal problem-solving experiments, reasoning, such as it is, operates as a means of finding paths to *instinc-tually given ends*. Biologically, reason is a servant to life. Let us recognize that in some sense the choice of the ultimate values with man, as with other forms of life, must be determined by biological purposes. It cannot be pulled by reason out of thin air. The actual supremacy of biological goals is patent in the individual's use of rationalization and doubtful reasoning in trying to save his own life and interests. But it needs to be recognized in a broader, more dignified and ethical sense, namely, in appreciating the proper role of reason in defining general human goals. Alas, in a more specific sense and through what psychoanalysis has recognized as the defense mechanisms, including rationalization, reason is not even a servant of inborn needs, but an abject slave, ready to live if necessary by dishonesty.

Apart from this questionable integrity, reason has failed because, without some illumination by science (or revelation), it has simply found itself — in and of itself — unable to set up a life goal. Even Plato and the classical philosophers who, in the first rapturous worship of reason, one supposes would have dearly loved to assign this privilege to it, failed in the main to find the precise goal it seemed always about to unveil. Before Christianity seized the emotions of the Roman Empire, attempts at the *tour de force* of developing ethics from reason led either to Stoicism or Epicureanism. The former died of emotional starvation and the latter perpetually degenerated into the goal of pleasure. The latter realized itself at best as the refined pleasures of a sybaritic life of luxury, wherever material conditions briefly permitted. It took less than a decade for the ceremonies set up for the Worship of Reason, at the first flush of the French Revolution, to fall into decay. In spite of all early enthusiasms reason persistently led to an insipid religion and fragile morality. The question remains: "For what are we being reasonable?"

Three brief instances of how confusion is added to the modern scene by "social science" movements in the name of rationalism must suffice for illustration — the Freudian impact on sexual values, the reaction of the "authoritarian personality" concept in psychology upon education, and the impress of Marxian economics upon social organization. What Freud actually said about the conflict of libido and superego in producing neu-rosis, and what the literati and even many popularly read psychiatrists (such as Albert Ellis), made out of it are, as Professor Mowrer has recent-ly argued very cogently (1960), two quite different things. Freud was as austerely aware of the dangers of rationalization as one would like a scientist to be: his followers were not. Wishful thinking on a mass scale quickly led to the self-styled "rational" — but actually rationalized — view that neurosis could be wiped out if people would get rid of their antique Victorian consciences and satisfy their sexual needs in an enlightened (and, of course, uninhibited) fashion.

Rather than master the far more complex therapeutic procedures which Freud and later researchers have indicated, many psychiatrists, professional and amateur, took to the simple practice of whittling away, in session after session, the genuine strengths of the patient's superego. The couch set out to cancel the choir. The result — as Frenkel-Brunswik showed (1954) in one of the rare follow-ups of a sufficient sample of "psychoanalyzed" cases — was the conversion of anxious neurotics into impulsive psychopaths. It may be true, as some leaders in personality research point out, that Freud himself had a distorted view of personality function because his evidence came too largely from pathology, but at least he respected the empirical approach. By contrast, the rationalist simplification by literary or journalistic intellectuals ended in a crude philosophy that he would not have accepted for a moment. Thus he certainly never proposed a modification of general cultural-moral standards purely in the interests of a diseased minority, but this is exactly what the "anti-inhibition" school of social reform has reasoned itself into. Essentially, psychoanalytic theory proposed a reconstruction of the faulty childhood relations of id and superego, not an abdication by the superego. Freud's view of the relative roles of the culture and those who cannot bear its burdens was clearly shown in *Civilization and its Discontents* (1930), and it should have been clear that the Freudian concepts envisaged culture as working toward a destiny of its own. The neurotic was to be helped by individual clinical therapy, not by asking culture as a whole to backtrack. The progress of the cultural pattern was not to be dictated by the neurotic, but by intractable and probably unpopular leaders gifted with genius and vision (as Freud indicated, for example, in *Moses, an Egyptian*, 1939).

Another instance of what simple-minded "rationalism" can do to values also sprang from an alleged social science concept, namely the "authoritarian personality" writings of Adorno, *et al.* (1950) at the end of World War II. In this case, unlike the Freudian, where a great man saw his views, in Kipling's phrase "twisted by knaves to make a trap for fools," it ended up by the followers being more scientific than the leaders, and eventually destroying by a weight of empirical evidence what was either deliberately or emotionally an attempt to dress up propaganda as science. (Adorno had escaped from Nazi Germany and his personal emotional reaction to the dictatorship began by confounding *authority*, as in science or scholarship, with *authoritarian* and really, with *totalitarian* practices, which are very different things indeed.) The rationalization that authority is bad, in moral or any other values, gave Adorno a free ride to fame on the wave of the perennial revolt of the young against authority. It overlooked that Nazism was itself a revolt — unfortunately a temporarily successful one — of the less cultured, lower-middle-class, immature youth against the established religious authority of Christianity and political authority of liberal Weimar Germany [4].

The third exemplification suggested above of a rational but not scientific attempt at seeking new values was deliberately chosen from economics, to balance two from psychology. However, since the Marxian formulation deserves more discussion, it will be deferred to the next section. It suffices that economists have agreed for at least fifty years that the Marxian system has the character of a rationalization, fitting economic observations to an essentially *a priori* premise.

It is not our purpose here to pin down the specific errors and trace the confusions in social and moral thought spawned by these instances, but only to point to the general character of the rationalist tradition. It is true that in these instances, as in all such fabrications over the last hundred years, there is some introduction of scientific data or theory, but only by way of lip service. For the writers are essentially concerned to apply "reason" to social affairs — by which is meant some *a priori* principle, in which they escape all the reflection on alternative hypotheses and all the labor of trial and error experiment necessary in investigating the real causes of the natural phenomena in question.

2.5   THE ABSENCE OF INSTITUTIONAL MECHANISMS SPECIFICALLY TO CREATE PROGRESS

In contemplating the turmoil of alleged progressive ideas one is reminded of "the tossing sea of steel" in Macaulay's description of the Etruscan army advancing upon Horatius at the bridge when "those in front cried 'Backward' and those behind cried 'On.' " That progress should be the accidental outcome of contending forces has been the rule throughout history. But with the embracing of progress as an ideal, in the last three hundred years (Bury, 1920), it follows logically that specific social institutions should be set up to engineer progress. Their task would be to examine and adjust those pressures which otherwise lead to social earthquakes and to direct the movement as smoothly as possible in directions scientifically evaluated as progressive.

It might be said that we have such a mechanism in political democracy, and so we do relative to the crustacean life of totalitarian dictatorships, but democracy can be merely adjustive rather than progressive and needs research eyes that it does not yet have. Chapter 9 here is substantially concerned with the planning of the evaluative and directive institutions that would accept progress as an integral necessity of the social organism. In this section it is proposed to take a preliminary look at the necessity for such organization, chiefly by noting the confusion and conflict which arise from the present lack of such organizations.

Some readers may wonder why the above attack on the inadequacy of the rationalist, *a priori* approach to social reform has been made, since at least it can be said that rationalists refuse to be bound by tradition, and consider themselves the forces of progress. Admittedly, in the eighteenth and perhaps the nineteenth centuries, the rationalists earned the gratitude of those eager for progress. But today the easy belief that we know just what progress is, and that it can be ensured by pursuing liberty, equality, fraternity, longer education and fuller stomachs is not good enough. It can easily degenerate, for example, into that satisfaction of the wants of the id which we study in more detail in Chapter 6.

Subjective, *a priori* goals are likely to omit much that is important. For example, the ideals of the French Revolution arising before Darwin, were necessarily pre-biological, and also had no regard for the energy and pollution considerations which the vast advance of the physical sciences brought about. (Yet they were carried over with no real change into the October Revolution of 1918.) But, above all, we must recognize that the purely rationalist approach is useful to criticize and destroy rather than to invent and create. Like the knife of a less skilled surgeon, rational argument may take away some valuable but not yet understood organ even as it removes some hated obstruction. (The destruction of religion in France in 1800 and Russia in 1920 may be such an instance.) In demolishing superstition (as did Voltaire), or vested political prejudice (as did Diderot and Paine), or long accepted axioms in the intellectual climate (as with Hume and Locke), there has always occurred some destruction (generally temporary) of other and vital values which reason alone could not justify. This serious damage is made greater by the common association of rationalism with the optimistic postulate that all change is progress.

By contrast the empirical, scientific approach to social problems or any other matter is capable of replacing old growths by new inventions. The steam engine, X-rays, penicillin, and so forth, were not the products of "reason" but of science, which uses reason but is much more. And similarly in social construction, investigation, discovery and invention of new devices that demonstrably work better is essential. These are not found by *a priori* fiat from some belief that to advance we have only to be more rational — in the speaker's own definition of the term!

Since all systems in time become obsolete and have to be removed, it would seem inevitable that governing authority must be periodically attacked. But this is far from saying that all attack on authority is progressive. Indeed, unless we wish to live in anarchy, authority in the abstract must endure. "The King is dead. Long live the King." Moreover, so long as there is social structure and human knowledge, there will tend to be the best accepted values in the one and the most mature and considered human judgment in the other. The question we shall raise later in this book — and it is a large and difficult question — is whether it may not be possible to build the research basis for progress, and the machinery for self-modification into the governing authority itself. Rigidity, for psychological and social reasons, is a part of all human behavior, but there are clear historical and contemporary instances of this rigidity being greater at the bottom than the top of society. (This is one of many reasons why the oversimplified Left-vs-Right formulation of political action needs to be recognized as obsolete.) The rigidity against which the progressive has to fight is in the future less likely to be found in the governing "authority" than in the thinking habits of the mass.

As a fact of history and biology it must moreover never be forgotten that although change is desirable, the chances of any random change being an *improvement* are decidedly less than fifty-fifty. It helps if we turn here to a well-investigated field of organic change: that of gene mutation. Of all mutations that occur in genes, the vast majority are deleterious. They go backwards to atavistic forms, or to new expressions that upset the harmony of the organism so severely that it fails and dies. Change *must* occur for evolution to proceed, but in any one step the organism must retain ninety-nine hundredths of its integrity if it is to avoid internal cancer or external defeat and collapse in the face of environmental demands. Similarly, the social body has, shall we say, a right and a duty to maintain the authority of large parts of its own "program" in face of "reason," though it should not and cannot do so in face of "experience."

But the effect of the alleged rational attack on authority *as* authority (not against specific, possibly fallacious views of authority) has been well illustrated in this generation as typically a growing contempt for moral authority *of any kind* among the general population, and a rueful and bewildered readiness of authorities, from parents to university presidents, to abdicate moral leadership. For example, there has occurred through rationalist arguments an abdication of values in sexual morality, in everyday civil order, in the punishment of crime, in defense of one's country, in relaxing what constitutes the harder, disciplined core of intellectual education, and countless other areas (Archer, 1958). If it be argued that there has been a substitution of new values, not merely an abdication, one can only point out that practically nowhere (as one would expect from the above stated nature of rationalism) has there been a construction of new moral *restraints*. In most historical periods of rationalist activity since the sixteenth century the social change has begun with what Crane Brinton (1938) called "the defection of the intellectuals" (meaning, of course, a section of the intellectuals rejecting their culture in favor of whatever revolutionary movement was popular at the time).

It will be discussed later, as a datum of social psychology, that these "intellectuals" are more frequently from literature and the arts than from the intellectuals of science. (Darwin, Thomas Huxley, Pasteur, Jenner, Einstein and others in science who upset the establishment were not primarily concerned to upset and were almost wholly concerned with reforming the scientific "establishment.") Lacking the reality-testing procedures of science, the literary or political "science" intellectuals are particularly prone to confound change with progress, and to take their departure in rational, logical argument from entirely *a priori* postulates, e.g., that pleasure is the logical goal of life, that all men are equal, that rivalry between societies is bad, that all strife is bad, that some men are naturally slaves (Plato), that the natural ecology of existing species should never be disturbed, that if it is medically possible men should live forever, that education can always be pleasant, that man is naturally monogamous, and so *ad infinitum* through the world of the "self-evident."

In the domain of ethics, which is our principal concern here, the effect of the literary intellectual leaders in this particular generation has been to persuade the educated man that the supreme sin is to take a "bigoted" stand on any ethical point, and that he had best let the words "right" or "wrong" fall into desuetude. A frequently associated assertion is that the humane person puts blame for criminal acts not on the criminal but on the society which brought him up. We are invited to cringe in collective guilt, when the most educative and remedial use of guilt might, for example, be (on the verdict of present psychological research) to seek to increase guilt in the criminal and the relatives immediately responsible for bringing him up. If society's sharing in the blame (but not the *complete* transfer of blame to society) were simply a more widespread recognition of scientific determinism, it would admittedly be to that extent an advance in comprehension of our social world. And it would be valuable to recognize that some forms of society are by this definition more guilty than others (crime is low in totalitarian societies such as Russia and China and authoritarian ones such as Italy). But the assertion of the guilt of society actually comes from the "reflex" liberal, responding directly with mixed-up humanistic and sentimental reaction, remote from any attempt to understand by intensive social scientific research what is actually happening. An empirical, deterministic approach would on the other hand, ask what the criminal does to the morale of society as much as it would ask what society does to the criminal. A scientific interest in understanding the proliferation of crime must consider a two-way causality. It must also include both environmental and genetic causes and consider the possibility that when society allows palpable genetic defects to perpetuate themselves in the criminal, it may be increasing criminal cultural values in society at large. In any case, if it were found that punishment discourages criminal behavior in the criminal individual and deters those on the brink of committing criminal behavior, scientific determinism would end by supporting the "illiberal" practice of blaming the criminal. The rationalist who also embraces science therefore does so at the risk that it may wipe out his fondest *a priori* "principle".

A particular development which almost inevitably grows out of rationalism in the modern situation is a belief in *inherent* ethical relativity. (A distinction has already been drawn above between this and *situational* relativity, in which the authority of some fundamental moral value is admitted and the variations are considered to be appropriate situational adaptations—different compass courses to the same port.) The result of accepting *inherent* relativity is of course, the weakening of moral force and authority anywhere. The rationalist encampment in the ruins of revealed religious authority thus creates in modern society a neurotic type of indecision on vital questions of ethical values. The culture scarcely dares to teach on moral matters. A Harvard psychiatrist recently summed up the problems of a large section of alienated students who came to him in distress by the remark of one of them "My parents never told me what to do." John Gardner, the Secretary for Education and Welfare, reminds us (1968) that "Every great civilization has been characterized by confidence in itself." In Western culture, and particularly among the "intelligentsia" of Europe, this confidence in its values is at a low ebb. As Barbara Tuchman, the brilliant authoress, stated in an address to the National Conference on Higher Education (Chicago, November 7, 1967, 22nd Congress) the missing element in our morale is the willingness to stand up and say "This is what I believe. This I will do and that I will not do. This is my code of behavior and that is outside it."

Unless the authority of a culture makes such definite stands, its value as an experiment is vitiated as in any sloppily carried out experiment. Fully recognizing that its moral values are not absolute, but only the best that can be divined, a society ought still to be uncompromising about enforcing them, though all the time intellectually examining them. One reason for the lack of respect for ethical values — and it is the problem this book seeks to solve — is that the moral truths adopted are patently *not* "the best that can be divined." Granted a rejection of subjectivity, and an attempt to find an objective and progressive scientific basis for moral decision however, ethical authority could become once more respected.

The young — rebels and non-rebels — are no less idealistic than in any other period of history; and they have more time and education to attend to these problems. They should be encouraged to attack them, rather than be fobbed off by safe and trivial distractions. At heart, most of them find the erosion of firm moral values painful and discouraging. The universities are justly criticized by society for not having constructively channelled this idealism by conveying such basic truths as that (a) human nature being what it is, people kick against necessary restraints as readily as against false ones, rationalizing that the first type actually belong to the second, (b) modern science should be capable, in principle, of guiding progress in moral values as positively as other fields of progress, eliminating the need to depend on the *a priori* assertions of revealed religion.

Thus the attack made here on rationalism is by no means an attack on reason, which is the indispensable right hand of progress against superstition, though relatively ineffective without help from the left hand, which is scientific experimental method. Nor is it a disparagement of the need constantly to question authority — personal and institutional authority as distinct from the authority of scientifically-tested ideas. But it is an assertion in accordance with what Becker (1932) well states, that St. Thomas Aquinas in the thirteenth century and Voltaire in the eighteenth (despite the rationalist advances claimed for the latter) *alike* depended on reason and on faith. Moreover, the basis in faith was no less important (though sedulously hidden) in Voltaire. In spite of science having introduced a radically new method, the bulk of reformist writers on social and ethical questions are still simple rationalists. If they decorate their arguments with references to Freud, or Keynes, or Margaret Mead, the outcome, claimed as a "modern scientific, intellectual, *avant garde* position," is grievously tainted and distorted by slavish adherence *to wholly arbitrary value presuppositions*. Almost without exception they are fashionable, mass values. As a recent biographer of Browning, commenting on the herd reaction of the contemporary literati against this then revolutionary poet, observes, "the intelligentsia have never been noted for firm individual independence of mind" (Ward, 1967).

To reach new values for social progress we are going to need the critical and creative independence of the scientific worker. And the individual alone will not do; for advance needs in addition, the massive support and organization necessary for an undertaking that is as formidable as research on cancer or any other highly complex problem in science.

2.6 SOCIAL CONSTRUCTION WITHOUT POSITIVE VALUE CONSTRUCTION

Among the existing methods of wooing social progress that have accounted both for some enlightenment and much seduction, fraud and confusion is that of describing Utopias.

These beguiling stories are generally imaginative and interesting, but also vague on basic principles and often quite unreal in assumptions. Uniformly, from Plato to Marx, Comte or Wells, the *moral values* fall short of being explicit, or, if explicit, the support for them is not stated. Thought on social progress has clearly been much affected by the Utopias of Plato, St. Augustine, More, Bacon, Campanella, Harrington, the Abbé Sainte Pierre, Butler, Morris, Jefferson (in a sense of practical projects), Bellamy, Comte, Wells or Marx, but few of them indeed state the sources of their fundamental values. Even the grand logic of the rationalists is largely absent from these creations, though some, as in Marx's leaning on a philosophy of history, Comte on a philosophy of science, and Wells on a view of continuous material progress, begin to approach general principles. In the main, however, they substitute only the purely personal predilections of the authors, or large fragments of existing moral values from inspired religions, for any explicit derivation of values.

Mostly they are concoctions put together by cooks rather than chemists. But though they also lack the real test of history which the politician (who is an experienced cook) faces, the check "Does it work?" has at least been carried out in imagination. (Though scarcely by an unbiased judge!) Whereas exponents of revealed religion, like Buddha or St. Paul, rationalists like Diderot, Proudhon or Voltaire, have simply taught values to men and left men and history to generate societies, the reformer who constructs Utopias has worked out the actual detailed blueprints for the social machine. In a few cases, like Owen and Marx, he has even been able to set it up in action.

To see what values can typically be extracted from this kind of construction, it would be best to take an example, say, Marxian Communism, Butler's *Erewhon*, or Wells' *Modern Utopia* and (as has classically been done with Plato's *Republic*) pursue an analysis. With space only for one—and that all too brief—let us consider the Utopia of Marxian Communism because of its current importance. However, in order not to put it at the disadvantage of all the travesties that men make when they reify ideal models (Marx's temperament would almost certainly have caused him to disown existing Communist governments), as well as to avoid current political emotionalities, let us consider rather what was drawn up in the blueprint *Das Kapital*.

It will probably be objected at the outset, by enthusiastic Marxians, that it is questionable if he should be classified among armchair Utopians, because he drew his Utopia from the "grand logic" of historical empiricism combined with deductive rationalism. Believing in inherent economic laws, he *claimed* that the creator of Communist societies had only to be a mid-wife to history. The Jehovah of historical process had somehow given Marx this assurance. As is well known, he nevertheless rejected the metaphysics of Hegel and the idea of history as a transcendental process of unfolding mind ("the philosophy of history is the history of philosophy"). He thus may seem to have accepted what was in conception, if not in method, a position close to our own; that the laws of natural science suffice. However, much of this appeal to science is shop-window dressing, as shown by his willingness completely to neglect the biological sciences, psychology and any social science but economics, as well as by the neatness with which his conclusions fitted his own emotional viewpoint. For example, his acceptance of socio-economic classes as the only true human groupings, neglecting those culturally-different, organically self-sustaining groups (of which nations and religions could be examples) which are *at least* equally important in history, betrays an extreme bias. Classes, nations, occupational and religious groups all have claims to being organic communities, and modern social psychology has something to say

(Borgatta and Meyer, 1956; Merton, 1957; Davis, 1949) about their relative structures, dimensions and interactions (Darlington, 1969).

The fact that Marx chose classes was determined partly by personal frustrations[5] and partly by a real historical upswing in their importance at the time he wrote. Moreover, if one accepted the central theme of the economic determination of history which Marx borrowed from Saint Simon, classes offered the best illustration. For the then galloping industrial revolution expressed itself most dramatically in class changes. The machinery by which the overthrow by the proletariat was supposed to take place is well known and needs only brief reference as far as present interests are concerned. The able and enterprising (as he admitted them to be) would produce and command the means of production, using the proletariat as a commodity. But technical advance would put the "small man" — the lesser entrepreneur — out of business, increasing the dispossessed proletariat, until property belonged only to the few. At that point the proletariat would take over, substituting control by a few politicians (elected, of course, to express the dictatorship of the proletariat) for that of a few great non-political managers. The assumption is that when state and big-business finally become one, the worker will be better off. In Communism, in practice as we know, the bulk (semi-skilled and unskilled) of workers have not become better off in the sense intended. The differentials in pay, privileges, leisure and status between workers and managers are now not significantly different in communist and capitalist countries. The main difference remains that the passing on of capital from generation to generation is in one case under a monopoly — a responsible and probably sociologically well-informed state organization but still an authoritarian monopoly — and in the other in the hands of freely endowing individuals, trusts, and independently endowed institutions. It may be objected that power in the hands of authoritarian "party officials" is not pure Marx (who talked of workers running their own factories and sailors their own ships) but a Leninist addition. If so, one can only say that Lenin was more perceptive of the inescapable practical conditions for navigating the ship of state.

Scrutinizing for implicit values one finds Marx proposing to remedy "the degradation of the working class," "economic slavery," the dehumanization of social relations produced by the industrial revolution and the growth of commerce, the exploitation and parasitism of one class in relation to another, and so on. There is no serious attempt to define what parasitism means[6]. There is probably the implication that the "greatest happiness of the greatest number" is desirable (but never spelt out or credited to the liberal circle which begot Bentham or Mill), and that individuals should have the maximum freedom compatible with the good of the state. Nowhere does one find arguments clearly referred to basic ethical principles, or an attempt explicitly to define the values of conscience that would require the political action indicated. A major biographer has simply said that the value judgments most widespread in Marx's writings rest on "an acute almost paranoid sense of injustice." But injustice has meaning only against a definition of justice. Is the positive statement of values avoided mainly because it would appear only as the uninspiring and banal conclusion: "All men should work, and for the same salary"? Certainly there is no detectable originality in the implied moral values, which for all one can discern in this ethical twilight *could* be just those of Christianity, or Judaism, or tribal authority, or Humanism, or even anarchy.

A political scientist might object that Marx saw as inherent in the historico-economic processes he is implicitly dealing with—even if he never brings it out—a moral system more permanent and independent than that of a particular terminal Utopia. But although Marx was much concerned with the class struggle as a process, and with revolution, emotionally, there is no extraction of basic moral laws therefrom. The French Revolution, for example, he saw as an encouraging historical precedent rather than an ideational growth—it was mainly a practical rehearsal and forerunner for the *final* revolution. With the downfall of bourgeois society the class conflict would disappear forever. A "rational" society would follow in which there would be neither rich nor poor, master nor slave, and in which the free man would "develop his capacities [for what?] to the fullest extent."

Thus, as we look at this most popular of the modern Utopias we are compelled to recognize that it contains no profound moral system for the guidance of humanity in its long adventurous pilgrimage. And the long term values, so grievously missing from the concept of the "Final Revolution," are surely what we need; for whether we start with modern history or the trivially longer perspective of ancient history, we have to recognize that the human cultural pilgrimage has barely begun. Instead of seeking the basic necessary values for the journey, Marx makes us the illusory offer of a "terminal Utopia."

As indicated above, an extensive examination of Marxian Communism is not intended here, and its particular technical weaknesses are not important. The intention is to make sufficient examination of some important typical Utopia to ask whether this kind of writing on social reform contributes to advance or to false goals. Our conclusion is that much the same defects disqualify most of these approaches, whether from the "left" or the "right" [7] variety of reformer; namely: (1) that they proceed to detailed social construction without any clear value construction; (2) there is no empirical, scientific examination of behavior genetics, learning theory and social psychology necessary to understanding group adjustment [8]; (3) they essentially involve the belief, from the infant period of social idealism, that we can seek a purely static "heaven" to be realized on earth. (Concerning the essentially static quality even of Comte's Positivist Church, Anatole France felt compelled to say "It assumes science to be [already] definitely constituted and [disapproves] of further prosecution of researches" (1920, page 103).) And Dickson (1969) appraising the imaginative and scientifically informed utopias of H. G. Wells is moved to conclude, "the heaven Wells dreamed of in 1900 bears a distinct resemblance to the hell imagined half a century later by George Orwell."

So it is from More's or Campanella's purportedly Christian Utopias to the latest of note, Marxianism, of which one of the greatest Marxian students (Berlin, 1963) well says, there is "no new ethical ideal in the system." Thus, in the last resort, both this lack of a method for seeking moral values and the denial of the boundlessness of the human adventure forbid enthusiasm for this approach. Utopianism remains essentially an attempt at social construction without research and more disablingly, without ethical value construction.

2.7   THE TREACHEROUS ALLOYS OF "SCIENTIFIC" AND "REVEALED" TRUTH

We have now made the rounds of the basic ways by which man seeks to understand and reach adjustment to his lot — essentially through religion, rational philosophy, and intuitive arts, the "ideational experiments" of Utopias, and science itself. A brief but fairly fundamental stocktaking has been made of the particular qualities of error and sources of confusion in most of them. Some readers may think this reconnaissance has been too long and that we should already be off on our main journey, yet to begin in Chapter 3. But sufficiently searching and carefully judicial processes must necessarily seem long. We want to be sure that a dependable perspective on the strengths and weaknesses of the current approaches has been reached before arming ourselves for our own expedition.

When all is said about the specific methodological defects of these approaches, one recognizes that the most dangerous confusions today do not arise so much from such *limitations* of historical method, intuition, rationalism, philosophy, etc., as from the *naive mixtures* of intuitive, revealed, philosophical-rational and scientific methods which public discussion now takes for granted. In recent years, as asserted above, the most insidious and disturbing examples occur in social reform that is offered in the name of science itself. Incidentally, this problem of adulteration is an old one, recognized by alert scientists since science entered society. As long ago as 1904 Max Weber made a strong plea to his fellow sociologists to make a clean separation of values and factors. Such expressions as "a well-integrated society," "socially desirable" or the more oblique "normally healthy" still abound in social science recommendations — in writings intended to be purely technical — without either definition or apology. More recently, Polyani (1958) has defended the role of intuition in "heuristic" empirical research without clarifying the difference between using intuition as an exploratory research *means* versus depending on it as an ultimate *evaluation* of a scientific conclusion. In particular the social utopianism just examined often shows an admixture — on the one hand — of good social planning, sound political craftmanship, and even objective investigatory techniques, and — on the other — a farrago of subjective, fashionable social values, purely personal convictions, and emotional prejudice. Most commonly these values are hidden in arbitrary personal digests of the ethics of certain revealed religions.

To understand the undependable and unknown alloy from which such popular thought and evaluation is made today, it is necessary to recognize that a hundred years of sharp and explicit conflict in educated circles between religion and science has not solved much but has ended in the indifference of exhaustion and the superficial amiability which can follow a truce. Thus the two most fundamentally opposed of the approaches here considered — revealed truth and scientific analysis — have actually become blended in such insincere compromises as Humanism, Unitarianism, and more massive but less explicit popular trends. To sympathize — if not to agree — with this current acceptance of a squalid incompatibility one must remember that for at least four hundred years there has been a crescendo of warfare in the minds of men between science and revealed dogma, reaching its most painful expression perhaps in the nineteenth century, in the suffering of such men as Darwin, Thomas Huxley, Charles Kingsley, Edmund Gosse, Matthew Arnold and their theological counterparts. In seeking to escape this warfare it is not surprising that some strange compromises were made and some inherently false positions accepted. These are psychologically more understandable if we do not forget, moreover, that scientists and religious men resemble each other in important aspects of personality — in the idealistic and ascetic character — more than either resembles the casual and "sensual" man in the street[9].

The more tolerant of scientists have generally taken the position that questioning the scientific correctness of *Genesis* or the historical reality of Christ leaves the spiritual truths untouched. They are content, with James or Dewey, to be pragmatists. But pragmatism is not the full illumination of empiricism, analysis and insight as required by science. The magnanimity shown by science in this area may be misunderstood and paid for dearly. It is one thing to seek to persuade — as we do below — that there is a temporary social need to hang on, in practice, to traditional religious values while the new house is being built. It is quite another to pretend that the two kinds of truth are not really on very different foundations, and that we have not entered on a problem of mixing oil and water.

Looking back on the lonely spiritual adventures and strivings that gave us Buddhism, Mohammedanism, Christianity, and many less developed religions, any sensitive and historically aware individual must, despite his rationalist inclinations, recognize that one looks back at mountain ranges of spiritual grandeur. It is likely that when the logical-empirical scientific creation of values has soared into its more amazing abstract calculations of the future it will still have to recognize that perhaps some more elevated but unconfirmed spiritual insights remain in the higher mountain peaks reached by the early religious pioneers. That is a question needing examination after the full tenets of Beyondism are developed in Chapter 4 below. Meanwhile, even the sharpest critic will agree that the authors of the great inspired religions appeal to what man senses to be the best in him. They attract most strongly the noblest individuals. But they comfort and sustain all. They have been indispensable props through dark and barbaric centuries, when centers of learning scarcely existed and the cultural foundations necessary for the systematic objective creations of science were hopelessly submerged in anarchy and brutality.

Let us also recognize that the practical acceptance of scientific and revealed truth as if it were a single welded whole, unquestioned as to its essential homogeneity, which we here consider a perilous mistake, has nevertheless been adopted by some of the greatest philosophers and scientists. This is a humbling and disconcerting consideration when we ourselves are venturing to sail out on a purely scientific search for moral truths. But it is no exaggeration to say that throughout the nineteenth century the great trail-blazers of science from Faraday to Kelvin in physics and from Cuvier and Darwin to Pasteur in biology, steadfastly adhered to traditional, "revealed" religious values. There were few real atheists among them and they preferred to believe that the great machine they studied was the explicit and magnificent creation of a *Deux ex machina*. Regardless of any philosophy, their emotions at any rate dictated to them that they should keep science and religion clear of the degradation of ultimate conflict. Even the far-ranging mind of Newton kept active scientific and active religious interests in two compartments, and some biographers believe that he considered his work in theology more important than his contributions to physics. But, except for the strange importation into biblical history of the methods of calculations he had perfected in physics — methods of calculation as remarkable as those by which he weighed the earth — his religious conclusions were based on the inspired work of the Bible. He followed revealed religion as devoutly as any Fundamentalist from the hills of Tennessee. Einstein is less explicit on religion, but in morality, as shown by his writings on politics, his ethics is taken over ready-made from democratic, Christian-Judaic tradition as unquestioningly as by any schoolboy. Bertrand Russell (letter to Gilbert Murray, in his *Autobiography*, 1968) writes "I am very anxious to be clear on the subject of immediate moral intuitions . . . upon which, as is evident, all morality must be based." His belief is clearly stated that "immediate intuitions are the only source of moral premises." This path leads to "knowledge, and the appreciation and contemplation of beauty, and a certain intrinsic excellence of mind" as the ultimate touchstones of moral meaning.

This compartmentalization of thought found prevalent in even the most eminent scientists and philosophers who were born as late as the nineteenth century is a remarkable phenomenon, worthy of intensive study by a psychologist–historian. One might speculate simply on the existence of a deeper childhood fear of the authority of established religion than the writers are consciously aware of. Or again, they may have had such an exacting respect for scientific precision that seeing no prospect of reaching into religion with this exactness, they readily accepted an habitual adjustment to a deep social taboo. It persisted in fact as a professional taboo in both science and religion, each of which forbade any common domain of thought. The split philosophy — essentially a modest two-way agnosticism — is intellectually more immaculate and socially more trustworthy than most of the compromises practiced today. Nevertheless, one must insist that we cannot have two truths. Neither the impressive standing of these men, nor the emotional comforts of toleration must deter us from attempting the scientific journey that may bring us to truths in both domains entirely through the gateway of science.

Actually, in the present times of confusion, the opposition at the popular level to the scientific path may arise less from traditional, revealed religious dogma than from the mystical equivalents of the Gnostic heresies that have often grown like tumors on the body of organized religion in times of decadence or disturbance. Such movements, breaking away from whatever rational integration has been brought into either science or religious dogmas, and claiming immediate illumination, are, of course, inherent in the whole intuitive approach. One result of the growth of science might be that its undermining of religion will lead to its sharing society with a berserk religious mysticism instead of an organized church. Superstition tends to spread in society from below upwards, and, as an expert in Greek history has pointed out (Dodds, 1951), in more than one society it has ended the intellectual and political realism of whole societies. The incongruity in the advent of superb scientific powers, e.g., in nuclear force, the creation of new drugs, and the spawning of material luxuries, in societies otherwise given to undisciplined emotional thinking, might easily destroy whole societies. For this and other reasons it is imperative to make the effort to put moral values and spiritual direction on a footing uniform with that of other knowledge.

Although most space has been given here to the precarious quality of the alloy of revealed religion with scientific method, the other sources of confusion here — utopias, rationalism, the arts, self-evident "political" truths — can create mixtures just as dangerous. Of the apparent totalitarian solidarity of Communism, the writer Kuznetsov (*Telegraph*, August 29, 1969) says, in regard to his native country: "As for the more intelligent, thinking people — here you have a state of chaos and great confusion in Russia. Some [believe in the possibility of] . . . a decent, more democratic, more liberal society, even though still ruled by the Communist party. Others pin their faith in science and the scholars — that they will become so influential in society . . . that they will be able to find some solution. Very many turn to religion, more and more everyday . . . . Nevertheless, the majority understand nothing and do not believe in anything."

One recognizes readily here that this gifted representative of the arts, as opposed to Marx, has nothing more objective to offer in values than "a decent, liberal society." But no matter whether motivated by inspired religion, or simple "human decency," the intuitive approach to truth in the last resort cannot be respected except as a tentative approximation and a temporary stop-gap. If the truths of the heart are real enough, surely they can eventually be made accessible and proved by intelligent explanation and demonstration? For, without the development, through psychological science, of ways of examining the validity of such convictions and communicating their content, the truths of emotional experience remain unverifiable, and constitute a coinage of low logical and educational negotiability. Even if the epistemological doubts about the foundation of truths of this type could be overcome, the special difficulty would remain that their acquisition must proceed by the costly and painful path of *individual* trial and error. This path has to be trodden afresh by each person and each new generation, with no guarantee that it will get further than — or even as far as — the last. From whichever standpoint one looks at the matter — validity, mutual communication, or the possibility of steady advance — the acceptance of a duality of truths is intellectually unsatisfying and socially dangerous. The numerous attempts to view science and revealed religion as "happily one" (from the early writings of physical scientists such as Newton or Lodge to the recent writings of some biologists (Ebling, 1969)) are well intended yet foredoomed.

Our hope is surely that the application of the methods of science to human affairs and the human heart, will eventually permit science to take over the second half of its total heritage, and yield a body of knowledge, such as modern psychology aspires to, in which "truths of the heart" can become explicit, open to proof, and subject to exact communication and teaching. Perhaps — though this breakthrough is not essential to our present argument — it may eventually be possible, by some miracle of neurological science or dynamic psychological analysis to transmit emotional experience and maturity as readily as we now transmit intellectual skills in the classroom. Even before that is achieved the educated man of today surely has ample evidence before him that science has answered far more fully than have other institutions the questions "What am I?" and "Where am I?" It is reasonable, therefore, to hope that it has inherently greater chance of more clearly answering the final question "What ought I to do?" Meanwhile, the brittle alloys of scientific and mystical, or dogmatic modes of thought which constitute the fabric of this morally confused society place us in extreme danger[10].

2.8 SUMMARY

(1) The continually progressing world view of science, and the disciplined thinking connected with it, have clashed with and undermined the

authority of religion. An unfortunate immediate effect, as this understanding spreads from a well-educated minority to society as a whole, is that the demolition of superstitions and of arbitrary dogmas has also upset the credibility and authority of the source of society's ethical values. In consequence, in Western culture[11] we are suffering some degree of what Gilbert Murray aptly described and documented, in the decay of Greek culture after Aristotle, as a "failure of nerve." Other causes — economic, political, social — may be contributing, but it is likely that we shall find the main root to be a confusion over values due to this collapse, combined with a failure to recognize that science is capable, in a radically new sense, of building up ethical values to replace what it has destroyed.

(2) Instead of looking to science, writers have turned in their dismay to almost every other available social institution. These we have examined as to the validity of their claims to direct progress — beginning with Rationalism and other philosophical approaches, from Plato and Aristotle, to the specific modern flowering in Voltaire, Condorcet, Paine and Rousseau, and so to contemporaries. Reason in the abstract is more effective in demolition than construction. Rationalism, e.g., in Locke and Montesquieu, sufficed to clean the site. But the scientific assumptions about human nature and social mechanisms, e.g., that man is inherently good and that education can make him politically wise, or that rational intentions can alone abolish wars, poverty and injustice, were crude and unrealistic.

A modern writer, equally fervent for the millennium (H. G. Wells, 1920, page 455) has stigmatized the encyclopedic rationalists' mistaken belief that "a sentimental and declamatory [approach]" is appropriate to deal with socio-political problems. For, in spite of their paying lip service to science (which was in any case poorly developed in biological and social areas at the heyday of the "Enlightenment"), they expected Reason alone to create new values. Reason can sometimes reveal and destroy irrationalities in existing systems. But too many writers on social problems still proceed as if logic without advances in human science as such, suffices. Nor do they recognize that the values in Rationalism rest on subjective, *a priori*, premises surreptitiously imported from the religions they seek to outmode.

(3) The Utopia-builders, such as Plato, St. Augustine, More, Bacon, Morris, Owen, Marx and Wells, and those political reformers acting as the empirical mid-wives of history, such as the Jacobins or Marx, contributed less to the creation of values than is commonly assumed. New values, if any, are with difficulty extracted from their ideal societies, whence they

have to be inferred as implications of political and other concrete rules. Mainly, however, they quite ingenuously imported with the status of axioms what they considered to be universal human values. These were commonly the unrecognized fragments from universalistic, revealed religions. This is why Communism, for example, has correctly been designated as one of several possible Christian heresies.

(4) The current tendency to turn to the social sciences for solution to social problems, though healthy in intent, is vitiated not merely by their crudity as sciences, which will pass, but by their systematically intermixing and confounding scientific chains of argument with unconsciously or naively introduced moral value judgments. The latter is a far more serious and dangerous defect. Remedying this defect requires explicit recognition of the problem and an agreement either explicitly to bring in revealed, dogmatic values or to take the step of seeking afresh for purely scientifically derivable values.

(5) If we are right in the argument beginning in the next chapter — that science is itself capable of deriving moral values — it may yet take years of brilliant and patient research to reach methodologically sound conclusions. During that time we should probably do well to lean temporarily on the ethical framework, though not the superstitions, provided by the deepest convictions of revealed religious authority. But this consent to an expediency is very different from deliberately maintaining that the truths of science and the inspirations of religion genuinely belong to the same method of truth-seeking, and are capable of amalgamation. For in the long run the alloy is a treacherous one.

(6) For reasons largely beyond our present inquiry, many great scientists and leading philosophers, while recognizing that these methods are different, have been content to proceed with a duality constituted by scientific empirical truth about nature and by intuitive, traditional religious truth about ethics. Still larger fractions of men engaged in political and public affairs, and of ordinary men in their everyday lives, consider it practical to accept the same basic inconsistency of origins. Ultimately this practice has consequences, however, as dangerous as those which spring from inconsistencies in other realms of experience, a fact which is becoming increasingly evident in some current deteriorations of morale. Although the difficulties are very great, we have to voyage in search of a new scientific, i.e., *combined rational and empirical* basis for finding ethical values which is uniform with our scientific procedure in understanding nature generally.

(7) Meanwhile, confusion is breeding a serious degree of social aliena-

tion which, from a glance at history, e.g., at the times of Diogenes and the Cynics, we know can be mortally dangerous. In one part of society we see a group with a growing loss of any sense of social obligation, a tendency to justify this indifference by condemning society, and a denial of moral values. In the more established part of society we meet an abdicating authority, avoiding conflict and (as Malraux (1949) perceived early after World War II) "doubting its own credentials." Without new sources of knowledge we do not know what progress really is and what values are sound. But having in these two chapters examined the available varieties and sources of knowledge, and the effects of following some of them, we can more confidently organize ourselves for the search for values.

2.9   NOTES FOR CHAPTER 2

[1] For example, economics uses reasonably refined mathematical models, but by failing to recognize that exchange behavior is a part of total human behavior, i.e., that in fact it is a branch of psychology, it remains bound to laws rooted in what are really local cultural conditions. Consequently, it is weak in predictions which involve the emotional behavior of man, even about everyday matters of interest to millions, such as the stock exchange! Sociology and anthropology have achieved order only as descriptive sciences. The fruitless flirtation of anthropology with psychoanalysis is fortunately over, and both of these studies are beginning to search for laws by more adequate statistical methods, but have far to go.

In sociology, the thin ranks of able researchers, hemmed between armchair, philosophical sociologist colleagues on the one hand, and short-sighted do-gooders with the scientific standards of social workers on the other, can advance but slowly. For the last fifty years it has shown a notorious bias against accepting the findings of behavior genetics and has tried to build a science of group behavior on a *tabula rasa* theory of the human mind, which was discredited soon after John Locke proposed it two centuries ago.

Although psychology has been methodologically more sophisticated it attracts, by its universal appeal, a body of camp followers and "non-investigators" more numerous than the professionals, whom the former substantially impede. But, alas, even the professionals have not succeeded by that test of the "sincerity" of a theory which exists *in its developing an effective technology*. Theories can always be made scientifically elegant, but more than scholarly "correctness" is needed, and here the physical sciences expose the poverty of social science by the richness and effectiveness of their own technological impact. For example, though learning theory has a vast technical literature, and certain social possibilities have recently been pointed out by Skinner, it has in fact not changed school teaching practices in more than trivial ways from where they stood generations ago. Similarly, personality theory, though potent in selection and counselling, and useful in clinical practice to those clinicians who understand it, is still in its infancy.

[2] As personal experience I recall that when several of my researches on the social class distribution of intelligence and personality, and the factor analytic definition of social status appeared around 1940, I was taken to task by social workers, academically labelled social scientists, for (a) supposing that social classes could have any "function," and (b)

imagining that they could have even any value as scientific reference points since they were temporary hangovers, due to disappear after the war.

[3] The range of meanings given to a word like democracy illustrates the radical change in method and concept necessary if political science is to become an exact science. Obviously it means something different in America, Australia, Britain and France, and still more different again in the U.S.S.R. and China. In the two last, for example, it is compatible with the dictatorship of an intellectual elite, but not in the former. In Britain and France it is compatible with marked social class stratification whereas in the U.S.A. and Australia there is an inclination to declare it incompatible with class distinction. And so on. The object here is not to pursue its meanings in themselves, but to insist (a) that one cannot use the term in exact discussion, and (b) that one *could* use what is conceptually intended by this term if dimensional analysis and quantification were introduced from social psychology.

[4] The particular instance of Adorno and Frenkel-Brunswik is of current interest in that it helped strongly to give intellectual, *avant garde* endorsement to the purely emotional discarding of a number of disciplined values which became extremely popular among half-educated students in the 'fifties and 'sixties. It filtered down from clinical psychologists concerned either to give parents an escape from the tasks of leadership or to browbeat them into positive abdication from setting standards. Like much rationalism—which nowadays is apt to claim some support from popular science in addition to the usual logical dependence on an "unanswerable" axiom—it was essentially a rationalization of id demands in the child. Socially, it developed a clever propaganda, using *ad hominem* instead of *ad res* arguments. That is to say, it arbitrarily defined the personality of anyone who believed in authority as obnoxious and psychologically sick, and set out to "investigate" it clinically. Like any *ad hominem* argument, e.g., Hitler's attack on the Jews, it could be more readily understood by the unintelligent than could any *ad res* analysis, and it had an immediate appeal to young persons irked by controls, and to maladjusted persons with a permanent allergy to authority or other realities of life.

As a researcher actively engaged, at the time of the above incursion, in more experimentally objective and statistically complex researches on personality, I may perhaps be permitted a comment which I believe is that of the scientific world today. A retrospect on what it contributed to basic personality research indicates that as a scientific concept the "authoritarian personality" was worse than useless. There simply was no such unitary dimension as that of an "authoritarian personality," attractive though this straw man might be for political incendiarists. The hundreds of articles and tens of thousands of hours of misspent research time by immature investigators thus stolen from possibly useful scientific work, stand as a monument to the gullibility that can hide in intellectual rationalism and a reminder that social science is not always science. In the end, this tumor within a science was self-destructive. As more experimental and psychometric evidence accumulated, it became evident that individuals with respect for authority, when compared with those who had not, showed numerous valuable personality traits, e.g., higher emotional stability, dependability in emergencies, unselfishness and so on. It also became clear that there is no unitary dimension, and that the whole issue of respect for authority versus being a law to oneself is an altogether more complex resultant, psychologically and socially.

As the evidence backfired, the enthusiasm within alleged technical psychology collapsed as suddenly as it had been stimulated. But the fact that the hypothesis proved to be scientifically defunct did not stop the literary camp followers and educators with a superficial psychological reading, from claiming scientific support for their crusades, much as they had done with Freud. Thus as far as the propagandists were concerned, science had served its purpose, and its scorched harvest fields could be forgotten. In both cases the incorporation of this pseudo-science in the so-called Progressive Schools movement made it possible to beat shrewder, common sense parents over the head with "well-known facts of modern psychology." Between 1925 and 1950 the flood of "psychologically enlightened" education fed by the turbid waters of a misunderstood Freud and an hysterical fear of the damage done by authority, burst out of the smaller circle of academic psychology and its camp followers into the prestigious journalistic intelligentsia, and soon all the world was in the sea.

[5] The writer has no intention to make any *ad hominem* criticisms of theories. And even psychological analyses of writers of the past, which help us understand the course of events, should perhaps be detached as in the present footnote comment from the main evaluation in the text. But no psychologist, Freudian or otherwise, can shut his eyes to the interesting probabilities, pointed out already by Wells, of the determination of the less valid twists in *Das Kapital* from purely personal experiences. As one who had given up his Jewish faith, and lived in four or five countries, it was natural that the Internationale should be his child and that he should become psychologically unappreciative of any functions of individual national cultures. Thus he could easily become convinced that the latter were on the way out in favor of class loyalties, in the very century wherein historians now see more clearly a definite growth in national self-determination all over the world. (Classes and nations are, of course, true rivals in the psychological sense that greater loyalty to one means less to the others, so that their synergies [Cattell, 1950a] are negatively correlated, which may account for their conceptual rivalry in Marx's mind.) The effect of this personal factor favoring the denigration of national cultural values was multiplied by the personal experience of a life of poverty amidst an increasingly prosperous middle class—the "bourgeoisie"—in which he was not accepted and which was indifferent to his views. It is not surprising that he was embittered, and that as Wells has said, the fulfillment of hostility in the destruction of the bourgeois class seems more important in Marx's writings than any concern with what happens afterwards.

[6] As discussed later (page 148), the differentiation of parasitism from symbiosis, in both the biological and the cultural-economic world, is a most difficult diagnosis. The fact that an inventor–entrepreneur earns ten times as much as an unskilled worker does not automatically make him a parasite, if in fact the enterprise which came from cudgelling his brains increases the real wealth of the community by at least ten times the contribution made by the unskilled. Marx's definition of parasitism seemed to be that the entrepreneur spent relatively little time in work; but the same could be said also of the unemployed recipient of public welfare, or the worker who does "make-work" jobs to maintain "employment" but offers no really wanted contribution. Until psychology discovers more about the relative "stressing" and fatiguing effect of various kinds of creative, routine, and emergency jobs, evaluation of contribution by hours of work on a punched time clock is quite misleading. However, admittedly the creative individual who is also conscientious would not want to cease work when his capacity to earn reaches some satisfactory level.

A psychologist interested in the personal roots of philosophical, conceptual positions, again cannot but ask if Marx's constant preoccupation with exploitation and "injustice" had some connection with guilt over his own thirty years of "parasitism" upon Engels and several other acquaintances, in the absence of his earning an adequate living to support his own family.

[7] Parenthetically, it should be evident that the "left" political Utopia is not singled out

here for criticism of its values and methods of reaching values more than equally popular "rightest" philosophies. The example could just as easily have been a Fascist system, as in the writings of say, Croce (1945). Indeed the position firmly taken here and already pointed out above is that the constant, facile reference to political or cultural "right" and "left" in the world's journalism is a quite unnecessarily crude, oversimplified, and misleading handling of a more complex multi-dimensional reality. But in so far as the dogmatic systems of Communism and Fascism can be placed at the ends of any meaningful dimensional polarity, the reaction to them logically to be made from the new social scientific development in Beyondism is emphatically "a plague on both their houses!" In the odyssey of real scientific social enquiry they are the Scylla and Charybdis, past which one must sail undeviatingly to more remote goals.

[8] As mentioned, present day Marxians, impressed by our modern social science, may wish to claim that Marx at least heeded empirically the historical process. But the spirit of science, following wherever the evidence leads, is just not in his writings. The only science considered is a questionable version of economics, while biological, social, psychological and genetic sciences are completely neglected.

As to values, these are partly personal and unexplained (except for the above) and partly a hodge-podge of revealed religious values current in Victorian England and nineteenth century European political strife. One might, of course, attempt deeper value inferences by attempting to find the outcome of politically dictated-vs-personal and independent saving and investment, or the family life effects of continuing-vs-not continuing the present laws on inheritance of property, and so on.

[9] Both share a spirit of superpersonal dedication to the task of maintaining truth. Indeed, it is no cultural accident that many leading scientists have been the sons of clergymen, and that the Puritan virtues are found more frequently in the family backgrounds of those who pursue science than those who pursue art, literature, business, journalism, etc. (Knapp, 1963; Roe, 1953; Weber, 1904, 1956). Biographies show that the scientist (like the early rationalist [see Becker, 1932]) has been almost as distressed as the theologian that the destruction by science of the dogmas of religion has been taken popularly as an opportunity for rejecting the irksome moral demands of religion. Many scientists have become uneasily aware that it is scientific developments that have in fact bulldozed the bulwarks of religion into a ruin. This the geologists did with *Genesis* and Darwin with the Garden of Eden. Freud joined in by arguing that the social and emotional importance of the Christian resurrection goes back to primitive totemism, and that the psychological function of religious ritual is to administer to the needs of a group obsessional neurosis. Seeing this they have, as scientific observers, justifiably pointed to the resulting real danger of a possible moral collapse.

The historians have also done their bit to demote religion, by shrinking the biblical world to a very local performance in the grand panorama of historical cultures. Therein we are shown King David, for example, as the unsuccessful chieftain of a minor hill tribe existing precariously on the outskirts of the imperial cultures of the time. And the historical uniqueness or even reality of Christ has been thrown in doubt by reports of *several* crucified redeemers, in that stormy period, borrowing what was already essentially a "Christian" doctrine already present in the Essenes, as recorded in the Dead Sea Scrolls.

[10] The actual danger can be variously estimated by those who watch the barometer of crime, drug addiction or youthful alienation. But there can be no doubt about the spiritual suffering of sensitive young minds from the farrago of conflicting values they are asked to

digest. For example, the writers for the man in the street characteristically speak of the Christian values of Western Culture, but many intellectuals already speak of this as the post-Christian era. Others attach themselves, as we shall see later, to a kind of diluted Christianity, vaguely designated as Humanism or simply Humanitarianism. Others believe that their main standards are best designated by "democratic values" that are in fact, neither a clear political conception on the one hand, nor a clear set of ethical principles on the other. Outside of Western culture still stranger and more unworkable amalgams exist. What are the moral values in the unhappy cauldron of Vietnam with its mixture of Buddhism, Catholicism, Communism, and older native religions? What are we to make of the recent survey of Poland which simultaneously reports (in different surveys) figures of 90% Catholic and 90% following that Christian heresy which is Communism!

The beginning of cross cultural psychological measurement (Cattell and Scheier, 1961) shows anxiety levels and neuroticism frequencies to be highest, and morale measures to be lowest in countries with the most incompatible mixtures of values. India, with more than five hundred languages and almost as many religions, has one of the highest mean scores on psychologists' measures of neuroticism, and, with Egypt, stands among the lower nations in scores on the morale factor. Yet these have had the advantages (to speak rationally rather than empirically) of the oldest and longest background of civilization. In clinical studies of individuals it is frequently seen that children, e.g., even sons of missionaries, who have been exposed to and dragged through many different cultural-ethical values end by being emotionally immune to all values and sometimes by becoming severely asocial or delinquent. Incompatibilities of value systems is undoubtedly one of the main sources of this cancerous disease of cultures, which even the ancient morale of the cultures which have had the longest time to get established, cannot, seemingly, overcome. Our hope lies in a rebirth of dependable values by the objective and universal principles of scientific research.

Meanwhile, the traveller's shrinking earth, and the spread of permissiveness in education, has increasingly thrust these inconsistencies powerfully into the bewildered consciousness of the average citizen and especially the young. Modern man is prone further to increase his bewilderment by belief that a solution lies in some nostrum which, by a sufficiently frantic shopping around among exotic or mystical doctrines, may miraculously appear. But in fact he is already suffering from spiritual indigestion, begotten of an over-rich cultural meal.

[11] We may find in the next twenty years that whereas the intellectuals' destruction of religious values, like the act of some dying Samson, carries down the pillars of Western culture with it, such societies as Japan, which depended far less on abstract values of the Christian-Judaic type, will continue with a temporarily thriving morale.

CHAPTER 3

The Basic Logic of Beyondism

3.1 THE BOND OF RELIGION WITH MORALITY, IN INSPIRED, METAPHYSICAL AND SCIENTIFIC PERSPECTIVES

The preliminary glance in the last two chapters at the present state of society suggests that at best our culture is in the doldrums, awaiting a fresh breeze, and, at worst, in a slowly sinking condition. The old values have failed; the new secular values typified in a Russell, Huxley or Sartre have led only to poor morale.

A basic scrutiny of the validity of avenues to truth has led us to the conclusion that much of the difficulty arises from dependence on methods less realistic and less subject to verification—as well as less organized constructively for research—than is scientific method. The attempt to maintain that the ethical insights of religion are still the best we can get, while admitting the obsoleteness of the religious view of the cosmos is increasingly breaking down with educated people. Through whatever avenue we get our understanding of the universe and its working, through that channel we must also get the understanding of ourselves. A coherent morality requires a coherent world view.

Now the new emotional values promised here cannot be conveyed directly to the individual as emotional values; they must grow from his appreciation of a new analysis of the world around us. What is about to be developed here can *lead* to a new basic life attitude and state of mind in the individual; but it is first to be derived from a cognitive view of the universe. The quick appeal of many religions is that they offer immediate solace to the individual, not bothering him with any complex origin, and

71

they then trust that many people with this attitude, when brought together to constitute a religious community, will "work out" new emotional patterns. This type[1] of purely need-satisfying religio-moral adjustment is open to all of Freud's criticisms of religion as being essentially only an alleviation of a neurosis.

It is true that some popular religious movements and philosophies today, including modern Existentialism, have also approached values purely in terms of satisfying subjective needs. It is also true that religions are properly concerned, on the one hand, with "a state of being," but they are also concerned with "works" and the advance of a society — the City of God. Theologians, as much as social psychologists, have recognized that these two are organically connected, but that in various climes and ages their functioning as an organic harmony has been rather fragile.

In maintaining here that a state of mind cannot be divorced from a state of society — a view which religious mystics, for example, contest — we encounter a principle deserving a little more explicit development. For it will be invoked here in several other contexts, and it has moreover the status of a primary postulate in social psychology. Just as the crystalline or organic structure of a physical object is found to be consistent with the individual properties of the molecules which go into it, so — as can be shown in actual group dynamics experiments — the "shape" of the whole group is a function of the form of attitudes and personalities in the constituent individuals. (This does not imply or require, however, the apparent *one-way* causal action that happens to exist in the present physical or physiological analogues.) Just as for each chromosomal structure given in the single cell there is an exactly corresponding total organism, so for each form of personal behavior and values — an average style of individual adjustment — there is a kind of society. On a miniature society of only two, but with delightful convincingness, Shakespeare shows us in *The Taming of the Shrew* the simple example that two people cannot hold together a marriage when each is in a constant blaze of angry self assertion. It is perhaps unnecessary with the sophisticated reader thus to emphasize that certain states of mind are prerequisites for a society to cohere *at all*. It seems desirable, however, to give in footnotes[2] a little more concrete social scientific support for the *principle of man-group equivalence*, stating that for each form of personal adjustment there is a corresponding form of society. The particular state of mind and moral values of the individual can thus on the one hand, derive from his view of the universe ("can," because much will derive reciprocally from his society itself) and, on the other, be consistent with a particular form of society—though, of course, as an average citizen, he may not have insight into all the connections.

Now the position we are about to develop, and which may be called Beyondism, for reasons that follow, begins with the acceptance, among other things, of the scientific view that mankind is in process of evolution in a physically and biologically evolving universe. It admits the possibility that the further evolution of his species may fail, but also that there is no inherent reason why the present stage may not be a mere first step in tremendous evolutionary advances yet to come. From adopting this latter hope, Beyondism reasons back by means of technical social-scientific arguments to defining what the moral laws for individuals and groups have to be to produce such evolution. And from this definition of individual behavior it comes ultimately to what the "state of mind" or personal values need to be.

Before proceeding to a more precise analysis of the logic and technical assumptions of this inference, it may help to see, as in any birth of ideas, what its ideational genealogy may be. In particular, what is the history and the past success of attempts to derive individual moral values from a pre-statement of a group goal? And what help to the ultimate clarity of this concept can we gain from analyzing the elements in the current Zeitgeist to which it is related or from which it is definitely different?

In the first place any origination from the typical rationalism of the French "Enlightenment" has been expressly repudiated. The demolition by the latter of religious "superstition" may have prepared the ground socially, and, of course, reason is an ingredient in the present development, but the other descendant movements from that particular origin are completely alien to Beyondism. Many writers, such as Voltaire, merely paid lip service to science and, in fact, rested their reforms on *a priori* human value postulates. Mathematicians in that movement, such as Condorcet, liked to think that science was with them, but mathematicians are apt to prefer their own frameworks to those of nature. Moreover, like Plato and Aristotle before them, the rationalists were impeded by lack of the sciences needed—the biological and social sciences—and an altogether excessive worship—common among philosophers, mathematicians and logicians—of *a priori* "illumination."

Since the position we are about to take will put much weight on the life and character of the group as such, and on the evolution of whole cultures, it is even more important than any repudiation of philosophical confusion with rationalism to make also a fundamental distinction from Hegel, Marx, and Herder as to the purpose of the group, and from Herbert

Spencer and Schopenhauer as to the character of evolution[3]. To the present writer—as to many modern social scientists—Hegel's writings about the group are too remote from quantitative observations and hence too mystical—and pretentious—to be acceptable. His philosophy explicitly claims metaphysical action—the effect of a cultural "spirit"—beyond the causal action known to science. In this respect it is akin to the Gestalt movement in psychology and certain Russian views on psychology and neurology (Luria, 1965) and in sociology (modern dialectical materialism). These, at best, describe the action of "wholes" in too mysterious a manner, and, at worst, succumb to teleological, extra-scientific explanations. Without entering on the more detailed empirical analysis of culture patterns to be presented later, let us agree that action of the total culture is real enough, but that it can be handled by factor analytic procedures and multivariate, emergent models in a scientifically effective causal analysis (Cattell, 1965).

On the other hand, there is one vital point made by Hegel to which many experimental social scientists still seem not to have responded— namely, that some kinds of scientific experiment can be repeated as often as one wishes, but history does her experiments only once, as regards any *identically repetitive* sense, and that these experiments are part of a unique, unfolding, irreversible stream of data. The difficulty of separating and checking general principles from data thus becomes greater in the social than in the physical sciences. And if we admit that some principles are never fully understood until they appear in their most complex contexts, and that history is a movement toward higher complexity, we may never hope to understand how to predict certain events from certain principles until the events have occurred! (Who would know the saltiness of salt in a universe in which sodium and chlorine had not yet reached combination?) Even apart from this principle, the predictor of social events faces the practical difficulty that the number of influences at work in any event is so great that the fastest computer might not be able to calculate the event as fast as it occurs. But, granted these limiting conditions, the behavior of social groups belongs to the same science as other sciences, and needs no Hegelian metaphysics or intuitions beyond science.

It is necessary to guard not only against the confounding of the Beyondist concern for the group with Hegelian super-individual values, but also of its concern for evolution with enthusiasm for the Nietzschean superman. The writings of Nietzsche are often such sheer rhetoric (and sometimes fine poetry) that the gross intellectual misunderstandings of evolution in his writings are overlooked. Actually if a poet of evolution is required, Tennyson is more lucid and more apt in feeling than Nietzsche. Nietzsche, feeling imprisoned by the wall of nineteenth century piety, perhaps aided, by his sledge-hammer blows, the necessary breaking of a crust of cultural custom. Thus the vitality and beauty of his poetry is undeniable, but, unfortunately, it embraces among other errors of scientific conception the notions (a) that the survival of the fitter is an affair of individuals, and (b) that competition and aggression are one. The praise of primitive aggressiveness does not belong to a realistic understanding of natural selection. If nature had shared this Nietzschean valuation two million years ago it would have produced a still larger gorilla and vetoed the small and shivering naked ape who became man.

However, we are not concerned with the details of these earlier, often misguided attempts to respond to evolution and to a more scientific world picture. A philosopher-historian concerned to find clearer beginings for our present development of ethics from science here will find them not in Hegel, Fichte or Nietzsche, but in Locke and Hume, and in Paine, Bentham, Mill, Comte, Haeckel and Spencer. What began with Locke and Hume has been called English empiricism. (As such it has been accused of distrust of pure reason when actually it respects reason to the point of wishing to protect it from merely *a priori* verbal logic.) It has stood positively by the tenets of scientific method. From this source, at any rate, springs the only attempt at moral systems—that implicit in Comte's positivism and that explicit in the utilitarianism of Bentham and Mill—which will help in giving the ancestral family atmosphere back of the present birth. They strike the essential note by bringing together, as here, on the one hand, a regard for the empirical laws of social behavior and on the other, an explicitly stated final social goal.

Accepting "the greatest happiness of the greatest number" as a reasonable goal for society, Bentham (1834) and Mill (1863) thus set out to ask the social scientists what moral rules and what social legislation would, according to the verdict of empirical research, lead to this goal. To avoid any false lead at the outset of this discussion of Beyondism, however, one hastens to emphasize also our *differences* from the above—while not overlooking what is at least a continuity. The *continuity* abides in the rejection, as a source of morals, equally of religious revelation (except as an historical make-shift), and of metaphysical, verbal-rationalistic systems. The *difference* arises in the present scientific ethics *searching in science itself for the goal*, instead of accepting the too subjective and spuriously simple formula of "the greatest happiness of the greatest number."

The utilitarian and positivist developments in ethical thought which gleamed for a moment in the nineteenth century, as a herald of that more complete integration with science that is now possible, unfortunately failed to survive philosophical criticism (as many essentially sound inventions have failed in their first form). A post mortem suggests that the utilitarian movement failed for four main reasons: (1) the unsatisfactoriness of the subjective, arbitrary and indefinite nature of the goal, (2) the unpopularity with people of religious values of the possibly *hedonistic* nature of this goal, (3) the neglect to include the unborn – the biological future – in defining as "democratically" satisfying the "happiness of the greatest number (present)", (4) perhaps, in suggesting (partly through its historical association with liberal thought) a greater degree of "laissez faire" in moral matters than experience of moral control indicates to be workable. In any case, after Bentham, Mill, and Spencer the next fifty years showed little follow up, except for Sidgwick (1893), and Stephen (1873) who clarified and criticized, but did not radically improve. Nevertheless, the general idea was handled, with typical caution by Darwin, with moderate enthusiasm by Thomas Huxley and Haeckel (1929), and (still as a general, not rigorously defined notion) as a reasonable, implicit assumption by countless progressive writers since, such as Shaw (1965), Wallas (1914), Wells (1903), Marx (1890), J. Huxley (1957), Russell (1955), and others. Having investigated some causes of failure in this first move toward scientific ethics, let us turn to a newer design for success.

3.2   IS EVOLUTION AS PRESENTLY KNOWN ACCEPTABLE AS THE FUNDAMENTAL THEME?

Morality has to do with goals, and if a desirable ultimate goal could be found for mankind it should be possible to hand to social science – the more mature science of tomorrow, if not the feeble infant of today – the task of finding what behaviors in individual men best bring us nearer to that goal.

The number of such possible goals – human happiness, progress, knowledge, self-realization, human perfection – is, of course, infinite, and many writers have chosen some of the more obvious or popular. It is evident that the goal must be defined in terms of mankind as a whole, for individuals might – and do – choose goals that conflict with those of others and therefore, lack universality, and individuals quickly pass while mankind alone can pursue remote goals.

Whereas the revealed religions, the rationalists, the Utopians, and even the utilitarians have chosen a goal by some arbitrary decision, the consistency of the scientific approach first thoroughly followed here requires that we *find* the goal, in nature, not merely manufacture it. It is in the discovery, by scientific investigation, of this goal that the appropriateness of the term Beyondism, for this system of ethics, will become apparent.

The major systematic discovery of science in the last century may be considered that of organic and inorganic evolution. Here seems to be the message and the meaning written still mysteriously but large across the face of our universe. That much remains to be interpreted is shown by the fact that conjectures about the real nature of evolution have reached a crescendo of debate. Is evolution always to "better" things? Is it inevitable or dependent on circumstances? Do we have evidence of one-way organic evolution anywhere than on earth? Can one make any objective distinction between the terms "change" and "evolution?" Does evolution have a variety of directions or just one? And so on.

Granted these uncertainties, yet it remains the one increasing purpose visible to us. And since man is one small instrument in the orchestration of this symphony his role may seem sufficiently defined. On the other hand, the possibility has to be considered that since he has self-awareness and self-will he is free to recognize and admit this grand purpose while declining to be part of it. Actually, as we scrutinize this latter possibility more closely below, it may turn out that the freedom of man to choose his "direction" is as limited as that of man trying to find his way out of the blackness of a cave. His self-will may have only the choice of committing suicide or living by the conditions of reality.

Leaving the first issue—the quality and nature of the evolutionary goal—to be illuminated as we proceed, let us ask first at this point only whether evolution is indeed the primary and supraordinate purpose increasingly visible to enquiry, and, secondly, whether it is an inescapable scientific conclusion that we must shape our human lives to it or whether, on the contrary, an additional act of faith is required. (Such questions as whether we are allying ourselves with a "benevolent" purpose come later.) Science is not unacquainted with having to make certain basic acts of faith in the otherwise sceptical and objective use of its procedures. First, there is Descartes' postulate of the existence of the investigator himself. "I am conscious; therefore I am," and, secondly, there is the act of faith of Hume (based as we have pointed out above, page 22, rationally only on probability) that we are not encaged in a Berkelian solipsism. It is an experience that the individual exists: it is an act of faith that a *common* world also exists. Some philosophers would argue that there are additional acts of faith, in accepting the usual epistemological foundations of knowledge. For example, is it faith or experience that there is *order* (in that natural law is never capriciously abrogated) so that "replicable experimental results" are possible?

In view of these one or two inescapable prior confessions of faith if we should decide that we have a *choice* of accepting or not accepting the evolutionary goal, one more act of faith in accepting it would not be outrageous to the scientific conscience. To accept what moral laws — and much else besides — derive from such a single act of faith would be much less arbitrary than making the numerous separate acts of faith required by the creeds of the great revealed religions. However, as I have suggested above, it may turn out that man has only the choice of living by the laws of the universe of which the evolutionary process is an inevitable part, or of refusing to live at all. Thus if it be called an act of faith to decide to live, then, accepting evolution requires an act of faith. But this scarcely has the usual meaning of "faith" when it becomes synonymous simply with existence versus non-existence.

Naturally one asks at this point "What might decide rational men *against* participating in the scheme of evolution — if necessary at the cost of disappearing as a species?" A first possibility is that men might react by something no different basically from the child's temper tantrum: "I cannot be immediately omnipotent so I will not play." A second and more thoughtful reason could rest on a decision that evolution is either indifferent to man, or evilly disposed to him. As man matures in thought he certainly becomes more aware of the possibility that evolution (or, indeed, the universe as a whole) is cold and indifferent, lacking in benevolence, and no more concerned with man than with an ant or grain of dust.

On the one hand, since love in some sense means life to us, we can surely take heart from the fact that the "machinery" of the universe *has* already produced man, and love, and understanding. There is surely at least the *probability* that it is engaged in expanding these qualities [4]. Incidentally, even if man desires to believe in some benevolent force with which he cooperates there is still no need to make God in his own image. Indeed, since it is the nature of man to cut down whatever attempts to usurp his power, the existence of a God so constituted would forever deny man growing into control of his universe.

But, if when all is said we remain with legitimate doubts that evolution is what we think it is, or doubt whether what we value as love *is* an increasing principle in the universe, *then the logical decision must still be to commit ourselves to evolution. For only by evolving in intelligence and knowledge can we reach the answer.* This we may call the "forced choice" argument, for it says that if we refuse the "act of faith" and prefer suspended judgment, then we would still make the same immediate choice (if we are to live and to know), as the person who accepts on faith. For him whose temperament says "yes" immediately the poet has spoken well:

> "Yet I doubt not through the ages one increasing purpose runs
> And the thoughts of men are broadened with the progress of the suns."

For the more sceptical there are still intriguing thoughts, urging him by a devouring curiosity to move forward. He may believe, as a perceptive scientist has said, that "Nature is probably not only stranger than we imagine, but stranger than we are capable of imagining." Consequently when we say the forced choice principle leads likewise to an acceptance of evolution we mean especially biological evolution, toward a brain capacity and structure capable of understanding beyond present limits. For mere accumulation of knowledge by scientific method at our present stage of evolution as a species is not going to be enough [5].

Thus, in sum, the only conclusion left to intelligent man is that our species has to accept evolution and strive for evolutionary gains. Before proceeding further on our argument it is interesting to note how new or different the goal of evolutionary progress is compared with those that have historically dominated human values up to present times. In Graeco-Roman times, and in the Orient until this generation, it is no exaggeration to say that *the idea of progress and process did not exist* (Bury, 1920). In Christianity, with its orientation to a future world, the emotional basis for progress was prepared; "the present is not good enough and man is imperfect." However, as sociologists have noted (Weber, 1904), it was only in the Protestant ethic that Christianity came near to translating itself into a philosophy of human progress on earth. There is a sense in which this social ethic may be said to have degenerated—despite freeing itself from "superstition," and moving to an objective basis—in Utilitarianism. For the goal of the latter and its contemporary derivatives, as we have seen, slipped into becoming merely Hedonistic or Humanistic. (If happiness or pleasure were in fact accepted as the goal, it might indeed be achieved by drugs, or implanted electrodes.) Carritt's (1928) criticism of the "summum bonum [as the] ignis fatuus of moral philosophy" is here well taken. Neither as the total good nor as the total happiness can the utilitarian goal be accepted.

Equally fallacious in the light of the goal we now embrace are the values of Existentialism, represented by Sartre's (1948) "Man . . . must count on himself . . . with no other aim than the one he sets himself." What he sets himself today—at least in the circles of the comfortable, self-centered encapsulated intellectual existentialist of our time—is a trivial aim indeed. On the contrary, the scientist who follows Beyondism finds his goal in his active, group-coordinated investigation of the universe. That goal is no passive acceptance of a tide of evolution. Rather it commands the Beyondist to navigate on that tide—to take matters into his own hands and experiment and strive. There is no contradiction between the inevitability of evolution for living things as a whole, and man's deliberately attempting to explore his own evolutionary potentialities. Will is a part of the natural process. Even so, this is not as an individual simply adjusting to his corner in the universe, as in Existentialism, but the positive undertaking of what is necessarily a *group* adventure. More emotionally akin to the Beyondism position is that of the theologians in several ages who were willing to consider an understanding of the universe a proper part of "natural theology." Here theologian and scientist are one in "trying to think God's thoughts after him."

This first and vital step in the Beyondist argument—the acceptance of the evolutionary goal—cannot be left without recognizing a deeper paradox which has long puzzled philosophers. Since Epicurus, men have asked how it can be that if some Supreme Being exists and is benevolent, he cannot keep evil out of his universe. If he is omnipotent and permits evil then he is not entirely good; if he is entirely good then he is not omnipotent. Analogously in the realm of evolution one may ask, if the ultimate outcome is pre-determined by a Supreme Being and immanent in the structure of the universe, why do we have to re-discover it by the long joy and suffering of trial and error? Surely we have to admit that the limitations of our present understanding are such that we are probably asking a false question. The enigma simply underscores our need to evolve in power of comprehension.

3.3   THE CHECK OF GROUP UPON INDIVIDUAL NATURAL SELECTION: COOPERATIVE COMPETITION

Acceptance of the goal of evolution requires next that we (1) seek criteria of the goal of evolution. What is evolutionary progress? (2) Consider the machinery by which it may be ensured. For the moment the first question must be considered answered by a glance at the tree of evolutionary differentiation, though we shall come back and examine it intensively later. Here the issue which confused Nietzsche at one extreme of politics and Orwell at the other, must be cleared out of the way before we can proceed without misunderstandings.

This issue is really bound up with the second question rather than the first. It asks "Should the machinery to aid evolution be set up to produce advanced individuals or advanced groups?" And since evolution uses two hands, (a) variation and (b) natural selection for survival, are these to be applied to individuals or to groups?

Much nonsense gets into discussions on these issues, largely as a legacy from rationalism and its revolt against group values. At the risk of being sententious one must repeat that no man lives or dies without affecting others; that virtually the whole of his mental furniture and dress is borrowed plumage from society, and that his fullest individual self-realization is impossible without society. As Hobbes reminded us long ago the life of the single individual "in a state of nature" is "solitary, poor, brutish and short." The supreme examples of individual creativity and originality, in both the sciences and the arts, are made by men who stand on the shoulders of their predecessors. The original scientist—without whom society would be poorer in countless ways—is especially dependent on an orderly, well-organized and technically highly developed society—extending far beyond that merely of his fellow scientists—if he is to make his creative contribution.

The proper reaction to the old red-herring "Is the individual or society more important?" is to waste no time on it, recognizing the question as an adolescent misunderstanding. It is as meaningful as "Did the hen or the egg come first?" A successful society must produce the creative individuals on which its success depends, and creative individuals have need to foster a strong and supportive society. Obviously every society must control, by conscience or force, anti-social or anarchic behavior, and its skill in distinguishing between creative "revolutionary" originality or mere benign deviancy on the one hand, and malicious individualist anarchy and purely destructive "revolutionary" hubris on the other, is a measure of its likelihood of survival.

At some very early stage of evolution, where no social organization exists, natural selection will rightly operate on individuals. In the case of man, as he differentiates by race and culture, natural selection has, with equal appropriateness, acted in groups. If we consider, as we must, that natural selection acts both on the genes and on culture [6] (though with some new laws added in the latter case), then natural selection of the group has priority, for only the group carries culture as well as genes.

Indeed, as we shall see, even selection at the purely genetic level is inadequate if handled only in terms of the individual, since its aim must be to evolve a genetic *pattern and distribution* within a group as such. In so far as it selects individuals, therefore, it must do so for their capacity — while maintaining creative individuality — to maintain viable groups. It is probable that highly capable individuals could be evolved by competition and selection purely among individuals — capable of tigerish ferocity, or of the courage of a Nietzschean superman, or of the passive cunning of an Orwellian welfare state social parasite. Yet no society fit to sustain intellectual and scientific creativity could be put together from such types.

The central principle to be borne in mind here is *that though natural selection must necessarily continue to act directly among individuals, the selection of individuals is always going to be checked and validated by natural selection operating among groups as groups.*

It will be understood that in the above use of the concepts of variation, natural selection and relative survival rates we are supposing a relatively sophisticated conception by the reader of what these Darwinian and Mendelian processes mean. (Such as may be encountered in Fisher's *Genetical Theory of Natural Selection* (1930), Allee (1938), or the excellent recent works such as those of Fuller and Thompson (1960), Vandenberg (1965), Lerner (1968), Darlington (1969), King (1965), and others.) Thus we cannot make such popular solecisms as assuming Lamarckian inheritance, or overlooking the joint effect of both birth rates and death rates, or ignoring assortative mating, mutation pressures, the genetic load, etc. Furthermore, we are for the time being assuming that there is sufficient in common to biological and cultural (behavioral) variation and natural selection for us to be able economically to speak of both together.

As to the last, it *may* be that when we are able *deliberately* to bring about social mutations, e.g., new political forms, which we believe to be progressive, the percentage of false, maladaptive steps is lower than the 99% (or more) usual in biological mutations. However, our awareness of what we are doing does not *guarantee* progressive adaptation. And, although we are right to continue attempting deliberate progressive adaptations, the first illusion we have to get rid of in this field is that we *know* that sending more people to college, or reducing income tax, or giving more (or fewer) people a vote, and so on, is "progressive." In the end the progressiveness of these measures has to stand or fall by their survival value for the group, just as biological mutations do. Rational rather than scientific defensibility means little. What is less rational than shaking hands, or refusing to eat pork, or being unable to marry a woman of the same totem? Yet such socio-cultural habits are endorsed by group survival and wiped out (with the groups that espoused them) when they devitalize the group. Cultural habits will also experience *learning* (as distinct from selection) reinforcement, by reward, when they are successful and extinction when unsuccessful, so that natural selection in the broader sense can operate on cultures *per se*, *without* biological selection occurring.

Probably most educated readers have been so taught natural selection, that they almost automatically think in *individual* selection stereotypes. And in recognizing that that is not the process here being considered they should recognize that survival considered between *species*, as mere aggregates, as Darwin and Wallace initially considered it, is also something quite different from the inter-group selection we are now talking about. Only comparatively recently, in the writings of Ardrey (1966, 1970) at the level of primitive behavior, and Darlington (1966, 1969) at the level of advanced cultures have social scientists begun to take a good empirical look at what takes place in inter-group natural selection.

On looking with this more sophisticated eye one is amazed at the length of time this inter-group selection process has operated powerfully in human development (perhaps a million years). Consequently, one should be prepared to find a substantial magnitude in the effects that might be assigned to it, and aware of the potentiality for new and important scientific generalizations in this area. Countless important genetic features of man, especially in the realm of behavior genetics, are clearly the consequence of a million or more years of nature's weighing of family against family, tribe against tribe and (in the last ten thousand years) nation against nation—each examined for its goodness as a single, total, functioning organism. Over these aeons tribes have constantly been biologically wiped out as tribes, and strong selections set in motion favoring particular gene frequencies and proportions. In the same process tribes have been culturally obliterated in the sense that their culture has either fallen to pieces or been destroyed, with or without simultaneous biological destruction, by human or other environmental genocide.

Looking through the window into early history presented by Greek writings, for example, in Homer, and in Thucydides' retrospective glance, the struggle among small tribes and states is perceived as being ancient and incessant in the period before the flowering of Greek culture. There can surely be little doubt that that investment in what must have been one of the most intensive group natural selection periods on earth was substantially responsible for the genetic capital which, in the comparative security of Athens, was finally expended in cultural advance in the narrower sense.

It is important to recognize that evolution—or Mother Nature if we would sympathetically personify it—has faced an exasperating problem at the point where inter-group and inter-individual evolution began to go on together. For the processes of within-group individual selection and between-group selection of social organisms are not merely potentially independent, producing different results, but—especially in regard to such vital traits as superego strength and self-sacrificing tendencies—systematically undoing each other. Certainly one can see that it is easily possible for the within-group selection of individuals, i.e., the relative survival rates among individuals, to produce genetic types and tendencies to behavioral habits highly favorable to selfish individual survival but in the end incompatible with the survival of the group. Thus, as an extreme and perhaps therefore artificial example, one can conceive a benevolently intentioned political and economic system that would in fact favor a massive production of borderline mental defectives or a commercial system favoring the multiplication of such selfishly individualistic but intelligent types that society would break down from absence of people to do the unpaid jobs. (The old civilizations of the Eastern Mediterranean, and finally Rome itself, finished up with more subtle merchants than patriotic soldiers.)

In short, evolutionary selection operating on the characters of individuals and upon the characters of groups are distinct mechanisms, and we should guard against slipping into any simple assumption that they automatically work in the same direction, and especially that group competition may be abolished because selection within groups will take care of the matter. A whole new scientific development is needed to structure what goes on in this interplay. But probably there is a cyclic effect in which individual selection for a time multiplies genetic types and social habits having no regard for group survival, followed by a periodic check by group survival which rubs out the unviable group. Certainly the probability should be recognized of what might be called *systematic malignant inter-individual selection* (henceforth SMIS)—malignant from the standpoint of group survival—which has usefully suggestive analogies to malignant cancer cell growth in the biological organism. (The term dysgenic is not adequate for this, because dysgenic should be a general term applying both to this within-group *and* to between-group dysgenic, i.e., anti-eugenic, processes.)

When we say that a society fails to survive under natural selection, we mean primarily and immediately that it breaks down as a group and its political organizations and cultural values fail to survive. However, in the past the biological, racial group concerned probably usually perished too (or was reduced in fertility by enslavement, as Darlington (1969) points out occurred to the negroes enslaved by the Arabs), and as certainly occurred for tens of thousands of years prior to recent times by the systematic practice of taking the inhabitants of defeated or broken cultures into enslavement. However, there is substantial evidence that even incomplete and *relative* breakdowns of a group culture have powerful biological-genetic effects, much reducing the population previously supported. Thus cultural and biological group natural selection are significantly positively correlated [7].

In this connection any naive conception of natural selection among groups as operating mainly by "international warfare" should instantly be abandoned. As far as any research exists in this new field, the best conclusion is that, in the first place, natural selection among groups works as powerfully through what we may call *ecological natural selection (E-selection), i.e., success or failure vis-à-vis nature itself*, in obtaining sustenance and guarding against natural calamities, as through what we may call *inter-group pressure selection (I-selection)*. Secondly, only a minor part of the *I*-selection seems due to warfare, and most to economic, political, and cultural pressures of the kind analyzed in more detail in Chapter 5 below.

So it is in nature *generally*, for that matter. The infinitely formidable and armored monsters of the Silurian period were not "beaten in war" by the squirrel-like precursor of mammals and man; but were beaten in *E*-selection by species more efficient in the art of living. Most popular discussion of natural selection—especially when it concerns groups and species— gets hopelessly warped today by thoughts of war, because of the dramatic obviousness of war. Almost certainly the relative success of human groups rest more on *E*-selection—in making a living from nature; organizationally, in arranging good education, resistance to natural catastrophe and disease, and, above all, morally—in producing a happy society in which to live and reproduce.

Parenthetically, it should be noted that our concern at this point with selection is by no means ignoring the equally important differentials in mutation, hybridization and other sources of creative change, including culture borrowing and reinforcement. These may be studied in textbooks of genetics, or culturally in Toynbee (1947) and Darlington (1969). The latter ascribe much cultural mutation to the friction of different cultures in mutual impact. For the time being we consider these mutation producing influences "held constant" while we deal with the selection which "weights" the changes produced.

As to the relative speed and power of action of within-group and between-group natural selection we at present know practically nothing, compared with what laboratory genetics studies know about such selections in colonies of animals, fruit flies and bacteria. But one can speculate that comparatively poor evolution of group-selection-dependent qualities has occurred in the last two thousand years, and that there would be even less if the world wide "hedonic pact" discussed below supervened. There have undoubtedly been periods, as in China in the Sung Dynasty, or Europe under the later Pax Romana, where inter-group natural selection vanished leaving inter-individual selection (usually largely by birth rate differences) alone active over most of the "known" world.

If Beyondist ethics is right in aiming at the evolutionary goal of producing *groups effective in giving creative scope to high individual intelligences*, any period or condition which is capable of maintaining *only* inter-individual selection, without any check by inter-group selection, is likely to be disastrous Individual selection alone might then produce geniuses but there would be no society to pass on their discoveries: dead societies tell no tales. We should expect that at the end of periods of such purely within-group selection any society would fall to pieces. At the very least it would fail to maintain the conditions for vigorous collective cultural action, as happened indeed, at the end of the sprawling Roman and Chinese empires.

Our conclusion from this section is thus that instead of fearing or avoiding the impact of natural selection we should welcome and embrace it. And, especially, priority should be given to selection among groups. A recognition of this I have called in earlier technical writings (1933a, 1938, 1944, 1950a,b) and referring to nations and other groups, a belief in *cooperative competition*[8]. That is to say, like players in some greater, more vital game than men usually play, cultural groups recognize that the maintenance of inter-group competition is indispensable to evolution, and they agree to cooperate in whatever rules are necessary to maintain it in effective action.

3.4   DEFINING EVOLUTIONARY ADVANCE

As we approach the position just foreshadowed—that the ultimate touchstone of moral behavior is how far it contributes to human evolution—the first question, temporarily put aside, becomes more pressing. It asks "What is evolutionary advance?" In the narrower, technical, immediately-foreseeable sense of the level of advance of a contemporary society we shall come to this in Chapters 4, 6, 7 and 8. But first we do well to ask what it means against the total back-drop of what is understood today about organic evolution.

There is still much about the dynamics of evolution that the biologist does not understand. Why do some life forms exist with trivial change for millions of years, while others are constantly altering? Why do the ecological challenges of one particular environmental change produce spurts in evolutionary adaptation while other climatic or radiational changes simply wipe out dozens of species? Above all, what do we mean by evolutionary *advance*? That it is virtually impossible to handle "advance" by any single "straight line" continuous direction has been generally recognized since Morgan (1923) pointed to the phenomenon of completely new *emergents*—qualitatively different configurations—in evolution.

In some general sense—and if all forms of life are considered as a single growing "tree"—biologists have no doubt whatever that they witness a progressive evolution. But in any single species they will admit that evolution may stand still or even in some sense "regress." Pushed to define what is "higher" some biologists will defend only the proposition that the "more evolved" forms show "more elaborate differentiation of organs." Huxley, for example, talks of "novelty, greater variety, complexity and higher levels of activity." Except for the last, these are not question begging and might be placed on a fairly objective and even quantitative footing. But are they only *later*, or also *better*? Is the digestive system of a cow more "developed" than that of a human because it has extra parts, and, in mechanical evolution is a propeller plane more advanced than a jet, for the same reason? What is hidden in this concept of organ differentiation is surely the more fundamental consideration that the more elaborate apparatus generally produces *adaptation to a wider range of circumstances* and thus better chances of survival.

A sure way of avoiding value judgments—but perhaps some valuable quantitative scaling too—is to say that whatever fails to survive is less advanced than whatever survives. Besides offering nothing better than a dichotomy this has several weaknesses. From the standpoint of practical use in deciding between the levels of evolution of two living organisms it has the defect that it cannot be used until one is extinct! Secondly, what survives and thrives in one situation may not do well in another. Some very "lowly" organisms that have found a highly stable and unchallenging ecological niche have survived, unchanged, where more developed organisms in more exacting environments have perished. And mere length of survival will not do. We have a conviction that the axolotl, which has learnt to live in caves and has lost its sight, has in some way regressed relative to earlier forms. And because I abandon my snow boots in summer and take to sandals, are sandals a more complex structure?

Inevitably one has to come to a formula which includes an evolution of the environmental demands as well as of the performance of the creature. The child who gets top marks in the third grade is not as "evolved" as he who gets middle grades in the twelfth grade. To reduce the length of exploratory discussion let us tentatively conclude that evolutionary level must be assessed through several expressions, as follows, among which, however, the most fundamental is the breadth of the environmental demand that is met.

(1) Among species in the same environment, those which fail to survive may be judged inadequate in some respect compared with those that do.

(2) Although organ complexity and differentiation is not primary, it is sufficiently correlated with superior function to be given some weight. For if other organisms could achieve the same adjustment capacity with simpler structure they would have replaced the more complex.

(3) Granted equivalent terrain, food supply, parasites and predators, the more abundant is the more successful.

(4) Most fundamental in the definition is the *complexity and breadth* of the environment to which the organism can adapt. Since environments are liable to change, and more species than are now living have disappeared through failure to adjust to changes in environment, ability to survive when transplanted is a measure of level of evolution.

(5) The more advanced is that species which is more aware of (understanding of) and able to control its environment. Purely cognitive "adaptability to new situations" is one of the psychologists best definitions of intelligence and refers to a capacity which is biologically one of the last to appear. Since such understanding and control is effective in ensuring survival across gross changes in environment (as in men on the moon), it comes close to being synonymous with (4), but can be distinguished as a measure of how *much change the organism produces in its environment.*

Let it be repeated that although the above reads initially in biological terms, yet by the time we come to man, with his complex communal living groups, it applies also to the cultures he has developed in those groups. For example, we may say that the more advanced culture is that which can survive, without breaking down, over comparatively severe changes in, and demands for new adaptation from its environment. Implicit in (4) and (5) is that where human societies are concerned part of their capacity to maintain insightful control of environment resides in their capacity to support individuals of higher levels of intelligence and education. Thus among men emphasis shifts to "degree of understanding and control of society's environment and itself" as the definition of the measure of advancement of a society.

The immediate reaction of the man in the street, and often of the artist and poet, to a definition of high evolution which amounts, operationally, to "being possessed of high intelligence" is often a scorn of the "egg-head" and an assertion that high human development is something more than cold intelligence. On more than aesthetic grounds, however, one may doubt that evolution is simply toward a cortex as large as a computer. The fact that survival is by *societies*, with all the demands for human interaction which this makes, is sufficient guarantee of a commensurate development of intelligent emotional life with sheer intelligence. Evolution so far, at any rate, assures us that growth of the cortex, as between man and animal, is necessarily accompanied also by growth in sensitivity of emotional response. Normal phylogenetic, like normal ontogenetic, development ensures a growth in richness of emotional life along with richness of understanding.

The possibility presents itself of a sixth component in the definition of evolutionary advance: that it is movement in the opposite direction from that of our past. Both the biologist (phylogenetically) and the psychologist (ontogenetically) are accustomed to defining *regression*, as return to an earlier stage, and to think of regression and progression as opposites. However, there is a trap here in that progress is by no means always in a straight line, and what looks like a backward step, moreover, may be only out of a cul-de-sac, or part of a plan of "reculer pour mieux sauter." This is a frustrating conclusion, because nothing provides quite so definite a reference point for forward movement as where we *have* been. Especially in human affairs even the least creative of leaders have been willing to accept this principle. They may be quite unwilling for example, to apply positive eugenics, but quite prepared to apply negative eugenics (reduction of the genetically diseased, the mental defective, the throwback, etc.) and in politics they are instantly prepared to abandon as "reactionary" any return to states of society such as slavery, tyranny, absence of education. Yet the argument that what is past is inferior to the present is a probability argument only. Anyone who, like a school child, believes that no previous generation could be as wise and well off as his own, is out of touch with reality. Many an educated man in an age of vulgarity has experienced what George Gissing expressed for all classi- cists in *By the Ionian Sea* (1956); a dream of living in Athens in the age of Pericles. While for others, it might be an advance to join Shakespeare's circle on the Banke Side.

A glance backwards to steady the course by the ship's wake may sometimes help the helmsman, but it may also put him on the rocks. Indeed, the problem of defining evolutionary progress into which we have so suddenly been pitchforked in the main course of our general argument is an extremely tough one. The conception of a consensus of some half dozen criteria which we have reached at this point is only a partial solution — though honestly founded in operations and as free as possible of subjectivity. Though the definition must be left here, as part of a first attack, we must return to give it more intensive consideration as we study human cultural and genetic advance more closely, for this is the king-pin of the argument on which, in practice, the whole derivation of moral law must depend. Yet, perhaps in the end, as in many scientific fields, decis- ions as to "progress" will have to be made in the light of higher and lower probabilities rather than certainties.

There is a sense in which certainty can be approached by using *histori- cal retrospect*. If a form of society has repeatedly failed the test of ultimate survival it is not a "higher form." But this helps us in current decisions only if (as discussed below) we can get reliable "medical" signs of morbidity. And the question then arises "For how long should one wait after a cultural innovation before one gives the verdict of success (no collapse) or failure (signs of morbidity)?" If no collapse occurs after innovation A, how long does one wait before starting another experiment B? In current real social experiments this period is not explicitly discus- sed or deliberately decided upon; it is fixed by impatience and historical accidents. But even if a reasoned choice were made it would unavoidably have an arbitrariness such as occurs when an experienced fireman decides how soon it is safe to re-enter a house after a fire. Granted acceptance of some suitable interval, social experiment has to become a succession of innovations made until a community collapses. At this point, if there is good recording and communication, the nearest relatives of that culture may wisely backtrack from the latest innovations, deemed to have been responsible for the failure, but retain the earlier mutations. This Western cultures did, for example, when, after climbing out of the Dark Ages on the back of the feudal system it nevertheless rejected government by hereditary aristocracy.

3.5   THE PLANNED BIO-CULTURAL DIVERSITY OF GROUPS IN THE GREAT EXPERIMENT

It is the naive belief of most "reformers" and "reactionaries" alike that they know *a priori* what is best, and also that they like the desired ideal state of affairs for the reasons they announce. With the latter—the reliability of insight—we are not for the moment concerned, though a common historical comedy is that in which the reformer gets sick to death of living in the society he has created. (This touches the issue we scrutinize later of how far reformist zeal is really directed towards an improved society and how far it is a psychological rationalization of poor ability to tolerate any reality demands whatever.) For knowing in some objective sense what is best, one must recognize that even when the argument for a particular social change does not rest on *a priori*, philosophical arguments, but is based on a preliminary study of social scientific laws and analyses, absolute certainty is still not possible. Wind tunnel experiments and calculations with models are made before engineers build a large and expensive plane, but the degree of success achieved is not known with confidence until the plane itself is flown.

So here, though we may define a more advanced society by certain criteria just discussed—central in which is superior potential adaptation to environmental challenges and changes—the innovation which Beyondism now requires is the inauguration of *well-recorded, comparative experiment, with whole communities.* Parenthetically, let us be clear that "advance," "reform" and "moral values" must necessarily be regarded as different facets of one and the same concept. When we say that moral values are to be determined as those which conduce to the evolution of society, as defined above, we are automatically asking that the direction of true reform be simultaneously defined. By the Beyondist approach progress and morality derive from a single definition.

The uncertainties, and the evaluations of probabilities about which we are now speaking, belong to a concept of organized social research which now appears as the central practical proposal of the Beyondist creed. What we may call the *primary research design,* to which we shall return from time to time, is aimed to get far more knowledge and direction from the principle of *cooperative competition* among groups than its haphazard present action achieves. If some hypothetical Great Experimenter could in some mysterious way implant in the hearts and minds of men a variety of diverse cultural aspirations to produce an ideal experiment, what would he set up? A social scientist will at once think of the basic plan in what researchers call a "factorial design." Therein each of several *cultural* variations would be combined in every one of the possible ways with each in turn of several *racial* variations. Further, *several* sub-groups or countries would be employed as instances to give due representation to each of these combinations. Taking measures on all of them, upon those indices of evolutionary progress defined above, the experimenter would then seek to decide which culture patterns and racial patterns *per se* are significantly more successful than others. Almost certainly he would also get the statistician's "interaction effects" since there is good reason to believe that some cultural aspirations and norms would more fruitfully fit upon some racio-genetic population distributions than others. Indeed, science needs to investigate as soon as possible the limits which exist for grafting certain cultures on certain genetic groups, or for developing certain genetic gifts within cultures of a particular type. For use of the factorial design above depends on the experimenter's freedom to put into practice most possible combinations.

Elements of this basic research proposition are explored further in later chapters, but here we must keep to the central fact that the variation and natural selection which now take place in our world haphazardly could, if coordinated world experiment became possible, be deliberately broadened in scope and more accurately recorded. This would permit both an increase in precision of the answers about the desirability of certain trends, and, for a given advance, an immense saving of human effort and suffering. The essential proposal is thus that experimental variation of culturo-genetic types of community needs to be encouraged, and coordinated, both toward greater comprehensiveness of alternatives and more systematic, objective comparison of results.

Many questions will at once occur to the thoughtful reader as he looks at this conception of an ideally planned group cooperative competition. How could the spontaneity and self-determination of individual culturo-genetic experiments be preserved while yet maintaining an overall world plan? How stressful should competition be allowed to become: Is there, if frustration produces pugnacity, a danger of suffering war? And finally one may ask "What actually *are* the individual societies—as socio-political organizations—that are considered to operate as the unitary organisms in this competition?" Are they countries, religions, classes, or some other aggregates yet to be developed?

Especially one wonders, "What should be the rules of such competition?" In working to the common end of improving man, how far dare we incorporate designs that risk the very survival of the individual group?"

The choice of progressive goals and the moral rules among the *individuals* sharing the group life of a given group are, according to the above argument, to be fixed by the group itself, in accordance with its own best view of its survival, and they are unlikely to differ much from current inter-individual ethics. But what does this plan mean in terms of the moral rules *among groups*? And what would be done if the authority in a society becomes so rigid as to decline advice to modify itself, in the light of demonstrated unsoundness of the changes and experiments which it is making?

The Great Experiment is a new world of possibilities in which no existing moral authority has yet developed the ethical rules, and for which no world organization or traditional political science has yet worked out the appropriate international law. Before considering so formidable a problem, demanding that the mind be very thoroughly washed of current prejudices, it would be best to exercise our methods on the problem of deriving the within-group, inter-individual ethical values from an evolutionary goal; for here the inferences are less conflicting with current moralities, and the connections more easily seen. Accordingly, Chapter 4 will be devoted to exploring the basis of inter-individual, and Chapter 5 of inter-group ethical laws. However, by the argument above for the primacy of group survival, the inter-individual ethical standards in a general way depend on inter-group laws, so in this Chapter, in Sections 3.6 and 3.7 which follow, that *general* direction of dependence will first be clarified.

The moral values shared by citizens which make for the survival of a group are in part the same for all groups in as much as the internal machinery of all groups must have much in common. In addition each has its own unique values as to what human progress means in terms of its own uniquely attempted experiment. Because of concern for its own survival every well-developed group in a Beyondist future will resort to its own social scientific research center to determine the former values, and to fix the associated internal legislation as accurately as possible. But in regard to the latter—the entirely *new* and, hopefully, uniquely progressive steps peculiar to the one group—investigation is likely to be less certain. One wonders both what the machinery for it may be, and how far the sovereign right of the group to its own spontaneity of choice is likely to be influenced by some world conception of what diversities of racial-cultural experiment are most needed. This latter question can be left for Chapter 9, on a machinery of world coordination, where it is argued that because of *the natural tendency to seek individuality of expression* the problem of getting all diverse "cells" in a factorial experiment voluntarily occupied is not so intractable as it may seem.

As to the formation of within-group values by the group itself the Beyondist position implies that new institutions of applied social science must perform in novel ways yet to be studied. Those institutions will be engaged in what might be called social *cybernetics* – the steering of social change. But although we have rejected as far as *final* validity is concerned such sources of progressive planning as utopianism, rationalism, inspired religion, or intuition based on scientific information the last especially is still likely to play some part. For in this realm of social experiment we are by definition operating with guesses and the group that would progress maximally cannot afford to delay all action until everything is proven to the hilt. But even in making an educated guess the work of the institutes which record, collate and digest data on the cultural and genetic health of the group will be extremely valuable.

Thus, for example, though any final – still less any crash plan for a simple minded – Utopian goal is absurd; "action research" toward "trial one step utopias" (if one wishes to call them that, as in "five year plans") may need to continue to be described, especially to the young, as intermediate Utopian goals. The hunch of Plato that "harmony of parts" is the social ideal, of Confucius that it is "stability under the Emperor," of Marx that it is a "final revolution" bringing a society without government, or the vaguer conviction of a Utopia in Russell that man has to climb "a difficult and dangerous precipice, at the summit of which there is a plateau of delicious mountain meadows" (1968, Vol. 2, page 35) may suffice to motivate the less farsighted. But it would be out of keeping with the analysis and spirit of Beyondism to pretend they are final goals.

Similarly, the "tension reduction" values (in the modern jargon) are likely to play an inevitable part in choosing "progressive goals," be they in the form of "the greatest happiness of the greatest number" or the Nirvana of Buddhism. And rationalist derivatives from *a priori* "self-evident" principles will also never be lacking. But the generation of new steps in within-group values by Beyondism principles needs more austerity, imagination, organized research, discipline and audacity than these. It needs austerity in expecting no "delicious meadows" in immediate utopias, but a road that winds uphill all the way. It demands imagination to conceive ideas beyond the stale alternatives of current political and religious values. It needs intellectual discipline, because the cause and effect relations which must govern choice of moral ideals, are only going to be reached by patient, organized, empirical observation and intricate mathematical analysis. And it requires audacity, as the first airplane, or heart operation or rocket to the moon required audacity. For in cultural and genetic innovation the risks of a disastrous mutation always exist, and in any case the recommendations for social values may be sharply different from the comfortable values to which we are now accustomed.

3.6   THE MORAL IDEALS OF INTER-GROUP COMPETITION

Some dependence on less-than-exact methods in reaching the choice of within-group moral standards in any one group can be risked by mankind as a whole. For after the individual group's gamble on its own within-group values there always remains the self-corrective mechanism available in the realm of inter-group competition. The general problems in this inter-group realm — so different from those in inter-individual moral behavior — will be introduced in the present section but expanded upon more thoroughly in a whole chapter (Chapter 5). (Just so the last section introduced us to the basis of within-group morality, to be expanded upon in Chapter 4.) But whereas the within-group moral values from Beyondism (at least in the survival rather than the specific progress area) turn out (though reached by a different method) to be reassuringly similar to the brotherly love of traditional religions, the between-group directives could be very different.

For the obvious first tenet of inter-group behavior, from the premise that evolution rests mainly on inter-group natural selection, is that failing groups should either be allowed to go to the wall, or be radically re-constituted, possibly by outside intervention. By contrast, successful groups, by simple expansion or budding, should increase their power, influence and size of population.

This is the logic of the situation, but it leads to conclusions that run counter to the habits of thought of the majority of people today. The result will be that for them emotion will add its lurid touches, and convert what has just been said into an alleged advocacy of a nightmare of ambitious group self-seeking. Finally, it will be dramatized that all this must end in a nuclear holocaust. Actually this conclusion is logically, politically and emotionally false.

It is logically false because the greater part of relative survival, among human groups as among animal species, hinges on "competition against nature" rather than against other groups. As already pointed out, that species survives which makes the best use of its food supply, or is better protected against the climate, or which has inhibitions against eating its own young, and so on. The precise relative importance of what we have called above "E-selection" (against pressures of the physical environment) and "*I*-selection" (direct interaction with other groups) remains to be evaluated, for different human situations and historical periods, but there is little doubt that the human love of the dramatic has overstated the latter. Furthermore, nations themselves prefer to believe their decline is due to a lost war rather than the more banal effect of poor economic and moral habits. (Indeed, they may be indignant, as France and Britain were after World War II, on seeing the defeated more prosperous than themselves.)

It is politically false because, war being, as Clausewitz (1943) said, "only an extension of policy by other means," the fact remains that political competition has more primary means. War is often a sign of failure in normally expressed competition, and an avenue that a good competitor will avoid. A far-sighted nation realizes that war calls for a lop-sided development, lamed by the weight of armaments, which, though supplying the capacity for a sudden blow, takes away normal political and economic long term development.

It is emotionally false because the concept of cooperative competition implies a brotherhood in a common religion of progress, in which real competition and objective comparison are an indispensable reality, but no cause for rancor. Indeed, cooperative competition, as Chapters 5 and 9 bring out more explicitly, is emotionally a very complicated balance, involving mutual assistance and shared hopes and strivings, along with inexorable regard for realities. It calls for pressures toward re-direction not unlike those in a parent bringing up a child, or in true friendship. However, in the greater space of Chapter 5 this proposition, on the question of risk of war, and the proposition that freer competition might actually reduce war, are more fully considered.

War is actually not the only threat to a widespread sympathetic adoption of the ideals of cooperative competition, or the chief obstacle to their working out most auspiciously in practice. A second danger is the tendency of competitors to fall into the "horse race" stereotype, i.e., to assume, under the hypnotic influence of concentrated effort, that there is only one direction that competition can take. In life, as distinct from the racetrack and the school examination room, there are many quite different directions in which one may succeed in surpassing in competition. But the hypnotic effect of the "tape" in the ordinary race, and a lack of courage in all people to diverge from the goal of the crowd *does* tend to cause conscious competition to increase rigidity. It takes some emotional maturity to buy railroads on the stock exchange when all one's competitors are caught up in "the rage" of buying industrials.

Actually, the forms of competition which decide relative group survival are diverse indeed. In order to get an informed overview we must, in Chapter 5, consider the mode of action—and the efficiency as an instrument of evolution—of each of such diverse inter-group activities as economic competition, political pressure, propagandist influence, migration and racial interpenetration, relative colonial expansion, cultural and moral advance, and much else.

In attempting to work out what the most effective rules will be in inter-group competition in these areas, to ensure productive evolution, one must constantly bear in mind that natural selection will act both on genetic make-up (race) and on culture (the totality of acquired habits and values). Social scientists are going to have to find out much more about both of these before inter-group moral rules can be reliably developed. Some complex mechanisms are certainly going to be found here, and about one such possible effect—indeed a high probability effect—we should take notice immediately. This effect we may call *culturally originated genetic selection*. It hypothesizes that the spread of a culture may virtually amount to the equivalent of a spread of a genetic group, even though there is no actual movement of people (and genes) from one culture to another. It supposes that a culture tends to favor a relatively greater survival of those whose genetic endowment fits the culture better. For example, a complex scientific culture, begotten of an intelligent people, will tend to favor, in another people which adopts it, a biological selection of the more intelligent. Hard-headed animal geneticists may think this fanciful—and admittedly its action must be slow and hard to prove—but human genetic evolution certainly has social features—and this is one of them—which the biological study of species as such never encounters. The concept will be developed further in a later chapter [9].

In order to keep perspective at this point let us remind ourselves that our main synoptic argument is that:

(a) A group which discerns the best internal moral values, and fosters them, has a greater likelihood of survival and success *as a group* in inter-group natural selection (which depends on both E and I competition);

(b) The laws of inter-group competition most suitable for ensuring this relative survival are quite different from those of inter-individual competition. They call for a detachment from attempts to shore up the falling walls of an essentially evil or ill-built culture that may be mistaken for ruthlessness or indifference. But it may be said straight away that this inter-group ethics does not condone resort to war and that it is anything but a license to rampant group individualism or ignoring the purpose of the community of groups.

(c) Though inter-group natural selection success will depend partly on the level of internal, inter-citizen morality in (a), it will depend additionally on (1) fortune in genetic and natural resource endowments, and on (2) the specific ethical values espoused by a group over and above these basic "group maintenance" moral values to which all groups must aspire. Mathematically we are saying that group survival is a function of three main classes of variables, issues of the soundness of internal morality with which we are here centrally concerned covering only one class [10].

No one steeped in revealed universalistic religions or brought up in the kind of education that underlies the studies often miscalled branches of "liberal thought" will expect freedom of natural selection among groups to be understood at first glance as a "moral ideal." Our Great Experimenter, with his overview of a factorial design of genetic and cultural variation, will not, however, let "competition" operate in crude and hectic, diseased forms, e.g., in parasitic decadence and treacherous military or economic onslaughts. Nevertheless, as the following section argues, conclusions are often popularly drawn on the other hand from universalistic religious teachings in the realm of inter-group morality that are quite dangerously wrong from a Beyondist standpoint.

3.7 MORAL LAWS WITHIN-GROUPS AND THE FALLACY OF UNIVERSALIZATION

Having looked at the logical inferences concerning *within-group* and *between-group* moral injunctions that follow from acceptance of the primary goal of assisting human evolution, it is appropriate next to glance at the problem of their reconciliation or integration. To the Beyondist there is no problem of reconciling them, for they have been reached with internal consistency from a single premise. They are nevertheless, *different*, not simply analogous or isomorphic, just as would be expected when the scientist applies the same laws of molecular behavior to, say, matter in gaseous, liquid and solid states, issuing in different descriptive laws for the three states.

But in the history of religious thought a misconception grew up in this area. It cannot be justified by logic though it can be explained by human nature. From all that anthropology can tell us, moral values of cooperativeness, restraint on mutual aggression, and the idealization of altruism grew up *within small competing groups* — of tribes and families. For a very long time those injunctions never went beyond the boundaries of the individual's own tribe. There was no "universalization" and, for that matter, little explicit verbalization of such loyalties that could have tempted anyone to generalize and codify. Obviously there is no ground at all for supposing that what we may call basic community morality sprang into being resplendent and complete, by the grand insight of a Moses, provided in every human tribe that lived in the last half-million years. These rules against murder, incest, theft, etc., must have grown gradually, by cultural natural selection, with some accompanying genetic development of sensitivities of conscience in the better surviving tribes, and have become coded far, far later in history.

Altruism and a regard for group values must unquestionably be considered an outcome of group natural selection — at least as regards anything beyond the care of an animal mother for her cubs. We shall not repeat in detail the evidences of such writers as Comte (1905), Haeckel (1929), Spencer (1892) and (more recently) Ardrey (1970) and Hardin (1964), presented in their different ways but pointing to one conclusion: that the goal of group survival would require and lead to the development of those same altruistic ethics as have been stated in the Decalogue and in such great universalistic revealed religions as Christianity, Judaism, Buddhism and so on. Groups survive to the extent that men love their neighbor as themselves, respect the rights of others, and are prepared to sacrifice their lives for the group.

It was a serious emotionally generated intellectual misunderstanding of Victorian times (which Nietzsche, despite his detestation of bourgeois values, ironically shared and propagated) that the Darwinian discovery of progress by natural selection meant a return to the idols of the jungle and "nature red in tooth and claw" (as, for example, Tennyson expressed it). Some misunderstandings of this kind persist to the present day. Dobzhansky (1962), for example, and many others with a professional training in biology that one would expect to aid their imaginations in overcoming popular prejudice, often come up with Judas denials of Spencer's "Social Darwinism" as when Dobzhansky says "A devastating critique of evolutionary ethics was given by T. H. Huxley in his famous Romanes lecture in 1893." I look in vain therein for any such devastating or even cogent argument as is stated, and find only the fervor of a man unable to separate his reason from his emotions, but this is at least better than some who seem unable to follow reason for fear of losing social approval.

Mostly the critics have in any case set up for burning in effigy a straw man which represents a complete travesty of the more subtle ideas of Beyondism, and even of the ideas of Herbert Spencer a hundred years ago. The basis for altruism is already genetically shaped in the social animals, as Lorenz (1966) abundantly demonstrates, but our argument here is for a more special development connected with the far greater advances in culture in the primates and especially man.

So long as priority in natural selection is given to selection by groups instead of individuals, as it has operated for at least half a million years with man, the characteristics which are heightened genetically and culturally are those of cooperation, unselfishness, and willingness to sacrifice for others. Parenthetically, the special further point must be considered later that progress in this direction has gone further culturally than it has genetically. The substantial lag that exists between the moral standards enshrined in culture and the genetic nature of man constitutes his special problem of "sin." But even the genetic part is far advanced in the primates and man compared to the repertoire of responses of the lonehunting predatory animals.

Without laboring this essential point let us in summary recognize that when differences among groups with respect to population intelligence level, size, resources, etc., are duly allowed for, the group with the greatest prospect of success is that whose members show the least selfishness. Without amplifying further, and admitting as above that other resources play a part, we must conclude that survival goes to these populations whose lives are given more freely to super-personal and community service. They are those whose emotional impulses do not run excessively to uncontrolled sexual-sensual and aggressive gratification, and who show neighborly altruism and cooperation in the best sense. It is no accident that morale and morality have the same root; for correlational research across national cultures shows that *individual* moral levels and *group* morale levels are closely bound (Cattell and Gorsuch, 1965). Of all factors contributing to group survival, preventing cultural breakdown, and avoiding dissolution into scattered primitive brutishness, that morale which goes with the virtues of unselfishness, considerateness, honesty, loyalty and love of one's neighbor is probably the most important.

The high coincidence of the ethical value systems of the various revealed religious sects and organizations with those which would be derived for within-group inter-individual morals purely for the purpose of ensuring group survival is striking. Many of the injunctions — such as the Jewish ban on pig meat, the totemic ban on inbreeding, the "Puritannical" ban (actually far older than the Puritans) on free sexual expression – would be unlikely to be reached by logical insight. Despite the complexity of many of these rules, half a million years of human group trial and error, of variation and selection by family, tribal and national groups, is surely a reasonable period in which to expect such complex creations to appear. Inspiration and "revelation" may have come either as a more conscious part of the trial and error or as a satisfying rationalization afterwards. But we do well to remember, as we dismiss the claim of revealed religion to a different kind of truth, that in fact, it has the pragmatic truth of a crude "social scientific research."

The problem that the social scientist faces today in attempting to develop Beyondist inter-group behavior moral values with maximum precision and objectivity is that history in the last two thousand years took the gift of pre-history from a million years and turned it, in the immaturity of conscious thought, into a perversion. The revealed religions were not content to stop at the point where they had appropriately developed within-group values. Instead they experienced a natural imperialistic urge to become universalistic. Ethics having evolved naturalistically and realistically, was next exploited rationally. Rationalism – short-sighted as usual – took the view that what worked within a group should work with all men. All men are brothers, said the great religions, and should drop their special value systems and group loyalties in favor of a universal citizenship. (Of course it was implied that citizenship was in the values of the particular religious cult concerned!) Thus ultimately it became rational for Christianity, and especially, Mohammedanism and Judaism, to put outsiders – obstinate infidels – to the sword in the name of universal brotherhood. Psychologically, we need nothing to explain this but the emotional imperialism which naturally grows in any intellectual system and in physical groups – that plus the gain in status satisfactions of priests and others which universalism gives. The familiar battles between church and state (from medieval Rome to Henry VIII and beyond) and church and class loyalties (in Marxism and Fascism) followed. Actually, the common sense and intuition of the intelligent man break through what he is taught in doctrinaire universalistic religions, and he patriotically reacts to the invading fellow Christian soldier of another country by denying the universalistic injunction "Thou shall not kill."

Anthropologists, historians and psychologists (McDougall, 1934, among the latter, has given us the most penetrating dynamic discussion) have tended to conceptualize the difference of value systems we are here discussing in religious group terms by distinguishing "nationalist" religions, such as Shintoism, Judaism of old, and countless patriotic organizational developments since the Renaissance, from "universalistic" religions such as Christianity, Jainism, Mohammedanism (and perhaps, one might add, modern Humanism). But there seems to be no thorough empirical historico-psychological study (Freud's *Totem and Taboo* being slender and incomplete) of what actually happened in the historical development of the latter from the former. For example, we know little about the conflicts and compromises that must have arisen at the dawn of history in terms of the natural attempt to extend within-group to between-group behavior. Nor has any detailed study been made of the various factors—psychological (as above), economic and historical—that would favor or inhibit the spread.

Regardless of mode of development, the universalistic religious expansion must unquestionably be regarded as a "heresy" from a Beyondist standpoint. It is a gross oversimplification. There *is* a brotherhood of man and a great common endeavor; but it is not one that can have any useful function if it denies the importance of cultural and racial differences. In fact, when given its proper expression it requires that all men cooperate to sustain and produce such differences, and give their lives to testing their validities.

There is in Beyondism a common glory of evolutionary endeavor, toward the wonder of a spiritual understanding beyond anything we now possess. Beyondism contrasts with the universalist religions, however, in *agreeing to diverge* in all values but this fundamental agreement to join in evolutionary movement. It expects men to adventure and explore in distinct genetic and cultural communities. Indeed, it asks them to stake their happiness and their lives on the divergencies, i.e., to stand or fall with the success or failure of their own guess at the future. And in periods of history where men are over-stressed and lose control—in fact in war—it still calls upon the moral man, in the words of Newbolt (1908):

> "To set the cause above renown
> To love the game beyond the prize.
> To honour, while you strike him down
> The foe that comes with fearless eyes."

For in this great game all are in the end struck down—men, cultures and races—but out of their endeavor comes the ever more comprehending future. The fact that all are superseded—that individual "failure" and supra-individual success are built into the system—calls for the emotional consolation which only a development beyond present religious values can give. As often as not it will be the fate of any man, culture or race to be anvil and not hammer in this creation of the future. These lost cultures and races are the wrecked vessels which succeeded in warning the exploring fleet of human cultures where the shoals were thickest. Perhaps, even the "barbarians" outside the wall, who, in ignorance or arrogance, do not share the universalist vision of Beyondism, also contribute.

Like any world view, a Beyondist moral system requires a world organization, and this development, appropriate to the more intricate relations among groups which replaces the false universalism of revealed religions, is discussed in Chapter 9. For clearly a Beyondist ethic has a positive constructive task to perform in the creation of culturo-genetic experiments and does not aim, as do universalistic religions, at one flat, grey, featureless homogeneity of domesticated humanity. For, to use the useful term of the physicist, such an end point in human homogeneity could be nothing but the running down of a spiritual energy in a final sump of entropy. Rejecting this Ghandian paradise of ultimate erosion to uniformity, Beyondism at the same time rejects at the opposite extreme of method one world obtained by conquest—the single political power aimed at by an Alexander, a Ghengis Khan, a Napoleon or a militant Communist group. The variety of independent, federated groups is the goal of Beyondism.

Central in world experiment must be a non-coercive scientific research organization. It will need the finest brains that science can produce, to plan penetrating comparative studies among the cooperating, competitive, divergent groups. It will be a deep well of exact information for all those groups who come to avail themselves of it in their own social planning. Nevertheless, let us not imagine that the reality of power will vanish. Planning is needed to avoid single, world dominating conquest, but force is a fact of nature, and it must be incorporated in what is perhaps inadequately described as a federated police force. In some moment of imbalance forceful domination will be attempted, by criminal or madman, and then, as Kipling described the inevitable (1940):

> "Once more the nations go
> To meet, and break, and bind
> A crazed and driven foe."

And so a center of research and knowledge can no more be left unprotected in a world of possible anarchy and violence than the human brain can be left without a skull. This acceptance of checks and balances on sovereignty does not mean acceptance of world monopoly, and subtle

competition" among groups. This is an agreement to go competitively in diverse directions for the sake of a shared purpose. What is required among individuals will follow from what is required among groups by that condition. However, although the virtues of particular within-group ethical systems are thus weighed for their survival value, the outcome of the natural selection process among groups is also determined by genetic and environmental resources unconnected with and extra to what is contributed by the present moral state of the group. It is an illusion apparently shared by dogmatic religion, Fascism, Liberalism, Humanism and most Utopian political systems that we can directly tell from our untutored desires what directions of change are progressive. Social scientific research can greatly raise the degree of certainty in choosing what is going to be progressive; but in the last resort the internal direction of social change remains an adventure and a gamble. Poorer methods than those of science, such as resting on Utopian guesses, may have to be tolerated for a while, though measurement, recording and rapid analysis will bring trial and error closer to scientific planning. But Beyondism differs fundamentally in not aiming at the static equilibrium of a Utopia or in putting faith in mere rationalism. It sees evolution as a continuous quest, and it recognizes that in the last resort what is progressive can only be defined either after the event, through the fact of group survival, or by less reliable indicators and predictors of evolutionary expansion or morbidity observable *before* the ultimate proof by viability.

(8) The problem of evaluating — without waiting to evaluate as *history* — the progressiveness and survival potential of an existing society is extremely difficult. A first expansion of the general principles stated earlier for evaluating evolutionary advance is briefly attempted here. At a level of probability one may generalize that features found more frequent in societies that are recorded to have failed are less valuable than those of societies still living (provided all societies compared have met essentially equal stresses). Similarly, elements that have lasted long in a society are more likely to be sound than those little tried by time. Additionally some indicators both of a general moribund condition in a society and of a positive effectiveness can be tentatively set up. Important among the latter is evidence of *capacity to adjust over a wider range of environmental challenges.* Thus in human societies a capacity to produce and utilize individuals of higher intelligence is one important objective measure of advance.

(9) The positive moral value indicated for cooperative competition should lead to a "grand experiment" of deliberately setting up bio-cultur-

ally diversified groups, along with a scientific monitoring system for evaluation of groups and exchange of information. This should set very few limits to the action of spontaneous variation, competition and natural selection among groups. The high percentage of failures inherently occurring in all mutations requires positive scientific measures to invent more viable variations and, at the same time to foster and supply the conditions necessary for effective evaluation for inter-group natural selection, in which one must realistically anticipate and accept a substantial number of "failures."

(10) The inter-individual moral rules which need to develop *within* groups are primarily (a) the "community laws" necessary for sustaining the life and survival of *any* group, and, (b) secondarily, some unique values in individual behavior specifically necessary for *each group's own experiment*. Historically it seems probable that the half-million years of inter-group natural selection has led to the present moral customs of revealed religions. They are followed by mankind often under the impression that they came by "inspired" and divine religious insight, without full perception of their rationale of action. Far from the action of natural selection leading to the aggression, cruelty and individual non-conformity, which the nineteenth century (*vide* Nietzsche) naively assumed, it is actually the source of altruism, self-sacrifice and the standard values of the Decalogue (or its equivalent in other cultures).

(11) The goal of evolutionary advance actually leads to three distinct areas of derived moral laws: (1) between groups, (2) between individual members of the same group, and (3) between individuals at large. The two first sets are relatively straightforward in derivation; the last has to encompass the complex attitudes and behavior of sharing a common purpose through a diversity of loyalties. It includes maintaining free speech, fair play, and mutual respect in completely antagonistic argument and action.

In the last thousand years universalistic religions have made the false (but naturally ambitious) step of arguing that within-group moral rules should be simply carried over to between-group behavior, and the interactions of all men regardless of their other affiliations. Beyondism takes issue with this, inferring from its principles that active inter-group cooperative competition and differentiation need to be maintained, and that the required balance of shared and non-shared values is more intricate than the emotionalism of revealed religion or the "obviousness" of rationalism suppose.

(12) Just as in historical religio-moral developments, the spirit and ideas of an evolutionary morality will need ultimately to become embodied

in social organizations. Chief among these needed creations is an international research planning and evaluating organization, which, though essentially only advisory, will need the political power support of a federation.

In this chapter the reader has nowhere been offered the detailed injunctions of a finished moral system, but only a rough map and a compass. It requires a substantial development of national and international social science research centers to reach practical, concrete ethical guidance.

3.9 NOTES FOR CHAPTER 3

[1] It is worthy of note that oriental religions, as well as the Indian shamans in their trances, and the whirling dervishes of Iran, have put greater emphasis on the subjective emotion of state of mind, whereas the objective, extravert West, in Christianity and Communism particularly, has put greater emphasis on service to fellow man and the maintenance of a healthily functioning society. Our contention, however, is that these are only differences of emphasis arising from differences of temperament and social need rather than indications of any true separability of religion into "types."

However, the beatific state of mind can *exist* only as long as it functions (except where societies charitably support schizophrenic or beatnik aberrations). The religious state of mind in a well-functioning society is essentially *peace* of mind, from accepting and expressing in life the moral imperatives. Other states of mind than this may be appropriate in an ill-functioning society, where state of mind and condition of society are locked in a sad cycle of cause and effect. For example, a state of narcissistic withdrawal may develop as "religion" in a social chaos where no "good action" *can* lead to good results. Thus Gautama Buddha, in his desire to escape from the misery of a confused society, other than by committing suicide, arrived at the goal of "a state of mind," which he defined as one of elimination of all personal desire. This goal—in many ways so similar to that of the schizophrenic—does not so much create a society as allow the religious individual to live in *any* kind of society and almost any kind of moral or political system. The spread of monastic withdrawal in the Dark Ages, though similar, offered the more constructive alternative of a morally functioning sub-society. Instead of disengaging a state of mind from community action, it disengaged a viable sub-society from the outer chaos. This *may* need to be done again at the present juncture of society.

[2] The person with free choice has a wide spectrum of religio-moral styles of living initially open to him—ascetic or sybaritic, criminal, drug addicted, schizophrenic, bourgeois, priestly, and so on—but only a small fraction are compatible with what, under some rubric or other, he views conceptually as a tolerable society (let alone the City of God). Most successful societies have had to stimulate fairly definite sets of values, such as, for example, have stabilized themselves in the Mediterranean Christian world, and its derivative colonies. These imply (1) a certain pattern of behavioral reaction, (2) a deep, permanent style of emotional adjustment, and (3) a state of mind.

A number of historical and philosophical debates have concentrated on the issue of which really comes first, causally, a state of mind in the individual or a certain structure and character in society, as discussed in note [1] above. In some cases it is obviously the former, as when Christianity began as a personal style of life, propounded by an individual, and undid the Roman Empire, shaping a society "nearer to the heart's desire." In others—and we will leave the economic determinist to cite his examples—the converse has happened. The material and political shaping of a society has apparently been responsible for the emotional adjustments that later became stereotyped in the molecules within it. Normally causality weaves both ways.

For our present pursuit, the causal direction is a less central issue, but recognition of the inner-outer *correlation* itself is vital. Later, in Chapter 6 it will become necessary to examine more sharply the claim which constantly recurs in history, that the state of mind in the individual is more important than "good works" and the structure of society. Something not far from this has been noted above, in Buddhism, and is evident in some degree in such diverse historical streams as the Christian mystics and the modern "hippy" taking his drugs. For the moment we shall assume that preoccupation with emotional state is an extreme, aberrant form and that in general the religious consciousness and the moral behavior of a group in a structured society stand balanced in importance, as face and obverse of the same coin.

[3] Modern followers of Marx will doubtless protest against classifying dialectical materialism with the tradition of Hegel, claiming that Marx rejected the metaphysical content of the latter. To a modern social psychologist he seems to be indulging in a "reaction formation," to distinguish himself relatively superficially from a tradition in which he had earlier been raised and from the main characters of which he never escaped. His argument—leading to "a final stage of history"—remains teleological, mystical, and bereft of that analysis into biological, genetic, economic and psychological causes which a strictly scientific analysis would require. As Darlington (1969, page 546) well says, Marxian inference "went beyond ordinary rules of scientific evidence. It had [supposedly] an ultimate inherent validity. It was dialectical!" and Hegel was convinced of "an inherent goodness of intellect," whereas Beyondism is content to see intellect as only one of several survival aids, to be evaluated by its survival contribution.

The distinction from Schopenhauer, and more clearly from Spencer—for Spencer was more clear—resides in the latter seeing life itself as creating evolution, whereas Beyondism sees life as geared only to continue, while the creativity comes from the impact between life and demands of the independent universe. If we wished to turn the microscope on evolutionary mechanisms *per se* we should have to give greater emphasis and consideration to the primary fact of "persistence" above (emphasized, for example, by Monod (1971) as "replicative invariance in DNA"). Living matter does not "aim" at evolution. Its property of primary persistence with secondary susceptibility to shifts is merely a precondition: evolution is forced by the impact of the environment on living matter.

Hegel and others of his persuasion are, of course, correct in singing hymns to the importance of wholes, and the Gestaltist psychologists likewise did well to shake the classical bivariate experimentalists out of their complacency, but one does not have to transcend science to do this. Wholes and patterns are being successfully handled by suitable mathematical models such as factor analysis, and their complex causal action is beginning to be understood. These distinctions need to be drawn clearly here because reviews by philosophers of the briefer, earlier presentations of Beyondism and "cooperative competition" (Cattell, 1933a, 1938, 1944, 1950b) which have heralded the present systematic presentation, have made the mistake of concluding that the present position leans on Hegel and Fichte. The wholes and patterns we recognize here are measurable and operate principally in the fact that evolution of the single organism is dictated by the pattern of community in which it lives, and by the teleonomic (purposive integration, not teleological or purposeful intent) pattern of coherence required in the genetic basis of the elements of its own behavior.

[4] Paley's (1802) famous "argument from design" included not only the notion that so marvellous a creation must have a creator, but also that we may consider the motives of the designer benevolent. However, one is not, logically, compelled to assume the creator to be kind. He might, if a peevish boy, create only to break or torment; but we have the rooted prejudice that a great artist rejoices in and loves his creation.

[5] The advance of the human species, in sheer brain capacity, during the next few thousand years may well make the difference between survival and catastrophe, since quite complex problems from the crowding of our own planet will challenge our intelligences. Such a demand for genetic selection in a comparatively short time can certainly be met, granted a readiness to re-evaluate our values in favor of evolutionary goals. Surely, if we do not perceive the whole plan now or know to what goals we are moving, the rational response is not indecision or fatalism. Rather it is the decision to progress — culturally and eugenically — toward levels of evolution where we may hope better to understand. For, let us make no mistake, it is biological evolution of mental capacity, and *not merely an accumulation of scientific research data* that is needed. The time must come — strange as the idea may seem, even to scientists — when, unless man himself evolves greater mental capacity as such, a "principle of diminishing returns" will show itself in relation to research effort. All that is discoverable and comprehensible at our level of intelligence may soon have been discovered and comprehended.

In education, psychologists are compelled to recognize upward limits to teachability. Nothing in the way of repetition, visual aids, verbal analogies, or anything else will enable a man of I.Q. 80 to understand Einstein's special theory of relativity — by any realistic, operational test of understanding. Indeed, we need not go so far; but can conclude that many millions living today can, by no lengthening of education, understand the principle of logarithms, the logic of Euclid's fifth proposition, the use of the subjunctive, or make appropriate distinctions in communication between such words as to repel and to repulse, extenuate and exonerate, humiliation and mortification, etc. (These are examples wherein persons of high and equal exposure to verbal education show systematic differences in percent of correct responses according to intelligence.)

In just the same way, outside culture, a point is reached where the messages written on the cloudy face of nature will span too complex a set of relations for us to read them. Our scientific instruments may record them sensitively and our scientists may debate the riddles offered, but perhaps in vain. There will always be geniuses, one hopes; but just as a genius among *homo neanderthalus*, or among the primates, might not, in our culture, get beyond recognizing one syllable words, so the genius of the next century may find himself unable to read any further in the open book of science. One hopes that scientific research will go on — indeed, on a far greater and more organized scale than we now manage. But it should be clear that when we speak of "progress" to that greater understanding which will tell us more about where our world is going, this ideal of progress is going to require both genetic and cultural programs. New mutations are needed beyond the genetic ranges which give us our present leading brains. Community commitment to biological evolution is thus the most basic — and more novel — step we have to take in answering the question of the nature of the evolutionary goal.

Thus, whether it be by the almost aesthetic argument that we must keep in harmony with that evolution which we find as a central principle in the universe, or via the more urgent and realistic argument of human survival, or through the simple logic that if one does not know whither the road leads, one should gamble on following it a little further to know better, one finishes by embracing the ideal of evolutionary progress as the most fundamental goal.

[6] The laws and mechanisms of natural selection are at present worked out far more adequately for biological than for cultural evolution, and cultural evolution as we have just seen, has extra mechanisms, e.g., culture borrowing, reinforcement and adjustive assimilation, not operative in biological evolution.

Enough is formally in common in both, however, for natural selection to be used as a concept for defining the main determiner of evolution in both. Cultural natural selection becomes an important topic for social scientific research. With special reference to history, such as Darlington (1969), Deutsch (1965), Merritt (1970), and Rummel (1966) are now giving it, the concept becomes vitally important to Beyondist understanding.

[7] When the Plains Indian said, as Margaret Mead records "the cup of our culture is broken" one recognizes that the biological life which it contains tends to run away too. Sometimes, as in this instance, the culture is broken by impact against a stronger more developed culture; but archaeology is beginning to suggest that intrinsic cultural decay, without outside pressures, can occur far more frequently than we have imagined.

If one is prepared to take the Old Testament as approximate history, it offers numerous instances, from Noah's contemporaries to Sodom and Gomorrha, of *biological* group natural selection being associated with virtues and defects of internal group *cultural* values.

[8] "Cooperative competition," as I have found from its earlier statement (1933), creates immediate emotional difficulties for those (a) who mistakenly assume that it connotes war, (b) who think of it as a team game in which their own individuality would be lost, and (c) who turn their backs in neurotic withdrawal, on all serious group demands.

Although it does not connote war, and is indeed a means of avoiding war by more freely expressing emulation in other directions, if war accidentally occurs we need not fear that it will "weigh" nations by performances totally different from those involved in other forms of competition. Indeed, it would be a serious mistake to suppose that, in the main, the more adaptive, progressive communities, are the countries that fail by the test of survival in war. When we document the sociology of war in more detail in Chapter 5, it is clear that both involvement in war and success in war are associated with, for example, higher levels of general education and with greater capacity to handle the problems of peace. There is a very high correlation of success in the peaceful arts of science, social organization, education and literary creativity with formidableness in war. It was the greater productive capacity and material well being of the democracies that mainly decided victory in the Second World War.

As to the narcism which sees emphasis on group performance as hostile to its individualism, one can only point out that several misconceptions are involved in it. First, few narcists realize how much of what they call themselves is actually group property. A good antidote to such egocentrism resides in the careful factual and technical writings of a sociologist like George Mead (1934) showing how extensively the mind and individuality of the individual is a loan from the group, commonly unconsciously plagiarizing individuals of the past. Often when borrowed finery is set aside there is little left but a "visceral individuality." It is an individuality which simply emphasizes wants, by subjective visceral choices, in an already almost totally supplied kingdom of cognitive ideas.

As to the third, much could be said on a form of withdrawal from group interests which is not introversion but neuroticism. A highly developed group form of this reaction is seen in the disillusioned Existentialism of Sartre — disillusioned because it starts with fundamentally wrong premises. These premises today are not precisely the premises of Kierkegaard, but Kierkegaard mixed with a decadent phase of French culture prevalent before World War II. In this and similar projections of neurosis into social philosophy nothing is left but concern for the loneliness and uniqueness and all-importance of individual experience. This system has been unable to spare any attention for either the group economic realities of Marxianism or the biological realities consulted here in understanding the adventure of groups. Consequently, in such philosophies the importance of *group individuality* and differential survival is either unrecognized or resented as an intrusion on the individual's essentially morbid degree of Existentialist introspection and narcistic self-concern.

[9] However, it can be briefly made clear here that the evolutionary actions in human groups is substantially different from that which the biologist is accustomed to think of in biological evolution. In the first place the role of natural selection among *organized groups* is very different from that among *species*, as commonly discussed in Darwinian selection processes in animals and plants. Species rarely show any species-wide organization, e.g., bird migration is in *groups* of a species but not including the whole species. The individuals in the species mostly survive *as* individuals. Indeed, as Lorenz documents (1966) most animals are more hostile to members of their own species than to others. Even when, as seen more in birds and lower mammals, groups of individuals combine, it is only in a quite rudimentary way compared with the primates, and, especially man.

In the highly organized cultural human groups considered here, the whole survival differential of group and group is tied up with the goodness of the internal organization *per se*. Because of the relatively high technical development of the modern study of ecology and the complexity of the dynamics of survival of species, there is a real risk of unconsciously leaning on the animal concepts there used and importing them, without the necessary changes, into the present discussion. Two distinctions are actually involved: (1) that in man the cultural differences in any case play a far larger role than the simple biological differences in determining survival, and (2) that the survival on biological traits alone is tied up more with biological traits that have social importance.

[10] However, these three classes are likely to be found by statistics to be interactive. The connection of the moral with the genetic is most likely to be overlooked by contemporary one-eyed sociology. But clearly the evolution of a powerful system of within-group cultural moral habits must proceed in association with genetic evolution of tendencies and sensitivities favorable to those values. This is usually completely overlooked in the average historian's account of cultural development (except in Darlington, 1969, McDougall, 1924, and a few others). Yet it is extremely important, as our discussion in Chapter 6 on the culturo-genetic lag principle shows. In clinical terms the implication is that what has been called the superego must have genetically as well as environmentally determined differences of level and force among groups.

The idea will surprise no psychologist, for every measured behavior yet investigated by nature-nurture variance ratios has shown some appreciable genetic component in individual differences—and therefore in differences of selected groups. (Thurstone's twin studies showed even a genetic component in so obvious a culture acquisition as ability to spell!) In any case there is direct psychometric evidence (Cattell, Blewett and Beloff, 1955; Jinks and Fulker, 1970; Cattell, Stice and Kristy, 1957) that superego strength—roughly "altruism" and "sensitiveness to obligation or guilt"—is partly genetically determined. This is what we should expect from its rise through natural selection. But the fact has consequences that have not been expected or discussed in ethical domains (*see* Chapter 6). For example, the re-establishment of a cultural ethical system, after some temporary breakdown of civilization should prove to be much much harder in some populations than others.

CHAPTER 4

The Moral Directives Derivable from the Beyondist Goal: 1. Among Individuals in a Community

4.1 PROBLEMS IN DERIVING OBJECTIVE NON-RELATIVISTIC ETHICS FROM STATING A FIXED GOAL IN A CHANGING WORLD

The basic principles for an objective ethics have been stated — scientific recognition of an evolutionary goal; a logic which compels its adoption; and social scientific research to determine the moral regulations within and between groups which best serve it. The necessary plan of this book is, therefore, now:

(1) To examine in richer detail the derivation of community values within groups, among citizens (this chapter).

(2) To examine similarly the second main derivation: that of the procedures and values appropriate for cooperative competitive behavior among groups (Chapter 5).

(3) To turn from ethics to psychology and ask how moral systems in general, and these new values in particular, conflict with or facilitate the expression of human nature (Chapter 6).

(4) Expressly to scrutinize the changes in values from those of the traditional revealed religions now directing much social action (since the impact of the new is likely to cause conflict in several areas) (Chapter 7).

(5) Correspondingly to examine the constructive new directions of action which a Beyondist ethics indicates in urgent current socio-political problems (Chapter 8).

(6) Finally to proceed to the new institutions and associated emotional satisfactions that concretize Beyondist values.

113

A revaluation which calls for a radical shift in the hub of a moral system is an immense undertaking. The fulcrum of the evolutionary principle and the lever of intensive scientific research (on a scale for which social science has yet to be organized) should nevertheless prove strong enough — given time. As far as the community ideals of behavior of man to man within a given society are concerned the principle has already been claimed that these are of two kinds: (1) the principles which all societies need to hold them together, for survival, which we may call *maintenance values*, and (2) the principles which express the *unique experimental values* of the particular adventure they have adopted. While a far-sighted society, well equipped with scientific research facilities, will definitely seek to check the survival value also of the latter, they are inevitably undertaken as a gamble, relative to the former. The former — maintenance morality — is a matter of world wide concern, common to every group, and a tremendous history of religious and political endeavor describes the attempts to reach the best behavioral rules.

It is probably safe to say that few societies have succeeded — in the blind trial and error of "inspired" religions — in reaching rules that are reasonably efficient for ensuring survival, and that still fewer have succeeded in keeping their populations observant of the ethical values that have been reached and clearly set out. The differences among societies in moral effectiveness have probably always been very great, and almost as great between different periods of the same society, e.g., between early and late Roman times, the Commonwealth and the Restoration in Britain, and so on. Unfortunately, until psychological measurements get set up in modern times whereby a graph of different aspects of morale and morality can be drawn over time, the guesses of "historical observation" are unlikely to lead to firm laws. It is no mere moralistic rationalization, however, that the periods of high morality generally come early in the history of civilizations, and that internally degenerate values supervene before societies collapse, though this does not seem a simple one way movement, or inevitable.

The problem of inferring, from the fixed goal of evolutionary survival, what the ideal within-group moral values should be has to take mainly three terms into the calculation. (1) A criterion term expressing the degree of success with which the society is surviving. (2) A term defining the level of moral behavior within the group at any given time, and, (3) a term expressing the natural resources of the group, in environmental wealth and genetic endowments of the people. A strong endowment in (3) has often hidden, for a time, the effects of a weak development in (2).

The very great difficulties in quantifying the first have already been discussed. In the first place it is two-facetted — success *vis-à-vis* the demands of the physical environment and success in enduring the predatory pressures of other groups. The complete breakdown of a society, in government (return to anarchy), failure to reproduce, return to a totally poorer economic level, inability to defend itself against even the weakest enemies, and so forth, is sufficiently unambiguous. But this evidence is too late for the society in question to readjust its moral values, though it may be good data for post mortem scientific analysis of moral causes and social consequences by the research institutions of the surviving societies. Even so, it is an insensitive, all-or-nothing type of evidence, and, as has already been suggested, moral research institutions need to develop more precise, refined, predictive measures on a continuum of health and moribundity as a means of expressing this first term; *the criterion term.*

Briefly to recapitulate the promising avenues for research on the criterion discussed on page 88 above, these are: (1) level of complexity of social organization maintained; (2) size of population per unit area maintained at a standard of living adequate for effective citizenship; (3) degree of control of the physical environment — against flood, famine, disease, earthquake, pollution effections, etc.; (4) creativity in demonstrating adaptability to even wider sections of the environment, e.g., colonizing climatically difficult regions, space travel, resilience to major natural catastrophes; (5) movement in a direction opposite to toleration of problems long suffered in the past, particularly the reduction of what we shall define as the welfare-economic-genetic burden.

Granted the possibility of evaluating the criterion — the vitality of progressive survival as a group — it then remains to find the equations relating this term to particulars of within-group, inter-individual moral values and practices. (The third term — the luck of endowment — being held constant, or — as we would say in correlational statistics — "partialled out.") A general character of these equations can be stated immediately — that they are likely to differ according to the level of development and state of evolution of a group. That is to say, the actual ethical ideal rules are unlikely to be immutable, but to change in the manner in which directions change as one aims at some ultimate goal from different positions reached. Again one must emphasize that the sanction for *positional relativism*, as we may call the above, is quite different from conceding *general moral relativism* in the sense used by some contemporary intellectuals, or in the sense in which anthropologists, say Mead (1955) or Benedict (1934), use it in describing diversity among tribal customs. In the latter the complete relativism comes from sheer ignorance of *any* rational target for behavior.

Concerning the actual scientific mechanics of deriving values from the prime evolutionary goal, our discussion (in so far as discussion is possible for a science scarcely developed) is postponed to Sections 4.3 and 4.4 below. Here there are prior and more general matters first to be clarified Thus whereas theoretically this approach is capable of shaping anything from the form of one of the ten commandments, down to a local bylaw or to legislation on some trivial form of social behavior, one may ask in the first place whether such detailed penetration of the culture by primary ethical values is ever desirable.

Socially, some reasonable compromise is indicated. The detailed injunctions of such religions as Confucianism, Mohammedanism and Taoism seem merely to have produced rigidity. The persistence of rigidity in detail defects the primary purpose of adapting and expressing the basic goal in terms of each successive current socio-historical position, which we have discussed above. There are, of course, practical reasons which sometimes make rigidity desirable. For example, human nature being what it is if people are given an inch they will take a mile. But scientifically, on the principle that an abstract concept is usually not fully defined until implications in many fields are established, it can be argued that such detailed working out of Beyondist principles is at least required to improve *definition*. Application may, in fact, raise new questions and add richer basic meaning. Regardless of whether the prime value is scientific or revealed, this is true, and it is sometimes validly objected to Christianity that leaving its basic values tied up in a personality — though *dramatically* attractive — leaves far too much to interpretation and projection. The historical fact that Christianity has succumbed to the fate of splitting into innumerable sects, of suffering down the centuries from a perennial disease of "heresies," and of being quite dubious even today on, for example, such a question as patriotic duty-vs-conscientious objection to war, shows that stopping with nothing beyond "general values" has its dangers. (Historical determinism recognizes, of course, that there are economic, temperamental and other causes for sectarianism *disguised* as ideological disagreements.) But it remains true that with sufficient original ideological precision these breaks could not have been defended. It is inevitable that the "flexible" religions will be tempted to retreat from thorny legal and contemporary political issues into "leaving the matter to a Christian conscience" or invoking "the personality of Christ." Though the traditionally devout will protest at this criticism, we must insist that moral laws and values cannot ultimately be left tied up in one personality — especially one only obscurely known historically and whose words permit countless re-interpretations. One must maintain again here what was said at the beginning — that intuitive interpretations lead to that slippery slope where all values are deemed "relative," and which descends into chaos. That danger is present from the start if the initial supreme value definition — from which more detailed conduct is to be inferred — is itself vague and non-operational.

While holding quite firmly *in principle* to deriving within-group morality from the relation of behavior to an absolute group goal let us nevertheless realistically admit that the task is a very difficult one. The unified social science capable of reaching ethical laws by logical and empirical procedures is, as just admitted, scarcely in its infancy. Awe-inspiring and overwhelming though the task may now appear for our undermanned and still methodologically groping sciences of psychology, economics, sociology, anthropology and quantitative history, the definition of the primary *evolutionary advancement* criterion roughly attempted above definitely has to be carried to far greater precision in application as soon as possible. Here we must recognize:

(1) That the higher order principles themselves, in *any* system, can rarely or never be fully fixed and understood by *verbal concepts and logic alone*. Their true meaning resides in mathematical and quantitative research and in behavioral formulations that extend into specific injunctions in specific situations. A value is an abstraction from particulars across a great span of behavioral backgrounds, and the particulars need to be both specified and measured.

(2) Unless the general principles have such conceptual precision that they *can* ultimately be accurately *interpreted* in rules of conduct in a given physical, social, and historical situation the hope of any *practically* effective guidance from such general principle vanishes.

These practical inferences from the principles in actual injunctions for behavior also imply moral feelings and needs. If someone wishes here and now to anticipate the "feel" of the emotional values which Beyondism embraces he can, it is true, let his imagination contemplate the adventure of man in his lonely and tremendous pilgrimage through the universe. But the spectrum of within-group civic, inter-individual emotional values which this implies in the effective dealings of man with man is something at present unstructured. It awaits definition by empirical research into the nature of the moral acts which engender evolutionary advance. This position incidentally, is just the opposite of that in such modern pseudo-religions as Humanism and Existentialism, in which what the propagandist assumes now to be appropriate human emotional values are made the basis for developing a social morality. In Beyondism man is asked instead to discover and develop a new emotional life appropriate to the behavior which a scientific examination of man's destiny in relation to the universe shows to be vitally needed. This logical situation in which a general concept is increasingly understood only as its practical applications are followed through is a research situation of everyday familiarity to the scientist, but sometimes baffling to the lawyer, the logician and the literary, verbal thinker. The scientist is accustomed to understanding a theory by what it predicts, and to continually refining and modifying its meaning additionally through seeing what it fails to predict. The logician, on the other hand, would require a precise and complete definition of, say, the atomic nucleus or the action of cholinesterase before consenting to discuss it. But the scientist knows that such demands for conceptual thoroughness at any stage of research advance are unrealistic as well as being prone to sterilize the intellectual processes of discovery.

Our purpose in this section has been to clarify the inherent principles in deriving moral values from a fixed goal in a changing world. But, as mentioned in passing, there are also emotional and "public relations" problems to be solved. Hitler in *Mein Kampf*, and emotional writers of very different persuasions elsewhere, have equally tried to argue that science is unsatisfactory as a repository of any popular faith because it is forever changing. That this evolutionary movement dismays the political absolutist is perhaps as it should be; but in the field of religion the call of the human heart for ultimate certainty is to be respected, for that is the meaning of religion. In the buffeting seas of immediate moral conflicts a religion should give the emotional assurance and guidance of an unwavering star. But this is what the basically fixed goal, for those with imagination to see the adjustment to each changing local scene in perspective, truly offers. Beyondism is a religion tied to only one dogma—and that a scientific one—from which the sentiments in all specific situations take their changing courses. These problems of emotional interpretation of a Beyondist world view must be mentioned here; but their full study belongs to the last two chapters. Let us not underestimate the propaganda problem, however, presented by the fact that traditional religions have long developed their passionate dramas, their tactful political affiliations, their mature diplomacy in international life, and especially their arrays of substantial emotional consolations. By contrast the bare framework of the

Beyondist principle, and the uncompromising, abstract search for truth in science will initially compete for popularity with difficulty.

4.2 EXPECTED DEGREES OF DETERMINATION OF WITHIN-GROUP BEHAVIORAL NORMS BY BEYONDIST PRINCIPLES

With this view of the essential situation in attempting to derive ethical standards and moral action from scientific research applied to a fixed criterion let us now consider the derivation processes themselves.

Two plausible misunderstandings likely to be encountered in discussion here need to be initially eliminated. They are actually technical, methodological issues in using the group survival (or evolutionary advance) criterion. The first is an abstract methodological point familiar to multivariate experimentalists, namely, that *all the variance in a dependent variable is usually not to be found residing in and tied up with the particular observed independent variable or cause that one has chosen to observe.* This may sound a remote and technical principle, but in the present context it brings out particularly a concrete point already made above: that the differences of nations in survival success (the dependent variable) are not tied up solely with differences in the relative soundness of their internal morality. Difference in resources, genetic and environmental (such as, in the latter, differences in land size, population, mineral resources, climate and alliances) also play their part. The equation for survival contains measures of all of them. By reason of accidental gifts of this kind the "wicked" may certainly flourish awhile, as the observers of biblical times noted. On the other hand, some modern historians quite ignore the moral, psychological forces we consider to be so important and see economic and other material factors as the sole "determiners" of history. They fall into a "univariate" explanation, and a wholly material one at that. But in revolting—perhaps with some justification—against the "great man" theory of history and scrupulously forgetting anything that such men as Carlyle said, they do not have to forget man altogether. For even the average man, with his hopes, his fears, his intelligence and his morale, is more than a set of economic statistics. Indeed, the sober argument of the social psychologist—although he keeps all terms in the equation—is likely to be that the genetic nature of a population, and, above all, its moral culture, are more important for survival than any accident of the resources into which that population is born.

The second possible misunderstanding to be headed off concerns the slip of substituting the term "government" for "group" or "group culture." In agreeing above that in principle ethical laws *might* be wholly expressed in government legislation we were not acceding to any proposition that governments normally succeed in doing this or that their life is the life of the group. The government and the group culture are normally distinct parts of some organic totality, as is recognized in the next paragraph, but it is at least an ideal that the legislation of a culture will come close to expressing that culture's moral values. And the beginnings of quantitative study of national cultures (page 133 below) suggest that the ethical values of a culture can usually be inferred, perhaps with some lag, from national legislation. Thus the distinction between government and culture is necessary for all future discussion here. Both the political scientist and the social psychologist habitually write about and clarify this distinction though substantial correlation between the styles of group behavior emanating from the two is nevertheless normally to be expected.

At this point those who like operational definitions for their abstract concepts may find firm ground in the concept of group *syntality* (Cattell, 1948, 1950a) which is a dimensionally measurable concept from the observed behavior of a group on many variables. Within this the distinction can be drawn between the *general population* properties (means on intelligence, personality, attitudes, etc.) and the *leadership* contribution to group behavior. Even at a simple observational level one can distinguish syntalities, as in dictatorship and primitive priest-king government, in which governmental and moral authority are in one, and others in which complex checks and balances exist among institutions with legally defined roles, and in which a group of peers living by agreed, explicit values and laws constitutes authority. Furthermore, psychological writings at least from the time of Freud, have discussed the corresponding differences in individual personality structures in individuals adjusted to these syntalities. The basic principle is that there tends to be an isomorphism between the dynamic structures and attitudes in the individual mind and the social macrocosm in which this microcosm dwells. For simplified discussion the dichotomy which Freud drew between the "horde" mentality, in which each individual depends solely on the external authority of the group and that society of law in which an individual acts upon a conscience introjected as an internal authority is a useful one.

In the former, control of behavioral morality is still externalized, the father figure not having been adequately internalized in the structure of a superego. It is then possible for the government, essentially as a dictator, simply to stand, in the emotional life of the individual, in the shoes of the father. There is then, in the corresponding group life no difference between the values of the culture and of the government at the time. No individual has the internalized values and emotional resources to maintain an independent conscience. Consequently no institutions or collections of individuals can stand against government, as a check and balance.

In the more developed "contractual" society among "peers" on the other hand, the psychoanalytic explanation is that there has been introjection of the father image, and the formation of an internalized system of values in conscience. Government can then only operate within limits of acceptability set by these internalized values and the laws which express them. (There is a tendency to think of this in political terms as an exclusively democratic concept, but oligarchies and conceivably other political structures could express the primacy of cultural values over the executive.) When a higher proportion of citizens remain at the horde stage of psychological development there will be a greater readiness to accept the government and the moral law as being one. When a higher proportion have developed superego formations the *law* becomes raised in emotional importance, above the contemporary government. Communist and Fascist states have accepted the former balance, making government and moral law virtually identical (regardless of whether dictation is by an individual or an elite group). Parenthetically, this may not always be their written theory, and Marx had almost an anarchist's dislike of government. But the dictatorship of the proletariat was by definition bound to express itself through a dictatorial government, as any realistic social scientist might have seen.

Democracy prides itself on putting the law above the state. It is important to note that in this it has been aided functionally and in historical development by the independence of religious institutions. Lest we in democracies assume too easily a moral superiority, or, at least, an unquestioned organizational superiority let it be noted that democracies have their real weaknesses, from which, from Athens to the French Third Republic, they have suffered severely. Among these drawbacks are: (a) the capacity of a dictatorship to hold its cards close to the chest, and play a single, secret consistent game, with rapid decisions in emergencies, and (b) the continual vulnerability of democratic values to the easy hypocrisy by which any arrogant escapist from social control and duty can claim that his "conscience" or his personal values requires him to avoid social obligations.

The issue of whether moral authority and the shaping of moral values resides in the government or in the vaguer entity we call the culture is obviously a matter for sharp scrutiny in any system such as we are now proposing where the criterion of inter-individual morality is the survival of the group. For, especially in emergencies, it would be easy for a government to insist that its actions, by definition, are the only moral ones. This is clearly not a correct understanding of what follows from Beyondist moral derivations, for governments come and go, and may represent the culture erroneously, whereas cultures endure, and continually grow. By the group we mean the cultural group.

In a horde mentality and emergency situation the dictatorship may be considered to have swallowed the church; but there is an opposite situation, which at first seems ideal, in which the church swallows the government as in the theocracies of ancient Israel, in the early Puritan communities, in Geneva, in the early days of Salt Lake City, and in a number of other societies.

The argument could perhaps be made that Beyondism is in effect a new theocracy with social scientists substituted for priests. Is this an undesirable state of affairs? If the culture is centrally represented by its religio-moral values, what is more Utopian than making the government an immediate servant of the culture? The answer, surely, is threefold: (1) That moral values in theocracies up to this point in history have come from revealed religion, and, as such, have been dogmatic, unchangeable, and partly erroneous. (Much of the food hygiene in the Jewish religion is surely inappropriate today.) (2) That revealed religions have considered themselves universalistic religions, whereas in as much as Beyondism requires variety from group to group, it cannot — in this within-group behavior — espouse the idea of a world wide sameness of values. (3) It is in the nature of progress that it requires discourse and calculation, with separate viewpoints officially propounded as in a court of law. The existence of some degree of separation of institutional authority, as in civil government, churches, universities, provides the necessary framework for the war of opinions within the individual mind.

Experience in history already clearly favors such division of community authority. Liberal and rational thought (except for the brief state worship of Reason in the French Revolution) in particular has brought arguments clearly on the side of having the "checks and balances" of civil law on the one hand, and an independent source of moral authority, open to discussion on the other. Indeed, democracy and the liberal writers, as exemplified for instance, by J. S. Mill in England, by von Humboldt among the Prussian liberal constitutionalists, and the designers of the American Constitution (with its independent Supreme Court) have tried deliberately to weaken central government, as an extra insurance for the safety of dissident opinion. They aimed to restrict the within-group action of government mainly to preventing anarchy or the harming of one set of citizens by another. This restriction on the extent of control by government, and the harm it can do, necessarily restricts also some of the good it can do. In slow, long term movements the latter defect may indeed be slight. But when the blast of war blows in our ears, the instant freedom of decision of a powerful government, or even a dictator, has its positive advantages.

Taking a look at history to the present, and stopping short of the new aspects to the question presented by the possibility of a scientific ethics, one must certainly grant that the existence of universalistic ethical systems, demanding, in the battles of church and state, a separation of civil authority and individual conscience, has been extremely valuable. It has firmly established in Western culture the principle of "rendering to Caesar the things that are Caesar's, and to God the things that are God's." If this institutionalized duality did nothing else, it could be supported as a useful part of the machinery of checks and balances in power that make for calm and far-sighted judgment in any group. But, as an aid to the individual psychology of development it has acted as an outer institutional support for encouraging the inner values of the superego – of the enquiring conscience. It has helped this system of inner values to be independent of, and a guardian upon, the necessary immediate and practical loyalties which the ego directs to the peer group or civil authority. The universalistic imperialism associated with churches may as we have argued, be philosophically an unfortunate error of over-extension. But at least by dissociating themselves from the political unity of the national group the churches even without universalism, staked out the free ground in which the rationalist, the liberal, and ultimately the scientifically-thinking proponents of "values beyond the present establishment" could live [1].

Now the innovation here suggested of deriving part of the moral values – that part concerned with inter-individual ethics within a group – from scientific research means that a newcomer among institutions – the moral research institution – has to be fitted into the above historical progression. Clarifying and researching upon what we have called the *maintenance ethic* (or *common community morality*) that is essentially the same for all groups, is unlikely – just because of being common to all groups – to be undertaken solely by a world organization. Like an educational system or a health service it will be better developed by some nations or religious communities than others despite the principles of, say, health, being much the same for all. The fact that the contrasting *specific community morality or ethical development values* are likely to be handled by the same research institution, seeking best to express the will of its particular group, makes it likely that just as in an educational system, the search for the best common basic principles in learning and the teaching of specific values will go hand in hand. So, therefore, both common and specific moral research are likely to be initially the job of each national institution.

4.3   THE PRESSING REQUIREMENT OF DEVELOPING A MORALS BRANCH OF SOCIAL SCIENCE

The difficulties that lie ahead—perhaps for hundreds of years—in deriving moral values by social scientific calculation from observation of effects on group progress and survival are so great that it would not be amiss to remind ourselves again of the basic importance of that enterprise. Only by this path can conscientious, dedicated men hope to regain the certainty and serenity of purpose that our culture has lost since the revealed religions were discredited by critical intellectuals. For the young who, as ever, are showing a lively interest in values, such research offers the much desired touchstone to distinguish between the mere selfish anarchy of revolt and the purposes of constructive revolution and evolution. And in general, it would lead to more vigorous, effective societies, less sapped by internal and external conflict, and better able to launch themselves on a common cultural adventure.

Nevertheless, for a long time that certainty is going to be relative, and the research answers merely approximate. At present the research data and methods are so almost completely lacking that the writer hesitates here even to illustrate the types of data and chains of reasoning that a research institution in ethics would handle. Perhaps a critical reader will be willing to grant that a modern committee of sociologists and anthropologists, given the criterion of group survival could convincingly demonstrate the positive contribution made by the main prohibitions of the Decalogue, e.g., against covetousness, robbery, adultery and murder.

But where behavior involves no obvious, strife-provoking interference with others, and no obvious sabotage of the health and welfare of society and its government, it would be quite difficult to prove anything to a committee of experts and still harder to a group of intelligent persons who for some reason have emotional antipathies to the moral values that seem to be indicated. For example, in the sexual field, and supposing "adult consent," are free love, adultery, masturbation, homosexuality and various perversions demonstrably reductive of group vitality? Is there no reproach attached to a man of wealth spending his money on fantastic, sybaritic luxury, provided he harms no one? What effect does full inheritance of parental property, as opposed to a substantially taxed inheritance, have on the survival strength of an economy? Is gluttony in an economy of abundance of any harm to society? In a society that seems able to support increased periods of idleness or non-productive recreation, (in the young and the unemployed unskilled particularly) is there anything "better" for social progress than continuing to expand that free leisure? Just how far, in matters that a man commonly likes to think "His own business" and in which he is obviously not *directly* affecting others, is social scientific moral research likely to be able to assert more positive, definite values? To evaluate numerous remote social consequences, and then to demonstrate that on balance the outcome is relatively anti-social or morally sound is a stiff task for research. But the history of science promises that it can be done.

For its hypotheses, though not for its methods, a young science of "consequence morality" as one might briefly designate this aspect of research, would almost certainly do well to turn – but without bias – to the hunches existing in revealed religion. Some of these biblical and other intuitions strike the man in the street as merely bizarre – as in the deceased wife's sister marriage ban. But even for some of these, and certainly for most others one might anticipate some support in a consequence morality based on examining community health and cultural survival. Yet others, like keeping one day a week for rest and worship (rather than exciting recreation) might be supported empirically by a finding that moral standards deteriorate less when such contemplation and reinforcement is practiced. Already some previously incomprehensible rules – like the ban on incest – can be seen in the light of a century of genetic research to be a true "consequence ethics."

To add to the urgency for shifting ethical values to a more scientifically demonstrable basis our modern society is now being besieged, at a rate of onset unknown in previous centuries, by quite new problems for which the old religions have no ready answer. For example, the increased knowledge of genetics just mentioned, as well as our statistical awareness of over-population problems, makes us aware that there is a whole area of behavior in begetting children – even legitimately – that was formerly no one's business outside that of the parents concerned, but which now looks very much a matter about which the community must introduce a consequence morality.

As to the more subtle issue of birth rate control to ensure progress through genetic quality, a line of brilliant and wise writers—Darwin (1926), Fisher (1930), Shaw (1949), Huxley (1957)—has long set out, to the satisfaction of intelligent men, goals and means in eugenics. But in attempting to do so in terms of the values of revealed religions such as Christianity they have been defenselessly driven back by reactionaries to a merely "negative eugenics," i.e., the elimination of only the most gross genetic defects. If society traced "demonstrable consequences" it would find in Beyondism a far more convincing argument for instituting a steady, monitored, cultural and genetic *advance* than is available in "revealed" ethics.

The onslaught of new problems that need a new ethics concerns not only our appreciation of much wider consequences of old forms of behavior than were seen before, e.g., those arising from unchecked birth rates or pollution, but also our exposure to a crescendo of new technological impacts. If one considers, for example, the bemused hesitations today over the correct position on the freedom of school children to smoke "pot" (marihuana), the absence of both clear goals and a sufficiency of *social* research in relation to them become apparent. (Resort to revealed religions suggests that Mohammedanism, with its dogmatic stand against alcohol and smoking would do one thing, whereas large branches of the Christian religion, permitting alcohol and having no stand on smoking, would probably open the door to the step by step transition which begins with marihuana and ends in heroin.) And so one could go on, through modern vagueness, permissiveness, and perplexity, from trivial matters such as the right of neighbors to defense against the new electronic capacities for amplifying noise, through the rights of disseminating pornography to those legally not adult, to such a profound issue as in the recent agitation against the death penalty for premeditated murder.

With both the ancient, perennial questions cited above, and the new problems just indicated, a very formidable list—not unlike a list of accumulated court cases in some overburdened judiciary system—provocatively awaits the new social science of Beyondist ethics. Without this, what really exists? The dogmas of revealed religions are irrelevant or unconvincing. The response to such questions by the social sciences as they exist today does not merit one's confidence any more. As we have seen they present an incongruous mixture of very limited experiments with a concoction of all varieties of inexplicit intuitive dogmatic ethical values anomalously embedded in attempts at scientific procedure. If the fundamental and radical concept of Beyondism should be accepted, namely, that a basis exists for deriving ethical values objectively from the criterion of progressive survival of the group, the unhappy state of the social sciences at this juncture would offer only feeble contributions. As we have seen, they are justly somewhat despised by the physical sciences, on the one flank, as lacking in "hard headed" methodology, and looked at with suspicion by the politicians, on the other, as impractically doctrinaire, and as the favorite thicket in which fanatics of various kinds can hide their true colors. Certainly many of the lesser sociologists, cultural anthropologists and academic political scientists can clearly be seen to have adopted revolutionary or anarchistic positions first and then to have dragged in the mere paraphernalia of science secondarily as a defense for their positions.

In any case, the lowly position, in prestige and grants, of the social sciences in such important organizations as the National Science Foundation is not due to accident. Some part of the trouble is the perception by intelligent outsiders that the social sciences have been unscrupulously used as vehicles for those insidious "contraband" values discussed above in Chapter 2 (Section 2.3). But the greater and less immediately remediable part is the realization by scientists themselves that these sciences have seldom got beyond a sheer amateurishness of scientific method and an uninspired choice of important topics. (Witness the fact that the first demonstration of effect of birth rate on population fluid intelligence levels has waited thirty years for a check.) Furthermore, this scientific inadequacy is evidenced, at least up to mid-century, by the fact that (particularly in cultural anthropology, sociology, political science and clinical psychology) the bulk of the writers and students have been obstinately reluctant to face the disciplines — mathematical and experimental — which the leaders in their fields have been trying to develop.

The remedy for this situation is not simply the pouring in of research money (though a radical change in level of endowment is certainly required) but rather a setting up of the basic guarantees just indicated, namely, (a) an habitual clear separation in publications of what we may call *effector science* from *values*, regardless of whether the later spring from revealed religious origins or the new branch of science required by Beyondism and which we may call *value science*. (b) A recruitment and training of social scientists which will leave them no longer partly peopled by axe-to-grind dissidents or drop-outs from the stiffer disciplines of the physical and biological sciences, and (c) the setting up of research institutions amply endowed for the enormous task of value science.

Far from being forever the "weak sister" among the sciences, the social science which studies morality is destined to be the greatest of all sciences. For the types of problem to be solved in deriving the morality of particular acts from the general principle of consistency with group survival, demand an imagination and grasp of complex method exceeding anything yet required in the physical sciences. Mathematics has been called the elegant queen of the sciences, and as in other complex sciences it is going to have as great a role to play in connection with the new science of morality. But if Beyondism is aptly so called, then this new science, with its rugged experimental problems, its supreme importance for human life, and its central integration of all sciences, may well be called the king of sciences.

Apart from its inherent technical problems, it will face social difficulties, because it will dislocate traditional ways of thought and excite jealousies, on the one hand, from entrenched agencies for moral value decisions, and, on the other, from those liberal, radical agencies which are accustomed to the idea that their dogmas already adequately represent "rational" and "scientific" approaches. As we have seen, the subjectivities and pre-judices in the axioms of the rationalist are if anything more fraught with error than the experiential and intuitive conclusions of traditional religions. In the radical rationalist approach one of the greatest dangers resides in the notion often loosely designated "action research." This is a still poorly thought out concept which confuses the progressiveness of science with the irresponsibility of minority government[2]. But it is only one of many perilous forms of charlatanism or sheer naivety that are likely to crowd alongside moral research on ethical values carried out in the social sciences with integrity and imagination. The social problems that arose from sources of this kind in the immature years of medicine were as nothing to what may happen from similar sources when social psychology struggles to found a science of ethics.

In summary, it has seemed desirable to make clear at the outset what value research — and social science research generally — is and is not. But a more detailed analysis of the difficulties which a Beyondist development is likely to meet in existing institutions, and of the current activities with which it should not be confused, is pursued in Chapters 6 through 8 below.

4.4   SOME FRAGMENTARY TECHNICAL BEGINNINGS IN RELATING GROUP VIABILITY TO INDIVIDUAL MORALITY

To ask that social science here and now show its capacity to draw a firm conclusion about the validity of some inter-individual moral behavior rule on a basis of its demonstrable contribution to group viability is to demand a miracle. Let us freely admit that as of 1973 any statement about derivation by truly documentable steps of scientific inference is a check drawn upon a bank of as yet non-existent data and principles.

Yet there are illustrative beginnings. The study of the dynamics of objectively measurable attitude strength and social attitude change is one such growth (Hendricks, 1971; Horn, 1966; Sweney and Cattell, 1962; Cattell, Kawash and De Young, 1972). The experimental refinement of concepts that once depended on psychoanalysis, such as ego strength, superego strength development (Gorsuch, 1965; Delhees and Nesselroade, 1966) is another. An approach which has yielded clear and widely applicable findings even though on a socially miniature scale, is that of group dynamics, in which the capacities of groups of six to twenty people to solve group problems, undertake common constructions, get along together, and so on is studied for various combinations of personalities, motivations, and leadership structures (Bales, 1950; Borgatta, Cottrell, and Meyer, 1956; Festinger *et al.*, 1950; Cattell and Stice, 1969; Gibb, 1969; Fiedler, 1965). But beyond these works specifically of the social psychologist one must keep in mind the need to integrate the areas of sociology, anthropology, economics, genetics, political science, human geography and quantitative history, as in the diverse approaches of such men over these two generations, as Sir Arthur Keith (1946), Ardrey (1970), Borgatta and Meyer (1956), Coon (1962a, b), Darlington (1969), Darwin (1871), R. A. Fisher (1930), Gibb (1956), Haldane (1925), Hardin (1964), Hooton (1946), Huntington (1945), Huxley (1957), Lerner (1968), Linton (1936), Deutsch (1965), Margaret Mead (1955), George Mead (1934), Mannheim (1937), McDougall (1934), Merritt (1970), H. J. Muller (1966), Needham (1929), Rummel (1963), Singer (1965), Sorokin (1937), Spencer (1892), Toynbee (1947), Waddington (1953, 1962), Max Weber (1904) and Veblen (1899) and some others. These men taken for illustration are, naturally, uneven as to agreed eminence and very different in approach. Few, indeed, bring any empirical evidence to the question and some leave whole areas untouched (as Toynbee, for example, leaves out all genetic determination). Their present professional variety is illustrated by the fact that one looks to a geographer, Huntington, for some most suggestive empirical comparisons; to an anatomist, Keith, for some of the most central arguments; to a botanist, Darlington, for the most subtle searches for social determinism; to a political scientist, Rummel, for some of the most imaginative mathematical treatment, and to some social psychologists, like Adelson,

Gibb, Gorsuch (*see below*), McDougall and Singer for the most concrete if fragmentary contributions of truly relevant information.

To make headway even with fragmentary beginnings it is necessary to have an adequate model by which to analyze data. In what follows we shall describe this theoretical model in its simplest form and then proceed to consider a first harvest of data that will illustrate an elementary beginning to the way in which social behavior and group survival need to be related. It has already been pointed out that *syntality* is an analogous concept for organizing an understanding of the behavior of a group to *personality* for the individual. When the most apt dimensions have been found by factor analysis for analyzing the behavior of organized groups, e.g., teams or nations, individual groups can be assigned a profile of scores, which truly constitutes the style of its culture pattern. (A culture pattern is a "type" among syntalities.)

It has been suggested (page 120 above) that the population average profile (*P*, below) is something quite different from the syntality, and that the internal structure of the group, *R* — the way in which people are arranged in institutions and roles [3] — is something different again. Now, condensing what is ultimately likely to be a most complex equation into its simplest schematic form we may say (*see* Cattell and Stice, 1969):

$$S = (f)P \cdot R \tag{4.1}$$

i.e., syntality is a function of population parameters, e.g., average intelligence and of social structure, *R*.

The dimensions chosen for *S could* be quite arbitrary, as illustrated in terms used to "place" a cultural group by Benedict (1934), Spengler (1928), M. Mead (1955), Toynbee (1947), and most anthropologists and historians in the past. Alternatively, they could be found as organic, meaningful dimensions as suggested above by the same factor analysis of a representative set of group behaviors as, when applied to the behavior of individuals, succeeded in giving us our modern understanding of personality and ability structure (Burt, 1940; Cattell, 1946, 1957; Eysenck, 1952; Guilford, 1959; Thurstone, 1944).

This objective, quantitative, mathematical attack on the structure and dimensionality of various kinds of groups has been pursued most systematically by Alker (Alker and Russett, 1965), Deutsch (1968), Cattell (1953, 1957, 1965), Cattell and Adelson (1951), Gibb (1956, 1969), Hadden and Borgatta (1965), Gorsuch (1965), and Rummel (1963, 1966). Such studies have typically taken for some eighty to a hundred countries the available measures on such variables as the community's average real income, the number of murders per million per year, the infantile death rate, the average length of school attendance, the expenditure per head on scientific research, the number of qualified doctors per 100,000 of population, the number of political assassinations per century, the number of times the country has declared war in a century, the number of Nobel Prize winners per century, and so on. A beginning has been made by correlation and longitudinal and cross sectional factor analysis in showing *how these variables are in fact connected* and in thus arguing to causal, functional relations between individual forms of behavior within a group and the total group performance. To those enamored of traditional theoretical positions in purely qualitative psychology, economics, history, etc., this may seem a too starkly empirical approach. But experienced researchers know that one of the surest ways to get theorists together from mutual academic isolations is to put their variables together, and theories should arise from perceived relations in the variables.

A simple principle about the life of national groups which quickly emerges is the relative constancy of values from one-half century to another. Changes of government, etc., as in Russia, by no means alter the main syntality scores. The cultural leopard does not easily change his spots. Another is an increasing support for the value of the formal model of resemblance of a social and a physiological organism. There is a constant feedback between the effectiveness of each organ and the total effectiveness of the body; the cells are to the total organism as individuals are to the social organism, and there are suggestive similarities of higher order abstractions such as immunological reactions to foreign immigrant proteans; the spread of the anarchy of cancer when immunological defenses break down; the maintenance of internal homeostasis by an internal (autonomic) nervous system and of external vigilance by a central nervous system; the arrest of peaceful, anabolic processes by epinephrin in "war"; and the normal onset of degenerative processes with the age of a culture. Any imaginatively conceived team of social scientists set up nowadays to investigate these causal relations in a social organism could well add a "living systems" physiologist using the concepts of J. G. Miller (1965)—and *all* should be good multivariate mathematical statisticians.

Out of such analyses, as illustrated by the work of Rummel and the Yale political scientists, e.g., Alker and Russett (1965), will hopefully emerge equations not only relating internal moral behavior standards to relative group survival (or freedom from high indices of moribundness), but also to the effectiveness of internal readjustments to external stresses impinging upon groups—from nature or from other groups. Probably a useful concept here will be that of *synergy—the total available energy of a group for group cultural undertakings*—and its relation to the life habits and values of the group members.

Novel though these existing attempts at integration across specialties are, they show clearly that a new body of theory is about to emerge which will be as different from current political science and sociology as chemistry was from alchemy. Actually, there are, of course, some recognizable precursors of the new concepts. Especially as we look at the new factor analytic "emergents" (Lloyd Morgan, 1878)—as in group *syntality* and *synergy*—and at their relations in experiments with small groups to the personalities and role behaviors of component individuals, we see parallels to qualitative precursors in concepts in political science, history (Darlington, 1969) and anthropology (Unwin, 1934). Definite analogues can be found in the human small group experiments, and possibly also in the careful ethological studies of Lorenz (1966) and Tinbergen (1951) with animals, to the notions we are here discussing in multivariate analysis of the large cultural aggregates we call nations. As to some specific theories arising with respect to the syntality concept, as expressed in equation (4.1) (page 130 above) it can already be stated that the correlating of social variables across nations yields indubitable evidence of the need for some ten to twenty distinct functional unities (Rummel, 1963; Gibb, 1956; Cattell, 1953; Cattell and Gorsuch, 1965), as listed in Table 4.1. Predicting the behavior of nations has been the perennial concern of historians (Roberts, 1941), but it is to be noted that a totally new, quantitative method, rooted in quantitative social psychology is here being brought to bear. At present social science has not unravelled the meaning of some of these dimensions—whence the crude descriptiveness of the titles in Table 4.1—but within them is contained a great deal of what one would want to say and could say quantitatively, about the "goodness" and general nature of a culture. Also, the "score" of a nation (literally a factor estimate from twenty or thirty variables) on every dimension provides a profile (illustrated on six dimensions only in Fig. 4.1) by which nations can be objectively grouped (by profile similarity coefficients) into "civilizations" (as Toynbee would call them). Thus America, Britain and Australia obviously belong to a family with a common profile quite different from that of the other six countries illustrated.

Furthermore, by what is known as the specification equation (4.2), an estimate can be made from these syntality scores of a given group's probable response or performance in any particular direction. Thus prone-

Table 4.1   List of Cultural Dimensions, Some in By-Polar Form

Factor 1:   Size
Factor 2:   Cultural Pressure vs Direct Ergic Expression
Factor 3:   Enlightened Affluence vs Narrow Poverty
Factor 4:   Conservative Patriarchal Solidarity vs Ferment of Release
Factor 5:   Emancipated Urban Rationalism vs Unsophisticated Stability
Factor 6:   Thoughtful Industriousness vs Emotionality
Factor 7:   Vigorous, Self-willed Order vs Unadapted Perseveration
Factor 8:   Bourgeois Philistinism vs Reckless Bohemianism
Factor 9:   Residual or Peaceful Progressiveness
Factor 10:   Fastidiousness vs Forcefulness
Factor 11:   Buddhism-Mongolism
Factor 12:   Poor Cultural Integration and Morale vs Good Internal Morality

(a)   Individual Profiles on Six Cultural Dimensions

High magnitude	High cultural pressure	High affluence	High conservative patriarchalism	High order and control	High cultural integration and morale
					Mean for all countries
Low magnitude	Low cultural pressure	Low affluence	Low conservative patriarchalism	Low order and control	Low cultural integration and morale

KEY: Australia — — —Britain - - - - - - - - - U.S.A. ———

High magnitude	High cultural pressure	High affluence	High conservative patriarchalism	High order and control	High cultural integration and morale
					Mean for all countries
Low magnitude	Low cultural pressure	Low affluence	Low conservative patriarchalism	Low order and control	Low cultural integration and morale

KEY: China ——— India - - - - - - - - - Liberia — — —

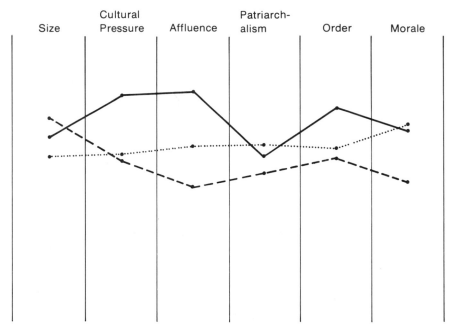

High magnitude	High cultural pressure	High affluence	High conservative patriarchalism	High order and control	High cultural integration and morale

Mean for all countries

Low magnitude	Low cultural pressure	Low affluence	Low conservative patriarchalism	Low order and control	Low cultural integration and morale

KEY: U.S.S.R. ——— Argentina - - - - - - - - Arabia — — —

(b)   Contrast of Central Tendencies in Three Culture Patterns

| Size | Cultural Pressure | Affluence | Patriarchalism | Order | Morale |

——————— The American-British-Australian Pattern
— — — — — The China-India-Liberia Pattern
·················· The Russia-Argentina-Arabia Pattern

The vertical scale is in standard score units for each factor, but is not set out numerically.

Fig. 4.1   Three Empirically-Discovered Types in Cultural Profiles.

ness to involvement in war is given by:

$$a_{iw} = 0.10S_{1i} + 0.58S_{2i} + 0.21S_{3i} - 0.44S_{4i} + 0.29S_{5i} \qquad (4.2)$$

where a_{iw} is the likelihood of any nation i being involved in an act of war (a_w); the S's are the scores of that group on syntality dimensions, 1, 2, 3, etc., and the coefficients before the syntality scores are what are called in personality "behavioral indices" (factor loadings). S_1 here is Affluence; S_2 is Cultural Pressure (*see* Table 4.1 and below); S_3 is Size; S_4 is Rationalism, and S_5 is Mohammedan Culture. The others have no noteworthy loading. (Values are reference vector values.)

Some of the dimensions in Table 4.1 (which must, as in Eq. (4.2), also be regarded as determiners in a causal, dynamic sense) are quite simple, such as: (1) Sheer Size, and its accompanying consequences. Others such as (2) Cultural Pressure — a measure of the stress of change and creativity that a culture is experiencing as shown in Table 4.3 — are much more subtle in action. Yet others, such as (3) Affluence and Education Level (a double name because these two ingredients go closely together), appear straightforward, yet require careful discussion (Table 4.2). Finally, (4) Level of Morale and Morality, the obverse of the number of crimes, etc., as shown in Table 4.4 comes very close indeed to that component of

Table 4.2   The Factor of Enlightened Affluence as a Dimension of National Cultures

Loading	Variable
(0.50)*	(High level of technological skill)
−0.73	Low death rate from tuberculosis (tuberculosis a disease of poverty)
0.70	Large gross area (contribution of large trade area)
0.67	High expenditure of tourists abroad
0.55	High real standard of living (judged by expenditures)
0.51	High real income per head
−0.42	High expenditure (all sources) on education (high level of intelligence)
0.40	High musical creativity (index of cultural interest)
	(More books read per person per year)
0.37	High sugar consumption per head (index of luxury expenditures)
−0.36	Low degree of government censorship of the press (liberality of education)
−0.27	Low suicide rate
−0.25	Low death rate
0.23	High ratio of exports to imports

Variables are here arranged in declining order of loading, averaged over two or more researches. Thus the first listed variables are more closely connected with the action or genesis of Enlightened Affluence than the last.

*Estimate from separate study.

Table 4.3   Connotations of the Factor of Cultural Pressure as a Dimension of National Cultures

Variable
High ratio of tertiary to primary occupations
High frequency of political clashes with other countries
High frequency of cities over 20,000 (per 1,000,000 inhabitants)
High number of Nobel prizes in Science, Literature, Peace (per 1,000,000 inhabitants)
High frequency of participation in wars (1837–1937)
High frequency of treaties and negotiations with other countries
High expansiveness (gain in area and resources)
Many ministries maintained by government
High creativity in science and philosophy
(High emotional complexity of life)
High musical creativity
High death rate from suicide
High incidence of mental disease (especially schizophrenia)
High horsepower available per worker
High percentage of population in urban areas
High cancer death rate
High divorce rate
More severe industrial depression in world depression
High total foreign trade per capita
Numerous patents for inventions per capita
Higher number of women employed out of home
More riots
Lower illegitimate birth rate

Variables are here arranged in declining order of loading, averaged over two or more researches, but for actual numerical loadings the original researches must be consulted. Thus the first listed variables are more closely connected with the action or genesis of cultural pressure than the last.

morality and morale in culture that we most need to study here. Tables 4.2 and 4.3 adopt the usual factor analytic convention of describing and interpreting the factor by setting out for each dimension the variables in the matrix that have experimentally been found most affected (loaded) positively or negatively, by that influence. In each one or two further expressions that are theoretically to be expected from separate studies, are added to guide further research, their speculative status being indicated by parentheses. In delving briefly into one of several complex methodologies — factor analysis — that the social scientist has to use, the reader unfamiliar with the mathematical background naturally cannot get the full import. But a general feeling for what the approach yields may be gained here. Each factor indicates *the pattern of expression of some underlying cause in cultural dynamics*, and, as a definite correlation pattern, offers potent guidance to new hypotheses as to its nature and origin.

The interpretation offered elsewhere for the Affluence-Education factor (Table 4.2) is that the real wealth of a country depends less on climatic and mineral resources than on education and intelligence (the issue of separating the two will not be taken up here). History presents many examples in support: when Germany was deprived of nitrates by the blockade, the chemist Haber drew them from the air; when the Dutch were deprived of their colonial resources, they took land from the sea; when the Jews re-occupied Israel, they converted it from a relatively desolate to a relatively well-to-do land. History will yield many other instances of "natural resources" being determined by intelligence.

The Cultural Pressure dimension (Table 4.3) is perhaps the most intriguing of these patterns for the psychologist. Several variables can be understood as indicators of what a Freudian would label "instinctual frustration" — from the regimentation required by urbanization, from what cultural complexity demands in education, and other mounting complications of life. This frustration expresses itself in three outlets: (1) aggressiveness, as shown by increased riots and wars, (2) neurotic internalized conflict, as shown by indices of psychopathology and suicide, and (3) cultural creativity, in part from that sublimation of drive which clinical psychology recognizes as a healthy concomitant of (1) and (2). Here is an illustration of the way nature may make a laughing stock of our *a priori* conceptions of social progress. For the nations of highest cultural productivity turn out to be those most frequently involved in expansiveness and clashes with other countries. (There is even a positive correlation of a people's frequency of involvement in war, and the number of Nobel Peace prizes awarded to its citizens.) And indices of community mental health are inversely related to musical and other forms of cultural creativity. We have much to learn. Such findings provoke some more sophisticated self-examinations, such as lead us to our argument in Chapter 6 (Section 6.5), that the aim of progress may not be simply to reduce human frustration and "tension" (i.e., not to bring about what some philosophies make synonymous with the "greatest happiness of the greatest number").

Finally, in Table 4.4, we encounter what appears to be the very essence of the internal moral level of a society, as pursued in our introductory discussion. (Here evaluated partly by indices that would be considered to reflect community or universalistic inter-individual moral expression

Table 4.4   Connotations of the Factor of Morale and Morality as a Dimension of National Cultures

Variable
(High standard of inter-individual morality)
Low frequency of syphilis
Low frequency of homicides
Low death rate from alcoholism
Low death rate from typhoid fever
More eminent men eminent outside the field of politics
Small proportion of births illegitimate
Better education and civil rights for women
Fewer families per single house
No licensing of prostitution
Fewer deaths from tuberculosis
No absolute restriction on divorce
Low divorce rate
Low gross birth rate
Low gross death rate

Variables are here arranged in declining order of loading, averaged over two or more researches. Thus the first listed variables are more closely connected with the action or genesis of morale than the last.

level.) The morality dimension identifies itself initially by its loadings on a variety of restraints on aggression and sex. But one soon sees by the visible breadth of influence of this factor on more remote variables how very pervasive level of morality can be in its effects on the health and welfare of society (Cattell, Breul, and Hartmann, 1952; Gorsuch and Cattell, 1967). Far more extensive research is now demanded on this basic influence, now that it is reliably factor-analytically identifiable (though not fully interpretable).

Parallel to these dimensions found in real groups (nations) investigated naturally, "in situ," there are supportive findings, as explained above, from small group (group dynamics) experiments, with experimental manipulation. Since these involve experimental conditions not as well known to the general reader as national histories and characteristics they cannot so well be briefly illustrated, but the general nature of the measured variables may be seen from their titles. For example, in studies of performances of 100 otherwise equalized newly-formed groups, with ten men in each, the "morale" of such small groups has been found to have no less than three forms or dimensions of expression, of which two are shown in Tables 4.5 and 4.6.

Table 4.5 Factor of Morale of Good Leadership

	Session 1	Session 2	Session 3
High satisfaction with leader*		82	78
High group unity*	82	55	81
High influence of formal leader*		60	70
High morale following dynamometer*†		29	54
High feeling of freedom to participate*	−35	−29	−43
High degree of interdependence‡	30	10	−02
Construction: Most planning done for early trial	03	−49	07
Small number of negative effectors*	02	−03	−52
Low "population" mean on Affectothymia (A) §	−04	−42	−12

*These are mean values of ratings over several group task performances made by all members of the group.
†A tug of war on an unpleasantly electrified rope—a situation which tended to break down morale and cause drop-outs.
‡Observers' ratings.
§Members are by nature unsentimental, forthright, self-sufficient, tending to aloofness.

The three columns of factor loadings on the right are given as they stand, for three experiments, without averaging, to show the degree of consistency reached. Note, however, that this consistency includes some thirty or forty *other* variables being consistently zero (nonsignificant) on this pattern across the three experiments.

The first (Table 4.5) has to do with a component in the "strength of a group" which arises from *goodness of leadership*. The hypothesis here is that in so far as the leader is able convincingly to interpret his leadership of the group's activities to more members as being expressions of the needs which originally brought them together as a group, the group synergy becomes greater. Additionally the correlations of leader success with personality characteristics of the leaders (not given in Table 4.5) show that he must also be perceived by the members as an ethically dependable person, capable of bringing out the best moral standards in the group.

The second (Table 4.6) is dynamically explained as the cohesiveness level of a group as it derives from the *immediate gregarious and mutual respect satisfactions of the members* in each other's company, and is thus a function of, among other things, their congeniality and mutual appreciation. A group, e.g., a recreational group, can often persist, in face of many difficulties (as shown by experimental manipulations), on the strength of this component alone. The combined result of this increase of group interest and avoidance of inter-individual conflict is a much greater capacity of the group to endure difficulties and to compete

Table 4.6 Factor of Morale of Mutual Respect

	Session 1	Session 2	Session 3
High degree of group organization*	42	61	66
High degree of leadership*	68	52	73
High degree of interdependence*	31	53	58
High degree of we-feeling*	50	66	51
High orderliness of procedure*	53	46	49
High mean on intelligence (B)	24	31	46
Much explicit concern with procedure*	25	56	68
Low group mean personality test measure on Protension (L)†	−85	−05	−13
Low degree of frustration*	−53	01	−04
High strength of motivation*	46	71	33
High freedom of group atmosphere*	08	48	04
Dynamometer: High level of non-shock pull	40	64	
High group mean personality test measure on emotional stability (C)‡	47	23	−06

*These are ratings across diverse task situations averaged for a set of observers behind a one-way screen taking notes on the group performances.
†In popular psychiatric terms this means that the individuals in the group are possessed of less than average paranoid, jealous, suspicious tendencies; a contribution to morale.
‡This means that groups composed of individuals of higher than average emotional stability (C, Ego Strength) develop better morale than those inherently emotionally unstable and neurotic.

The three columns of factor loadings on the right are given as they stand, for three experiments, without averaging to show the degree of consistency reached. Note, however, that this consistency includes some thirty or forty *other* variables being consistently zero (nonsignificant) on this pattern across the three experiments.

successfully with other groups, as shown by the positive correlations (not all shown in Table 4.5) which it has with speed on group construction, ability to win a tug of war in which powerful electric shocks are dealt out equally to both sides (Dynamometer pull), and other measured "group dynamics" behaviors.

The group dynamics research with small groups has also permitted more definite findings about the relation of specific within-group, interindividual behavior rules to group success than is yet possible with large cultural groups. These findings extend over the relation of average scores on various personality and motivation factors to the total group performance, e.g., lack of sentimental emotionality in morale in Table 4.5; of leader and hierarchical structure to performance; and of styles of mutual perception and response to group performance. For example, it is quite clear that higher intelligence, higher superego strength (conscience) and lower neuroticism—calculated as averages across the "citizens"—are related to more successful performance by the group in *most* performance and "survival" areas[4]. As Table 4.6 shows, "Morale of Mutual Respect" is raised by choosing group members low in L factor (low paranoid, non-litigious or non-anarchic), high in intelligence, and high in C factor (emotional stability and realism).

4.5 THE NECESSARY EXTENSION OF WITHIN-GROUP MORAL CONCERNS TO GENETIC FUTURES

The correlational analysis of world data on the relative behaviors of groups, resulting in the emergence of concepts such as Morale, Cultural Pressure, and Education-Affluence above, as well as the very successful prediction of the behavior of small groups from the personalities and group rules involved, offer hope that social science is on the brink of being able to prove at least some connections between group survival and specific inter-individual moral values and inter-individual behaviors.

While these empirical advances are being made, the scientist concerned with deriving within-group, inter-individual moral behavior rules from group survival conditions will also be thinking ahead and asking whether there may not be altogether new areas of within-group behavior the connection of which with group survival has never been observed. One suspects that such moral tillage of fields of behavior now left wild, which the more successful groups of the future will recognize and bring under cultivation, include reproduction (that childbirth is more than the concern of the married couple involved), mass media communication (that like education, the mass media owe a duty to positive standards), pollution of the environment, rights to migration (that habits of poor morale shall not infect populations of high morale), as well as quite a variety of economic customs and regulations now treated as if they had no ethical consequence.

A brief exemplification can be given by examining the first of these possible extensions, that to reproductive behavior. Here, although Beyondist values in general are already approached by those of many revealed religions, we see a definite gap. Humanistic, rationalistic reform values have also failed to touch this area. Prior to 1880 and the beginning of Darwinian penetrations—virtually all writers on social progress and moral ideals (including, for example, Rousseau, Marx and even Mill) consider only the economic, educational, criminological, and other cultural reforms which seemed to them to constitute the totality of social progress. The whole of the biological half of human affairs they largely ignored. One may speculate that this omission was an unavoidable part of the hastiness of a short-term, pioneer perspective, or a more systematic consequence of a rooted prejudice of liberalism against (a) biological inequalities, and (b) the idea that the reasoning man *himself* needs to be reformed as a reasoner and a biological being. Progress, however, is always both educational and genetic, and both must therefore be brought under moral direction. Even when it is recognized that genetic change is an integral part of social progress, the writers on social progress have often sought to evade the issues involved in recognizing that this involves both personal rules of morality and new demands for morality of public action. In particular, it involves clear thought and action in the field of what since Galton's time has been known as eugenics, i.e., the task of improving the next generation's genetic endowment. Obviously, exploration of what our responsibilities should be in this area, and of the social action and legislation necessary to express them, becomes one of the first respects in which a Beyondist ethic is required to transcend what most existing moral systems have to offer.

Behavior geneticists have already abundantly demonstrated that all behavior differences so far investigated are *partly* determined by genetic individual and species differences (Vandenberg, 1965; Fuller and Thompson, 1960). Consequently, whatever behavior better adjusts a society for survival is to be achieved partly by genetic selection. Although no individual can be morally responsible for his own genetic make-up, he has major responsibilities for that of the next generation. This extended responsibility of the individual devolves from the group's responsibility, for the criterion of group survival and success connotes that certain directions of genetic movement can also be considered ethically desirable or undesirable. For example, there is substantial heredity in intelligence, and, if intelligence enhances group survival, moral behavior would be such as will increase the community mean level thereon. And since there is also a genetic component in sensitivity of conscience (Cattell, Blewett, and Beloff, 1955), if individual altruism is desirable for group survival then it is correct to recognize that the *genetic* constitution itself of one society could be said to be at an ethically more satisfactory level than that of another. This will be a much debated extension of the meaning of ethics.

The addition which these considerations bring to current thinking on cultural evolution is that if a culture is committed to certain cultural developments, it must, if it is serious about those values, commit itself equally and correspondingly to a certain set of directions of genetic, racial development. If the indications for group survival were that a nation needed to bend its culture in the direction of habits of, say, (a) a greater restraint on emotionally impulsive behavior and (b) a generally higher level of mathematical understanding for a complex social technology, then it would follow that it needs also to advance its genetic evolution in the direction of increasing the chromosomal substrates of higher C factor (cortical control of hypothalamic, emotional activity) and higher general mental capacity (g factor, in technical measurement terms) (Cattell, 1971).

That culture patterns and racial patterns tend to change in a coordinated fashion will generally be conceded as fitting scientific principles and common observation (Mather, 1953; Darlington, 1969). With the canoe culture of the Polynesians, for example, there has developed a physical build and constitution particularly apt for survival in tropical seas, not surprising in islanders whose forefathers survived according to their capacity to reach land from frequent off-shore mishaps. It is also noteworthy that some of the best mathematical abilities come from the Middle East, India and other roots of very old urban civilizations, where presumably a living could be ensured by success in accountancy and business, and where selection presumably has operated over more centuries on such capacities than in other parts of the world. A rich array of diverse and interesting examples from general history of genetic developments being tied to cultural developments is given by Darlington (1969) though, of course, virtually none can at present be subjected to rigorous quantitative treatment [5].

As to the importance of the principle that both the genetic *and* the cultural characters of a group contribute to its survival there can be no doubt. But penetrating research has yet to be done to find out with what frequency, and in what manner these interact. Interaction could be in two directions: (1) The growth of a culture pattern, occurring primarily through historical and geographical circumstances, could then act selectively on the genetic pattern of the race supporting it, producing appropriate genetic distribution, and (2) new genetic patterns, produced by mutations and hybridization effects, could in turn favor particular cultural creations [6].

Clear cut examples in human history of genetic mutations being followed by cultural innovations are, of course, not easy to demonstrate because no geneticist was at hand to record with modern means of observation the given chromosomal changes. It is probably from the field of animal experiment that we shall obtain the first clear and quantitative demonstration of the principle — in relation to what slight "cultures" can be set up in animal groups. The data of Scott and Brace (Cattell, Bolz, and Korth, 1972) on measured untrained *behavior* of various breeds of dogs shows that it is possible to sort the dogs, unseen, largely into their biologically different breeds *by the measured behavior alone*. In short, what dog breeders did to separate by physical appearance (breed requirements) has sufficed at the same time to separate them in behavioral tendencies and capacities. Further, the work of Tryon (1940), Rundquist (1933) and others on rat breeding by behavior selection, and of Dielman, Schneewind and the present writer attempting to impose different learnt "cultures" on rats shows emphatic "cultural" differences between different genetic "strains" (races) in the same species. These are sometimes dramatic, as when in our own experiments the maze-learning "culture" which we planned to teach to different rat strains proved actually not to be acquirable (in any measurable time) by one (hooded rats) strain, while Wistar and other strains learnt with quite different speeds.

The importance of genetic levels, e.g., of intelligence, in acquiring different cultures is unquestionable. But the acceptance by society of a moral obligation arising in regard to such facts, and its inauguration of social machinery for the necessary, technologically complex "genetic engineering" are distinct further issues. In eugenics as it has so far been understood, the improvement of some single trait, e.g., health, intelligence, resistance to particular diseases, elimination of some specific pathology, has been the limit of the concept. In what, as a more generic concept, we are here calling *genetic engineering* (or socio-genetic engineering) there will be a technical aim of producing special *distributions* and new *constellations* of genes. There are already instances known to geneticists, where, for example, it seems advantageous to the group as a whole to tolerate the existence of a disadvantageous homozygous ("pure") necessary form of some gene for the sake of the more than compensating advantages present in the corresponding heterozygote. It is just possible that the continuous genetic production in society of schizophrenia-prone individuals despite their constant elimination by the poor reproduction rate of fully affected individuals, is due to such a mechanism. In such instances the likelihood of the disease is very high in the recessive homozygous form, which is eliminated, but the genes persist in the population because of certain probable *advantages* to the heterozygous cases. Sickle cell anemia which brings severe disease in a homozygote and relative immunity to malaria in a heterozygote is a more widely known and better documented example. Since by the nature of population genetic laws the heterozygote cannot be retained in the population without permitting the recessive homozygote also to appear, the technical and ethical problems of socio-genetic engineering are challenging. The argument—a very speculative one—for not seeking to reduce the frequency of criminal or mentally disordered types on the grounds that the same genes in other combinations and patterns might be valuable, is one we shall discuss later. Meanwhile one asks, "Can society morally allow individuals to suffer to this degree for the sake of the whole community?" This issue is obscure, but two other issues that are allowed to confuse discussion on genetic engineering should be handled forthwith. First, the word race is appropriately used for a population pattern of genetic endowment statistically significantly different from another. (The word gene, incidentally, comes from the Greek word for "race.") But it may turn out that under such a statistical definition we are dealing with two very different kinds of origin. One type of racially distinct pattern may arise largely from geographical isolation and inbreeding, and have little relevance to the modern world, while another may be the result of millennia of cultural selection, and be relatively important for cultural direction of development. The "geographical race," largely the result of climatic adaptation and genetic drift, may turn out to be only trivially different and perhaps irrelevant to the differences in cultural experiment we are now considering. On the other hand, especially in the experiments of the future, the creation of new races by cultural and deliberate idealistic genetic choice is likely to become increasingly important.

A second source of confusion over genetic matters is the "insuperable objection" to genetic improvement sometimes vociferated "that no one knows what improvement is and no one has the right to direct it." What this overlooks is that precisely the same problem exists in the parallel field of cultural improvement, where no one uses the argument as an excuse for doing nothing. The direction of genetic as of cultural improvement is always only an educated guess—akin to action research perhaps, but convertible into a true experiment. And the decisions of a society can be in genetics as democratically based as its cultural decisions (assuming we are speaking of democracies). The same machinery as guides society's decisions on its educational and ethical values can guide its genetic values. The different issue of whether we at present have good enough technical knowledge to carry out the desired genetic engineering is more appropriately taken up in Chapter 6 on practical, current socio-political impacts of Beyondism.

Yet another problem which challenges society to acquire a conscience on genetic relative to cultural activities lies in the possibility that a spend-thrift society can "warm its house" culturally for a time by burning up its household genetic furniture. For example, for the military survival and cultural predominance of French culture around 1800, the finest physical specimens of two generations were drafted into the Grande Armeé of Napoleon. Patriotic frenchmen since, including de Gaulle, have concluded that this lowered the average stature of the French (and, less demonstrably, general vigor, visual acuity, etc.) and have deplored the genetic cost of that vanity. A slightly less obvious example is the sacrifice of genetics to culture (in the intellectual sense) which seems to have occurred in the middle classes of Europe over the last two hundred years. Throughout the Victorian period, in order to be free of domestic chores and economic demands restrictive of cultural "self-expression," the more gifted treated themselves to pianos, studios, and salons instead of children. These societies created a fine sparkle of culture, but did so by unnecessarily and immorally using their genetic endowments for fuel. Only recently have there been attempts to measure such effects sensitively enough to record short period changes (Cattell, 1938, 1950a; Fisher, 1930; Muller, 1966; Lentz, 1927; Higgins, Reed, and Reed, 1962). For effects glaring enough for all to see (though mixed with other mechanisms), as in Egypt or Greece, the historian has to observe over two or three thousand years. Most primrose paths of this kind are fairly obvious and could be corrected as soon as the culture gets the slightest insight into, and conviction about, the principle that morality needs to be extended to include such genetic foresight.

Genetic management is needed also in regard to hybridization, by controlled immigration. If two populations are about equal, e.g., in intelligence, but different in type, hybridization, as Kretschmer (1931), Darlington (1969), Dobzhansky (1960), Dunn and Dobzhansky (1964), Muller (1966) and others have brought out, can be a step in a possible improvement that may be permanent. On the other hand, with less fortunate conditions it can be disastrous. After the "luxuriation of the hybrid" there may come, except where stringent selection eliminates the equally numerous unfortunate combinations and retains only the effective, a long period of cultural eclipse. Finally, in the complex relations among genetic elements, there is the possibility, admittedly hard to evaluate and demonstrate but theoretically clearly possible, of what must definitely be recognized as a parasitic relation. Therein two distinct genetic groups could live in one culture, one being parasitic on the other. This is given separate discussion in Section 4.6 below.

In summary of this section, the alternative of action or inaction with respect to extending moral concern and values to genetics can take three possible forms:

(1) Doing nothing. Here societies would take steps which (one hopes) would advance culture but leave genetics to look after itself. This is the story up to the twentieth century, and may well account for the wavering rise and fall of cultures, such as Gibbon (1910), Spengler (1928), Toynbee (1947) and others have studied.

(2) Taking steps to ensure that genetic selection favors those strains succeeding in the culture, so that the genetic pattern closely follows and aids the development of the culture pattern. This has many virtues, but "culturally-led eugenics" has one weakness: that should culture cease to be provoked into changes, or move into some cul-de-sac, the genetic pattern will similarly stagnate. This adaptation we have called in another connection below "genetic adjustment to the cultural cocoon" (page 165).

(3) Perhaps with increasing guidance from "chromosomal surgery" (or genetic engineering), eugenic or socio-genetic planning can be undertaken to produce racial developments *per se*, judged to be advantageous without regard for their fitting existing cultural directions. This is "venture research" in which the very nature of the group sub-species would be altered by laboratory-created novel sets of chromosomal instructions, or by very deliberate selection and hybridization of genes already in the population. For example, chromosomal engineering might present the technical possibility of producing temperamentally more self-sufficient, less extraverted types, despite society being actually composed and led by political types inclined to spend much time in "sociability." To the extent that all venture research (see discussion of this concept in Chapter 9) is partly a step in the dark it would call for bold experimenters.

In view of the recent enthusiasm, begotten of the remarkable advances in "chromosomal surgery" for what Lederberg has dubbed "euphenics" — the deliberate manipulation of genetic material to produce desired change — a longer distance perspective needs to be asserted. Such direct gene substitution could profitably be undertaken to reduce a number of well-known specific disabilities, such as phenylketonuria mental defect, diabetes, etc. As Spiegelman, its chief discoverer, says, it is good for obvious defects only. Anything more subtle than planning to remove gross physical defects awaits as much research advance in psychology, medicine and, especially, social psychology, as has been made in molecular biology. It is in fact a *far* more complex undertaking than substituting genes if one is to discover the full individual and social, cultural life consequences of a gene substitution. The "phenotype" is not just one effect, as, say, in polydactylly, but an extremely complex group resultant. Furthermore, as discussed elsewhere in this book, the ethical question (*see* Ramsey, 1970) will come up as to how well the genetic engineering is directed to the *specific* values of the group. Even in the technical field there can be disagreement. For example, chromosomal changes might be found to increase sheer mental capacity, but some might argue that this will not then increase creativity — though Spearman (1930) and Burt (1940) could in this case produce evidence that it does.

Clearly the new demands introduced by a Beyondist morality into inter-individual dealings and social obligations call for both positive eugenic pursuit of the culture (Type (2) above) and the attention of practical politics eventually to genetic engineering (Type (3) above) with the general aim of increasing man's potential to command a cosmic environment yet imperfectly contacted. This, however, is only one instance — though an important one — of a needed increase in the realization that the survival criterion will markedly extend the demands on inter-individual morality into new areas.

4.6   THE ELIMINATION OF PARASITIC BEHAVIOR AMONG CULTURAL INSTITUTIONS AND GENETIC SUB-GROUPS

A definition of ethical behavior within groups which takes general group health as its criterion must at some point handle the phenomenon of parasitism. Parasitism, which runs through the biological world, and constitutes a special problem for any argument connected with evolution, needs far more subtlety in its definition when applied to human social groups than is at first apparent. All ecological balances are "parasitic" in the sense that one species or group would suffer momentary problems of adjustment if another species were removed. Men are parasitic upon cattle and cattle upon grass — and vice versa. If the term is to be useful it must be distinguished from symbiosis — mutual usefulness. Without space for indefinite philosophical refinements, let us say that where (a) one takes far more than it gives, (b) depends wholly on the second, having no other means of support, and (c) perishes when its host perishes, we have a symbiosis of a parasitic kind.

Now within a society the normal condition of its coherence is that all are necessary to the life of each. But parasitism can be defined as a relative matter and, in approximate terms, half the membership may be conceived as giving more than it takes and the other as taking more than it gives. By definition the latter section is parasitic. To the extent that it endangers the life of the group it is immoral or less moral than the other part. In the broadest sense parasitism and relative immorality are synonymous.

It is the inescapable tragic element in the long process of evolution by groups (which largely determines the basic evolution of the individual as a physical and ultimately spiritual organism) that when a group perishes, the altruistic and the parasitic go down together. Setting aside, for the moment, the human sense of injustice and futility which this begets, the important fact is that this greatly slows down evolutionary advance by reducing the efficiency of inter-group natural selection. To the extent that parasitisms can be avoided in competing groups, the whole inter-group process of selection is heightened in efficiency and rendered more reliable.

Mankind has justifiably become almost paranoid about parasitism. It is justifiable because survival depends on vigilance. Groups have to be sophisticated about parasitism just as the small animal in the jungle has to be infinitely suspicious about every strange sound or shape if it is to survive. There are, consequently, constant allegations of parasitism against classes, races and professions. The Greeks were suspicious of leaders becoming tyrants; the undeveloped countries pillory "colonialism"; Marx saw the managerial, entrepreneurial class as parasitic on manual workers[7]; Voltaire and Rousseau thought the black crows called priests were the chief parasites; Hitler and the instigators of Russian pogroms accused the Jews of parasitism; suffragettes (and "women's lib" today) see men as parasitic; various groups regard the Rhodesian whites as parasitic on the colored, and some regard the American colored as parasitic on the whites . . . and so on.

Parasitism and the suspicion and distrust it engenders, whether it be by man or tapeworm, has always been an unpleasant topic. It would be comforting to believe that when real it is self-terminating. If marriage becomes one-sided in its satisfactions, will this not lead to divorce? And if the worker is under-rewarded by the manager will he not go elsewhere? If the anthrax germ kills an animal will it not perish with it? This optimistic view is untenable, especially to a man trying to get rid of a bad virus cold in the head! A failure of the group or organism is doubtless a final solution and also a clear proof that the symbiosis was parasitic, since it did one partner no good.

A classical simple experimental study of parasitism is that set up by Mowrer and the present writer's son in which two rats are put in an elongated cage with a bar-press at one end which releases pellets of food at the other. One rat will discover in the usual operant conditioning fashion that pressing the bar yields food which he can get at the other end of the cage. But the second rat discovers that by staying at the far end he can get the food before the first rat (the worker or entrepreneur, according to political taste) is through with his job. What then happens? Usually rat 2 gets satiated and eventually permits the enterprising "discoverer" rat time to get himself some food. It is very rare for him to learn himself to operate the bar. The society persists as a somewhat unsatisfactory welfare state. But it can also happen that the goodwill of the operator extinguishes — he gets insufficient reinforcement for his efforts — and in that case, worker and parasite perish together.

If we think of the institutions of organs within a society constituted by business, industry, defense, religion, science and education it is probable that some severe parasitisms have existed among them at various stages of history. Obviously the Christian religion contributed, in the early Middle Ages, far more to the health of the societies in which it worked than it received in return. Civic order and the strength of armies gained from the morale it created, but the "feedback" from the communities aided was for long so weak that only vows of poverty and celibacy by the devoted few made this service workable. By the time of Henry VIII it is just possible that it had in turn become parasitic. Karl Marx saw it in yet another way, when he claimed that "the opium of the masses" enabled management to be parasitic upon the workers.

In perhaps no age up to the present has society as a whole been anything than parasitic upon education, i.e., it has drawn economic and other gains from education without having the wisdom to support it up to the point of optimum return on investment. It took the performance of the Prussian troops in 1870, raised on a new system of free and compulsory education, to persuade hard-headed governments of neighboring countries that this was a desirable cultural innovation. But in the last one hundred years the biggest victim of parasitism has been science. Business has grown rich on it; the worker has experienced undreamt of increases in real standard of living from it; while leisure, travel, and luxury have increased for all.

If the average man is asked why he lives better, with far more leisure, and longer, than his great-grandfather in a similar occupation, his complacent smile will usually lead on to an uneasy, and vague assumption that he must be a better man than his benighted ancestor. Intelligence tests show that recent generations have not gained, and perhaps have even fallen, while in terms of political management or moral fiber our society has made no radical gains. The benefits have demonstrably sprung from the work of a quite small scientific elite. Their dedication, hard work and habits of objective, disciplined thought are so remote from the average man that when these men bring their thinking to bear on socio-political problems – as the Nobel physicists and chemists Soddy, Pauling and Shockley, for instance have done – the average man's sentimental feelings are outraged. In any case neither the average man of democracies, nor the aristocrat or priest of older societies are prepared to reward science as it has rewarded them. In a few countries like the U.S.S.R. and U.S.A. a higher percentage of the national income is fed back into science than in the bulk of the world's countries (prepared to be parasitic in science and in medicine upon the former). Yet it is still done so grudgingly (as in the moon program) and inadequately, in terms of how large the returns are, relative to those from other investments. However, we can be confident that in the course of inter-group natural selection the countries that learn the morality of taxing themselves more for science than for more obvious things, will prevail.

The fact that parasitism requires complex considerations in its definition, and that many accusations are wild, does not mean that it is not a serious problem, or that with sufficient social-scientific attention degrees of parasitism could not be more accurately quantified. A social scientific "cost accounting" analysis will doubtless develop in connection with Beyondist morality. It will take a long and far view, over time cycles and area (the embryo is not a parasite). A perennial issue in this field, leading to constant dislocatory strikes and lockouts, concerns the relative earnings appropriate for different occupations, and this gets crudely polarized as between "workers" and managers (or "capitalists"). An initial naive conclusion that constantly sets social brush fires alight around the world is that if person or institution A is better off than person or institution B, in the same community, A must *ipso facto* be "exploiting," i.e., be parasitic upon, B.

In examining this, let us be clear that we are not asking whether, on some particular ideal value system or another, two people in the same community should have a particular difference of living resources. (Nor are we falling into the belief that binds the naive of all the earth that differences are synonymous with injustices!) The question stops short of that and simply asks whether, if *no* re-distribution from the equilibrium point naturally reached takes place, the differences in earnings may directly represent differences in absolute contribution to community wealth. Let us imagine an experiment in which a not very bright and definitely lazy people are living at a low standard of living x. There arrives in their midst a set of intelligent and planful strangers who, without altering the number of hours that the population is accustomed to work, begin to make a much better utilization of the resources. They so dedicatedly plan their own lives, that they live at a real income level $4x$, while the natives are raised to $2x$ from a former level of x. History suggests that it will now almost inevitably happen that the "natives" will claim they are being exploited by the newcomers since they live at half the material standard of living of the latter. Despite their absolute living standard having actually been doubled by the arrival of the strangers, history tells us that they may indeed, either expropriate the "managers" or destroy them. Emotional reactions, chiefly of envy, cause them to be more concerned that the newcomers are living at an income twice their own, than that their own income has been doubled. This allergy has repeatedly caused societies to fall back to lower standards of living.

Generation after generation we see the young and naive (and the old and envious) distracted from the real issues of social progress by this particular red-herring derived from a misconception of parasitism. It is noteworthy that Communism, starting with the notion that all must be subordinated to the welfare of the manual worker, was soon compelled, in the interests of rewarding production and maintaining real standards, to readjust by raising the rewards for good management. Indeed, the earning rates of managers stabilized at much the same ratios to semi-skilled and unskilled earning rates as in capitalist countries. In the case of the scientifically gifted and productive individual, Communist Russia seems to offer even greater rewards, relatively, than do capitalist countries. (Parenthetically, this argument does not involve at all the question of *inherited* wealth, where Beyondism would have to take the position that it should be phased out at the same rate as that with which genetic eminence disappears, in families, which is fairly fast.) In any case it is clear that an evolutionary morality definitely repudiates Jaurès' socialist slogan "From each according to his capacities; to each according to his needs," though it would support the charitable provision of basic life *securities* for other reasons. A more intensive examination of the economic implications of an evolutionary morality is made in Chapter 8, Sections 8.2 and 8.3.

The basic objection to parasitism as we have seen, is that it renders the already difficult problem of ensuring an efficient inter-group natural selection still more prolonged, painful and wasteful. It dirties the test tubes of the experiment. In its broadest sense parasitism can be between individuals, between institutions, and between cultural customs and habits, and its effects can operate on both cultural habit and genetic elements. Social science should therefore devote intensive study to determining not only the morality of standard popular habits and values, but also of institutional structures, to determine that each is "pulling its weight" in terms of the viability of the group in cooperative competition. Precisely parallel arguments hold for the genetic substrate of a group, wherein genetic parasitism, running longer from generation to generation than the cultural habit parasitisms, need to be brought out and remedied.

As to the last, it is consistent with the general deduction of within-group values that genetic considerations require an adjusting of the eugenic conditions *within* groups to align with those demanded by survival *among* groups. In the last two thousand years of history — that of pre-scientific civilized cultures, dominated by universalist ethics — there have obviously been no effective attempts whatever (except for a dawning awareness as Rome declined) to discuss or bring about that alignment. At least in certain cities in the U.S.A. today the survival rate (birth rate minus death rate) and the economic incentives (cost of a free, state-educated I.Q. 70 child versus the family college expense of an able I.Q. 130 child) would, if continued long enough in time, largely replace the average and brilliant by sub-average intelligences. Though the possibility is fortunately at present still remote and hypothetical, there can be no doubt from our growing insight into group dynamics that such a replacement would sooner or later cause a breakdown of society as we know it. (A primitive subsistence society of poverty, disease and anarchy could, of course, persist. A discussion of likely changes in a society with falling I.Q. has been given elsewhere [Cattell, 1933a, 1937a].)

The parasitism problems with regard to aggregates of genes are even more subtle than those of cultural habits and institutions. Harmful recessives are constantly present as a "genetic load" in the populations of all countries and can survive — essentially and by definition parasitically — by association with other genes. A deliberate and potent eugenics program would aim to extirpate these, not hide them. As an example of the complexities, however, we may take the issue of operating in the same society with genetically different sub-groups. Let us take a case where one type has genes more favorable to high intelligence and the other to resistance to malaria. A society composed of the first type might succeed as a society by virtue of its gifts of intelligence (and malaria deaths need not reduce the total population, granted an appropriate birth rate). On mixing the two, however (in a malarial environment) the differential in immunity endowment to malaria would result in the intelligent maintainers of the culture being completely replaced by lower intelligences. Or again, in a welfare society, any tendency of a group to a birth rate less controlled by social standards—and this normally happens with the less intelligent and the less temperamentally foresighted—will result in that genetic sub-group inheriting the society. Such a society by its cultural momentum may stand for long, but eventually just like a great ceiling beam, which is eaten from within by death watch beetles and no longer what it appears to be, it crashes and has to be replaced. Thus it is a general rule that inter-group natural selection *eventually* takes care of a faulty within-group natural selection. But it is a slow and costly safe-guard and the aim of within-group moral scrutiny should be to avert these crashes by a regular cost-accounting for parasitisms, cultural and genetic [8].

4.7 THE RIGHT AND DUTY OF A SOCIETY TO PURSUE ITS OWN CULTURO-GENETIC EXPERIMENT

The hope of evolutionary advance lies in bold and well-directed experiments in human social life. Hopefully, these experiments (as suggested in the initial overview given by Chapter 3 and the suggestion for organization in Chapter 9) will occur in some comprehensive framework agreed upon or at least acceded to by a federation of powers on a world wide scale. To wind up our consideration of the ethical position concerning within-group, inter-individual behavior, it is necessary to turn now from what we called the *common* moral values, necessary *to maintain a group as such*, to the *unique* moral values which follow from the specific nature of the experiment planned by each group.

In the framework of Beyondist ethics such experiments are both a duty and a right; but in our present crude beginnings of a world society, as represented by U.N.O. and U.N.E.S.C.O. and certain would-be universalist religions, there is much confusion about these rights to unique community values and virtually no conviction about the duty to develop them. There is confusion, for example, between the fact that each culture should be prepared to accept another's right to its own peculiar culture, and the right which each culture nevertheless has to "lack enthusiasm" for the particular directions of development in another culture.

A glance at history—and in fact, especially at recent history—shows constant political, propagandist and military attempts to interfere in the internal values of other countries. Democracies seem to see it as their right to upset Fascism or Communism in other countries — and the behavior is reciprocated. Mohammed upset the religious values of other countries by force; Britain forbade the ceremony of Suttee in the Hindu religion when it controlled India; the U.S. beams "subversive" ideas by radio into communist countries, and so on — the tale is endless. Whether this internal attack is a legitimate branch of international cultural competition we shall discuss in the next chapter. But if it is, it should be carried out without the hypocrisy of our expressing moral indignation that it occurs.

This hypocrisy operates both ways. In U.N.E.S.C.O., for example, there seems to be a social and intellectual taboo on men frankly avowing that they prefer their own race and culture. It is an expression of the power-seeking of a short-sighted committee that they impugn this right of racio-cultural self-determination, while abrogating to themselves the right to interfere in the internal cultural and government affairs of individual countries. If a really wide spectrum of forms of government deserves to be tried, then fascistic and communistic governments are as necessary and desirable as *experiments* as are various democratic and theocratic governments. It cannot seriously be argued that they should be discontinued unless, in addition to our emotional aversions, we can assemble social scientific proof that they enter the class of what we have designated moribund cultures. And in that case, if the social scientists of the countries concerned are convinced, they will themselves take remedial measures.

The spirit of brotherhood which is part of the cooperative competition central in the Beyondist ethic should express itself in a "wish you well" blessing on the racio-cultural course adopted by each group, even though it is very divergent from that of the viewer. This liberality should nevertheless be accompanied by each culture having a shrewd bet that certain courses are wrong and a firm intention not to contaminate its own experiment by embracing into its system the racial or cultural emphasis of another. Two problems arise here which education in Beyondist principles needs to overcome. First, it is necessary to control the instinctive aversion to foreign ways of doing things, rationalized as moral fervor, which we have just discussed. This requires recognition of the difference between *common* and *unique* inter-individual morality, and avoidance of imperialism in the latter, based on false standards of "universalism" in ethical values. Beyondist principles should never allow one to lose sight of the fact that the required cooperative competition between groups has to be both racial and cultural, and that the within-group ethics have to adjust to this. Parenthetically, "racial" has little relation to races as they presently exist, which, as mentioned, are largely geographical accidents, but refers to the improved genetic variants more deliberately generated as described above in relation to explicit ideals of genetic pools.

A major problem in developing an internal group ethics that is workable is likely to arise in connection with handling envy, if wider considerations should require that society have a rather large range in genetic endowment. Envy is a problem also in regard to uneven distributions of wealth, leisure and other "goods," but there it is more amenable. For as Lloyd (1971) points out in his very interesting psychological analysis of the last days of Christ, envy of personal superiority in natural gifts, e.g., in intelligence but above all in character, is the most intractable of social problems, and, according to his analysis tends to bring out manic and even hostile defenses in the benevolent leader. Most religions have considered envy and jealousy as evil impulses needing control, and Christianity has placed it among the deadly sins. But Communism adjusts by bending to it, mainly, it is true, in the realm of goods.

If, as is possible, Beyondism should require as a condition of the most effective group survival potential that considerable variety and range exist in genetic endowment within a group, then it cannot entertain that envy and destruction of excellence which has lately been construed in some societies as a legitimate expression of democracy. Actually, when we come to immediate practical steps (Chapters 8 and 9), Beyondism suggests that since there is a sense in which a democracy depends on births as well as culture, it might be very desirable to *decrease* the intelligence range (in the cybernetic revolution) — but if so by breeding out below an I.Q. of say, 90, rather than above 140. Yet in other genetic characteristics the call might well be for greater range.

Just how a moral goal and an educational system to approach it, can be set up to handle envy, cannot be our concern here, where the concern is with the derivation of the group's inter-individual ethical values themselves. (Later chapters handle specificities.) But it is worth discussing in itself, as an illustration of the derivations of values, what a shift to Beyondist goals would require. Certainly it is easy in some moral system to become morally indignant at the "injustice" of another by sheer luck (as far as he, but not his parents are concerned) inheriting, say, greater mathematical ability, or more handsome features, the capacity to get along on six hours sleep a night or some other gift effortlessly received from nature. A political assertion that all are born equal is not a bad salve for our feelings. A religious conviction that all souls are equally significant in the sight of God is a more fundamental gift to perspective. A less intelligent, but unfortunately more widespread nostrum for envy than either of these sound principles is the notion that "environment can do everything" and that "There, but for a mistake by those who brought me up, go I." Great though the virtues and gifts of education are, they do not include the questionable virtue of being a panacea for the imaginary ill of inequality. For an effective education, applied to talents with sufficient intensity, merely magnifies the effects of innate differences. And with respect to improving the *average* performance of mankind the widespread assumption that education is our most optimistic creed, while concern for genetics is pessimistic, is a complete misunderstanding. We should be thankful that our psychological make-up is rather obstinately determined by heredity. At the least, genetic laws tell us that one or two generations of abstention from formal education (such as have occurred more than once since medieval times even in European countries) will not throw man back to the ape. And on the positive side let us reflect that a genetic improvement in the average inheritance of, say, intelligence, or immunity from diabetes, will accomplish once and for all, in one generation, what otherwise has to be maintained by painful, expensive environmental efforts repeated generation after generation [9].

As indicated in our first encounter with genetic engineering above, the social and scientific mechanisms *per se* must have their discussion postponed to Chapter 8. Here the issue is simply to realize that the *unique values* in a society will include a *positive, unique genetic target*, and that far greater mutual tolerance and respect are needed among groups in relation to these divergent goals.

At this point the group aiming to improve its genetic status encounters the objection that although applied genetics has made enormous and convincing improvements in cattle, race horses, corn and peaches, it does so because there is a consensus of expert opinion as to what is wanted. Who would have the presumption, we are asked, to say what constitutes a good human being? Yet the very opponents of *eugenics* who use this argument proceed unquestioningly in *euthenics*, e.g., in such matters as education, living standards, e.g., for public welfare, and in the cultural and moral upbringing of children—to set a definite target toward which *education* is to aspire [10].

If a society has the right to shape the young citizen toward a socially approved ideal, as God permits it to see that ideal, then that right extends to genetic as well as educational improvement. Or, in other words, and by a less traditional, metaphorical definition of the sanction, the Beyondist ethic of *community adventure* commands a community to clarify and aim at its ideal as much in racial as in educational, cultural matters. A possible source of the relative hesitation to act as deliberately in the former as in the latter is that educational steps are viewed as reversible, whereas genetic steps are not. Actually, this is a matter of degree; neither the misbegotten nor the miseducated individual is fully "reversible." In genetics our steps certainly need more consideration and are closer to a precipice, but since in any case steps have to be taken, i.e., we reproduce, it is immature to shut our eyes and leave all to chance. A second do-nothing argument (i.e., let accident prevail) is that we should not act "until we know enough." To this the answer is that genetics is even today no mean science; but that even before a single textbook on genetics existed breeders and horticulturists produced immense developments in domesticated animals and plants. This becomes a special case of the principle of "venture research" discussed more fully as a concept in Chapter 9 and elsewhere, pages 167 and 433.

Admittedly, if there were only "one world" and the choice of genetic direction had to be made once and for all, for the whole of mankind, it might reduce us to paralytic hesitancy. But, by the Beyondist design, all societies are experiments, deliberately varying in their ideals, in the interests of allowing natural selection to work among them in a grand design. With hundreds of alternative developments being made each society may try out to the full its own cultural and genetic variant without a sense of final risk for all mankind. But with this right to choose there goes also the obligation to make the most thorough trial of the preferred type. The answer to the corn and race horse argument (and to Aldous Huxley's *Brave New World*) is that no one knows what the best type of man is, culturally and genetically; but that to find out we have to experiment. The criterion remains in the inter-group natural selection domain: it rests on the judgment that the best type will produce, for the space of the given era, a more successfully surviving society.

It was perhaps scarcely necessary to stress in the introductory argument above that aiming at a genetic target does not mean aiming at a single racial type. A culture is not concerned with an ideal man, but a fine variety of men. Socio-genetic guidance has to design ranges and patterns of genetic distribution fit for whatever complexity of culture seems likely. While the statement that "It takes all sorts to make a society," is usually a resigned confession of submission to a mismanaged, degenerating population, yet, it is true in a different sense that given freedom of choice, any intelligent community planner needs deliberately to encourage a specific optimum in diversity of types (*see* Beadle, 1963; Spuhler, 1967). By optimum is meant optimum for the survival of the group, by meeting the varied cultural needs of a well-functioning society. Though it is easy to see this in principle, no social scientist would deny that it still remains an extremely complex technical question as to what degrees and kinds of genetic variety would be expected to maximize evolutionary progress.

If sufficient emphasis above has now been given to the importance of a pattern of diversity, let us nevertheless recognize that the crux of the planning issue can most simply be illustrated and discussed by asking how the *central* tendency in the target is to be decided upon. Certainly crassly to encourage the largest *possible* genetic variability in each group, with no regard to a central tendency, would be absurd. In the first place, that would tend to negate the design of having significant experimental differences—relatively—*between* groups. Furthermore, as far as history and group dynamics research indicate, it seems likely that an optimum diversity can easily be accidentally *exceeded*, both in terms of cultural habits and of genetic variability. The ancient Hebrews recognized this truth in the legend of the Tower of Babel. One suspects, however, that when genetic science addresses itself to this question it will, in the interests of having a genetic reserve to ensure survival in the face of a great diversity of environment challenges, ask for a greater variety than the designer of a good *culture, per se*, would perhaps want. For otherwise there are several ways in which too great a span of types upsets the morale and effectiveness of a culture in its existing life.

Granted that in principle each group must pursue a distinct socio-genetic experiment, and that goals are to be chosen by the group itself, one may ask, since the idea is new, precisely what combination of socio-political machinery and technical wisdom is anticipated as the effective means. The investigatory and executive machinery by which *genetic* decisions are made need in principle be no different from that by which the parallel *cultural* decisions are brought about. In a democracy, discussion will shape the moral ideals of the freely developing citizen. But since—as in principle a Beyondist must—we consider that all types of political organization ought to exist among the societies of the globe, then we ought consistently to expect that the genetic decisions will be made in each by the same kind of group authority as makes its educational decisions[11].

The coordination of eugenic and euthenic ideals could occur, as seen above, in two ways: (1) By genetic change following and fitting to the culture (in which case *eugenic* and *dysgenic* translate to "culture-genetic-positive" and "culture-genetic-negative" respectively), or, (2) by the bolder path of directly inducing that genetic change which promises greater powers, followed by observation of the type of culture that this produces. In either case, culture and race will tend to move into a positive conforming relation provided internal group selection is operating well. For example, it is conceivable that a great cultural interest in athletics will lead, by marital selection, etc., to genes for powerful physique. Conversely, it is hard to imagine a spontaneous or induced genetic variant which produces fine singing voices not leading to a cultural development of musical interest in choirs.

The technical resources for a society's genetic campaign are (a) mutation, (b) hybridization, followed by (c) natural selection, which latter can be effected either by (1) differential birth rate or (2) a differential death rate. Space forbids discussion of technicalities here, in spite of the need to dispel myths and misconceptions which clutter popular discussion, but further attention is given in Chapters 6 and 7 and in a note [12].

However, one technical point is especially relevant to the social and moral considerations with which we are here principally concerned. It is that in the future a scientifically controlled increase in mutation rate is likely to play a far more important role. What seems to negate this approach is that for every mutation produced that is favorable, about 1000 appear that are bad. This fear of a "genetic load" of regressive mutations, especially through accidental nuclear radioactivity ironically seemed to blind Hermann Muller (1953), the discoverer of the effect, to its probable usefulness [13]. He sees only the danger. The level of "mutation pressure" (however induced) that a society can sustain depends on its readiness to face a high birth rate and a correspondingly high selection rate (in nature, by death: in civilization by birth control for defectives). Civilized societies, long cushioned against such fierce selection, would probably declare the attempt to handle an increased mutation rate by decidely increased birth and death rates impossible. Indeed, in rejecting it they are likely to declare it ethically wrong though what is more certain is that it would be economically prostrating.

Fortunately, science is already offering a promising alternative. It seems likely that the majority of the undesirable mutations could be caught shortly after conception, and aborted. In plants, Dr. R. G. McDaniel recently pointed to a means ("mitochondrial complementation") of finding which hybrid in corn is a possible improvement without waiting for full growth. Human medicine is already able to recognize genetic failures of certain kinds in utero — by amniocentesis — and is avoiding births of the grievously malformed or maladapted. Such developments are needed, for even with the present natural mutation rates any society committed to small families and a virtual arrest of natural selection would otherwise be heading for a slow but catastrophic degeneration. (Reliable estimates give twenty-five percent of all conceptions as failing to develop to a fitness to survive at birth, fifty percent of these by chromosomal defects.) If the well springs of genetic advance available in stimulated mutation and chromosomal engineering are to be used it will be necessary to have an increase in conception rates accompanied by a monitoring of gestations. Ultimately the task becomes a very complex scientific divination, because it is not a question simply of eliminating physical defect and physiological abnormality but also neurological deviation incompatible with a healthy social life.

The parallelism of the mechanisms of cultural and racial self-determination as they concern the *unique* part of group values is less complete than the parallelism for maintenance values. Genetic adventures need greater time to ripen and show their qualities. This marked difference in tempo in the rhythms of variation and selection in the genetic and the cultural experimentation almost certainly has extremely important consequences, not yet worked, for the interaction of race and culture. For a population of a given (diverse) racial composition there is probably a virtually infinite series of possible cultures — though different from the infinite series possible for another. One consequence may be that an appreciable chance of a best fitting culture being found exists, and that the next move is then up to genetic change. If so the latter is more important than we realize. The kaleidoscope of cultures must not then blind us to the possibility that it is the genetics of the population which defines what culture will ultimately be most stable.

An undoubted common feature, however, to genetic and cultural movement is a two-phase rhythm — that of admixture and solitary digestion. In culture, Toynbee (1947) has talked of a desirable rhythm of interaction with other cultures and "times for withdrawal and consolidation." The lesson which Darwin taught biologists from his observations in the Galapagos Islands — that isolation of some kind is a necessary condition for new species formation — has still not become part of popular thinking. But, for those who can read, it is written large across the history of human racial advance that there is a grand rhythm of a phase of hybridization followed by a phase of seclusion and inbreeding. The aim of the latter is to produce a clear predominance of the particular new genetic mixture pattern that is proving most advantageous. In a country such as America, which has had perforce to praise the melting pot, there is perhaps a tendency to forget that a melting pot needs also to boil on its own awhile to remove the dross[14]. On the other hand, a misapprehension of a different kind may be impoverishing the genetic experiments of countries like Sweden, Spain, Egypt, Iceland and Israel which have developed relatively pure strains without much recourse to hybridization.

Much remains to be discovered and thought out regarding the institution of the unique genetic developments of societies and their coordination with unique cultural developments. However, the argument from an evolutionary criterion of morality is clear: that the consciences of men in communities must be concerned not only with the ethics by which all societies are maintained, but also with the values unique to the cultural and genetic enterprise of the given society.

Along with the right and duty of a society to pursue its own culturo-genetic experiment goes the need to develop research and executive institutions to monitor such developments. A view of the political and scientific nature of such within-group moral guidance centers—moral cybernetics institutes if one will—is developed in Chapter 9. They should suffice to ensure such continuous, well-informed progressive movement that revolution will be swallowed up in evolution. Societies will need to develop a sensitive socio-biological information, computing and research organization as different from anything we now have as the nervous system of a higher vertebrate is from that of a jellyfish.

Before leaving the conception of a scientific research organization within each group devoted to this purpose, one should, however, note that those scientists will have to be as proficient in deriving inter-group as inter-individual behavior values. For the best direction of evolution in any one society will require an alert observation, measurement and analysis of the course of all groups, to get the benefits of a *comparative* science of evolution.

4.8   SUMMARY

(1) The aim of the present chapter has been to explore the derivation of within-group moral values—the values necessary to group maintenance and progress—within the framework of the evolutionary position stated in the previous chapter. Any group will have certain values that are shared with all groups—called common maintenance values—and others—which may be called unique community values—peculiar to itself. The former are concerned simply with keeping *any* group functional as a group. They are the non-relativistic moral values deducible from the fixed

goal of group survival. The latter are concerned with advancing the group in the special experimental direction it is choosing to explore.

One should guard, however, against the mistake of conceiving the latter as "relativistic ethics" in the currently used sense of subjective values, culturally local and unrelated to general principles. For the unique parts of the within-group values are still deducible from an attachment to the goal of human evolution. They vary only as the bearings of different ships headed for the same port vary. The human need for ultimate constancy is not denied by a scientific ethics.

(2) Since the desired laws of inter-individual behavior are those which insure greatest group viability, the necessary first step, before social scientists can set out empirically to discover the laws which maximize this criterion, is to define and illuminate the criterion. Some five largely independently determinable criterion measures are suggested (page 115) for evaluating groups on a continuum between moribund, unlikely-to-survive, and highly viable cultures.

(3) The present level of technical development of the social sciences is pathetically inadequate for understanding far-reaching cause and effect relations between individual behavior and group survival. They are thus at present incapable of determining in anything but a most approximate manner the requisite laws of individual moral behavior for ensuring high viability. Nevertheless, there would probably be a reasonable adequate consensus from experts in the existing, non-quantitative social sciences, notably history, that already natural selection among groups, i.e., a prolonged practical application of this criterion of group viability, has been the means of generating and maintaining of tolerably effective inter-individual moral rules[15]. Even at the animal level (see Lorenz, 1966; Tinbergen, 1959; Eibl-Eibesfeldt, 1970) the behavior in humans that is in our sense ethical is paralleled at an instinctual level. In man, prior to the present Beyondist plan to recognize and derive by research ethical laws from group functioning, moral values came by "divine" revelation bequeathed to inspired religious leaders, such as Moses, Buddha, Mohammed and Christ. A very large number of such inspirations must have occurred, and those that have survived have been shaped by further trial and error, followed by natural selection maintaining the groups with more felicitously adaptive belief systems.

Both by study of these animal and early historical adjustments and by simple inference along Beyondist lines it is obvious that the central teaching on within-group, inter-individual relations is the importance of love and treating one's neighbor as oneself. The innate loneliness of the individual is bequeathed him as a guarantee that he will seek the oceanic experience of love for all his fellows – the subjective experience of the objective truth that only the group can be immortal. The fact that science seeks ethical "rules" need not and must not blind it to the fact that these are only a guiding framework and that the breeding and teaching of spontaneous and positive love, sensitivity and altruistic enterprise is also a defined requirement of the group criterion.

(4) The insightful and precise relating of the behavior of individuals to the performance of the group belongs to the future, and depends on genius in the development of experimental and quantitative social psychology. Two basic existing approaches – one with small and one with large groups – are briefly set out here which aim at (a) a precise and structurally meaningful quantification of a group or its culture pattern, and (b) an attempt to relate the patterns to the characteristics of population behavior and role structure. In small group dynamics experiments dimensions called morale of leadership and morale of congeniality appear in the area of interest here. In large, national cultural groups, out of a dozen descriptive factorially independent dimensions, three seem particularly relevant here: cultural pressure, affluence-education, and group morality level. Each is expressed in diverse measurable group behaviors and the measures permit drawing a *syntality* profile by which a culture's pattern affiliation can be calculated. The *synergy* of a group is defined as the dynamic part of this syntality.

(5) Although our concern cannot extend here to the technicalities of social psychology a gross statement of the core model is relevant to other arguments. We operate with a two-stage model in which, first, the syntality dimension vector, S, is considered derivable from population characteristics, P, and the group structure and resources, R, thus:

$$S = (f)P \cdot R \qquad (4.1 \; above)$$

and the viability (survival potential) of the group is in turn estimated from the syntality, S, and the environmental conditions (implicit in b values) expressed in the simplest linear, additive model thus:

$$V_i = \sum_{x=1}^{x=k} b_x S_{xi} \qquad (4.3)$$

(*See* specification equation on page 135.)

In as much as the k factors in syntality, S, include also such endowments as natural resources and factors influencing survival which are not morality factors, survival is not wholly determined by internal morality (measured in population behavior, P) and cannot be taken as an estimate of survival without partialling out statistically the scores on the other factors, such as ability traits in the population, generous natural resources (in R), etc.

(6) A requirement which *Beyondism* brings out clearly, and which is neglected by existing within-group ethical systems, is that behavior affecting the genetic make-up of the group comes quite as much under moral law as behavior affecting the culture. Genetic make-up can be changed in a culturo-genetic positive system simply by letting the culture select for closer genetic adaptation to itself. But it seems desirable also, in the interests of avoiding stagnant equilibrium of encapsulation in the "cultural cocoon," not merely to let culture lead genetics, but to adventure directly in genetic engineering. In most contexts the use of the term race here does not refer to existing races, mainly products of geographical isolation and selection (Coon, 1962a), but to groups carrying new divergent patterns of gene distributions produced by either cultural selection or deliberate genetic creation.

(7) Cultural and genetic *inter-group* natural selection, as will be more clearly evident in the next chapter, is inherently likely to be more prolonged, vacillating and inefficient than selection among individuals (in a steady environment). If a million years has been necessary to bring some finish to physical man, perhaps three million years of group natural selection will be necessary to develop social man. One of the chief reasons for this is the possibility of within-group, inter-individual parasitism, which is a special form of negative morality, in the parasite, and reduces the efficiency of inter-group selection. Although a reliable definition of parasitism is subtle, it can be made. An important aim of deriving an objective assessment of the morality of within-group, inter-individual behavior is the efficient elimination of parasitic and criminal behavior. Without this the tragedy of the worst pulling the best down to destruction with it becomes a wasteful, endemic cause of breakdowns in culturo-racial experiments.

(8) Both in cultural and genetic advance, the means are (a) production of variability, i.e., the trying of new mutations or borrowings, (b) hybridization, i.e., the trying of new combinations, followed by (c) withdrawal for consolidation, and (d) the elimination of faulty varieties that do not aid group survival. The rationalist reformer is quite apt to shut his eyes to the last requirement. But by the fact that more innovations, genetic or cultural, are bad than good, this elimination has to be severe. Science is likely to provide means, for those countries that avail themselves thereof,

whereby individual genetic mutations can be more rapidly tried without costly increases in birth and death rates. There is also the possibility of reducing loss by trying out culturo-genetic mutations in small experimental groups. Incidentally, equal importance is given throughout these considerations of evolutionary inferences to cultural and racial effects, but because the latter have been grossly overlooked in many sociological texts it has been necessary to give somewhat more detailed explanation to them here. In their reciprocal relationship the fact that cultural mutation and selection takes place at an altogether faster tempo points to consequences, not yet closely worked out, in the form of the culture tending to adjust more thoroughly to the genetic possibilities.

(9) Regardless of the means of evaluation of inter-individual, within-group, common and unique, cultural and genetic values, it is, in the light of Beyondist ethics, the right and duty of each group dedicatedly to pursue its own variant. A rhythm of "hybridization" (cultural and genetic) and withdrawal-with-consolidation in regard to the new pattern has been and needs to be characteristic of the evolutionary process for groups. However, this is a necessarily slower rhythm for genetic than cultural experiment. Societies are justified, despite supposed "liberal" arguments, in screening at their borders, in order to maintain the integrity of their own culturo-racial experiment except for deliberately planned hybridizations and borrowings. For this purpose, and to get skilled guidance on both moral and unique cultural values, it is likely that social science research centers for moral cybernetics will be set up in each group, additional to international comparative research centers. But discussion of their relation to existing scientific and government organization is deferred to Chapter 9.

(10) The explicit and constant reference to group survival as the criterion of morality of individual behavior must not be misunderstood as making the group more "important" than the individual. Individual and group are links in an endless causally interacting chain, each indispensable to the other[16]. It is from the mind of the creative individual, reacting to the situations created by the group, that the group alone draws its capacity to live and grow.

4.9   NOTES FOR CHAPTER 4

[1] The imperialisms of universal religions of course differ. For example, some demand a brotherhood of man which precludes loyalty to *any* one nation and specifically forbid combat on behalf of such. Christianity has been both subtle and, as some think, vague on this issue, as witness, for example, its lack of clear support for conscientious objection to

war, and the uncertainties as to whether the rules of sound inter-individual behavior are intended to be carried over in exactly analogous way to the dealings between groups. (Indeed, in admitting "just wars" it has even recognized, though never very explicitly, that different rules may apply to war with other than Christian states.) But, it is quite clear from countless journalistic and poetic laments about the "hypocrisy" of great nations that the popular interpretation of universalistic ethics is the naive one that *inter-individual and intergroup rules must be the same*, e.g., that the greatest love and expression of moral altruism would be for one country to give up its life for another.

That such injunctions have never been seriously regarded by intelligent leaders, and that, for example, the Christian countries have for five hundred years been the most successfully aggressive, does not eliminate the danger of universalistic ethics yet being interpreted in what, from a Beyondist standpoint, is a completely non-adaptive and fallacious generalization. In general, any basic principle is being followed with greatest internal consistency when it actually expresses itself not with mechanical rigidity but with obvious modifications of form adapted to changing basic conditions. One would, therefore, *expect* that the single-minded goal of service to evolutionary progress might require fairly different rules of conduct between individuals, on the one hand, and between groups, e.g., national or religious groups, on the other. That is why we have not risked for a moment the assumption that they are the same, but are examining them separately; the inter-individual rules here, and the inter-group behavioral inferences in Chapter 5.

[2] A sharper definition is becoming necessary for so-called "action research." One must distinguish first, genuine research which happens to use social data *in situ* instead of in the laboratory. It does not manipulate, but lets events take their natural course and by superior and general multivariate statistics teases out the natural causal connections. This is an important and recognized branch of research method (Cattell, 1966, page 31) and if in the present context it needs any designation could be called *site research*.

A second meaning — and one central to the Beyondist concept of experimental societies — is one that might be called *venture research*. Here, without any possibility of laboratory solution, an actual community voluntarily agrees to undertake a social experiment, which is both (a) an experiment in values and (b) an experiment in mechanisms, and *is completely recorded and analyzed just like a laboratory experiment*, the laboratory being the world.

At present a third variety which is being called "action research" (as far as I can tell from the mercurial writings of Lewin[1948]) has been espoused especially by such groups as SPSSI (Society for the Psychological Study of Social Issues). Most unfortunately these groups have tarnished action research by using it as a vehicle to bring about in the name of science, certain social deviations depending entirely upon their own subjective values. (For example, community housing changes have been brought about in groups too poor to complain or for it to be demonstrable whether they are willing or unwilling participants.) Neither the ethically necessary *voluntary* action, nor ((a) above) the explicit statement of basic values involved, nor ((b) above) the exact scientific follow up of results essential for true *venture research* have been maintained. The impression is unavoidable that the underlying aim is the propagandistic one of bringing about a change, and that the conditions necessary to a true research are quickly abandoned, substituting a dubious "pragmatic" judgment, and leaving a largely irreversible change. By the current definition, Herod, Pizzaro and Hitler among others, were engaged in "action research."

Possibly the term "action research" can live down its early history and re-define itself as *venture research* as now defined above. But substantial house-cleaning is needed by some of the groups concerned, whose political coloring is far brighter than their scientific standing.

For without immaculate ethical standards, explicit statement of the experimental values ventured in the experiment, and rigorous experimental follow up and analysis of multiple consequences, such undertakings can bring the whole idea of social experiment into disrepute.

[3] The sometime extreme importance of R is evident in such concrete instances as a Mohammed, a Pasteur, a Hitler, or a Napoleon, where one man's contribution to P is quite trivial, whereas his contribution to S is considerable, because of his military, religious, political, or scientific leadership role. (At the literary level, the point that history is more than a sum of average men is made by Carlyle's writing on heroes. But the opposite literary emphasis, in Marxian economic arguments or Trevelyan's social arguments for the massive effect of the mean P is also respected in this formula.)

[4] The technical point illustrated above and by the specification equation (4.2) supports the statement earlier that morality is not the sole determiner of group survival. If the specification equation for survival under some particular stress were written, it would probably include loadings on several other dimensions besides that in Table 4.4. Morality is not everything.

Here, in a more technical footnote, we can also take note that the general calculation from inter-individual behavior characteristics to the estimation of group survival is a two-stage matter. It requires (1) an estimation of the dimensions of syntality for any group from the characteristics of its behavior and its population, and (2) the further discovery and application of the importance (factor weights) of these dimensions *among groups* to estimate the ultimate criterion of survival. (Or whatever premonitory signs of failure or success are to be used.)

Thus, conceptually and mathematically, the criterion of survival is an independently existing and measured *third* term, i.e., beyond internal individual population measures and beyond the resulting group syntality measures. But, as scientific calculation is often more subtle than verbal logic, so there is a sense in which the understanding and even definition of this criterion is tied up with the syntality dimensions. The first tentative empirical correlations of syntality with survival measures may increase our understanding of how "successful survival" needs to be defined. Possibly, for example, much of the "survival across changing environments" will prove to be so immediately predictable from just three syntality dimensions—the "education-affluence," "cultural pressures," and "morale dimensions"—as to be virtually nothing more than their sum.

Certainly, the social scientist is strongly tempted, as he perceives the nature of these particular syntality dimensions to venture the *a priori* "value" judgment that the creative adaptability shown in high "cultural pressure" scores is immensely important to any group that would survive in a changing world. He may also offer the suggestion, with some confidence, that a high score on the "education-affluence" trait has been vital. (For example, in our recent complexly organized conquests of space.) Finally, as he considers the various aspects of a culture positively influenced by high "morale-morality," he is likely to have a strong hunch, even before empirical evidence arrives that a high score on the morality dimension is necessary for almost any great coordinated endeavor that a group might want to undertake.

[5] An instance of some relevance to political cultures suggests itself to the writer in the observation that a seemingly temperamental readiness to adapt to orderly authority coincides with the area—Italy and Spain—in which the firm control of the Roman Empire endured longest—over enough generations perhaps to produce selection in favor of increase of an "authority trusting" mutation. These racial mixtures also have, at least in their own

native areas, a low crime rate. (The large scale massacres of the Spartacist revolutionaries alone must have eliminated a substantial fraction of authority-intolerant temperaments.) Conversely, the thousand years of democratic organization of Iceland has, according to Huntington's data, produced a very high degree of orderly independence of thought and individualism in that country. The case of Germany might be cited as anomalous, but in an area with a long tradition of patriarchal authority natural selection might well have favored authority-accepting sons. That genetic selection through culture may have produced such results is at present a speculation. But with present advances in measurement of personality traits such as superego strength (G) and dominance-submissiveness (E) it becomes susceptible to a research check.

[6] An appreciable fraction of sociologists, professionally accepted in their field, have not only underestimated and neglected but positively denied any role to genetic, racial differences in behavior. Social psychologists with some biological education, on the other hand, see the flimsiness of this position. Perhaps the simpler but still organic field offered by agriculture is one in which the issues in the more complex field of social psychology can be clearly illustrated if not solved. Obviously environment is there as important as heredity, and nothing will grow in a desert. But if one asks in fact what, in our time, improved fertilizers and improved breeds have done for output, the latter is at least half the story. Leading agriculturalists, e.g., A. H. Boerma, Director General of the FAO, point out that improved seed varieties have done more than anything else to increase the world's rice and wheat output.

By contrast sociologists and historians have oscillated with obvious unreliability between extreme views. Historians like Hegel and Toynbee see history largely as cultural impacts; Darlington goes deeper beneath the cultural appearances. In sociology it is no exaggeration to say that as far as recognizing genetics is concerned we are now stuck with a whole generation of brainwashed students. The art of creating smoke-screens ("learning and genetic tendencies are so interwoven it is impossible to assign distinct weights to them"); the attempt to use equivocating hybrid terms like "ethnic group" when "race" is meant; and the use of Billingsgate scurrility ("racist") instead of careful argument whenever careful analytic evidence is put forward for genetic contributions, characterizes this generation. The treatment of Jensen's and other evidence on racial intelligence differences by the SPSSI, Kleinberg, Hirsch ("the notorious Berkeley white supremacist A. R. Jensen") and others, and of Coon's scholarly work on racial differentiation by Montagu, Dobzhansky, and others sufficiently records the climate in which research has had to work in the 'forties and 'fifties of this century.

The proper revulsion against Hitler's nonsensical genetics no more justifies continuing to pollute science with politics than politics justifies Lysenkoism in Russia or such persecutions as those of leading scientists Jensen and Shockley in this country. It is refreshing to find in an up-to-date volume of world history (Darlington, 1969, page 547) a willingness at least to challenge this last generation's rabid environmentalism with the view that "it is the [genetic] inequalities which create advances in society, rather than advances in society which create the [class and cultural sub-group] inequalities." As suggested above, culturo-genetic direction of action must remain today a shrewd guess; but it is a real scientific advance from the position of, say Kleinberg or Montagu, to entertain "genetic-to-cultural" causation, intellectually, as a research possibility. However, to restore a more comprehensive approach among certain groups of social scientists alienated from biological science (notably sociologists and cultural — not physical — anthropologists) after these two or three generations of miseducation is not going to be easy.

[7] A view of inter-class parasitism with more historical perspective is that of Darlington (1969, page 372) when he commends the wisdom of the Emperor Constantine: "Constantine, in moving his capital from Rome, escaped from its parasitic population, that is to say, from the whole of the plebs and from most of the patricians."

[8] The principle that parasitism should be reduced wherever it can be clearly established is none the less vital in importance because it can be established only with probability and because the political and ethnic worlds get uncontrollably emotional about such issues. For example, on the issue of discriminatory barriers against free migration, the common argument from "compassion" (often hiding a manufacturer's desire for cheaper labor) is that people in areas of poverty should be free to migrate into those which are succeeding in keeping a good living standard. Instances currently generating heat are Jamaican migration into Britain and Asian migration into Australia and South Africa. Since poverty is not accidental, but determined, there is an appreciable probability that the poverty-producing habits will be imported with the people.

The particular quantitative psychological and economic analysis of the present instances is not available, but the social *rights* issue is clear: that one does not migrate into the host's house unless invited. Against this right of the host it may be objected that the claim of having occupied an area first—which is brought forward in, for example, Australia, Ireland and Israel—cannot be considered absolute if such occupation was in fact, a meritless and ancient historical accident. The right to expansion discussed in Chapter 5 negates the idea of boundaries fixed for all eternity regardless of good stewardship. However, the loud attack by supposed "rationalists" on the Australian policy of keeping Australia predominantly Caucasian and restricting Asian immigration, overlooks complex realities and moral values. In any case, we lessen the chances of patient research on such issues by an emotional shouting of "racist" (on the one hand) or "traitor" or "ignoracist" (on the other).

However, caution is called for when we do not yet know enough about the limiting and modifying effects of differences of genetic temperament upon a culture to risk the loss or perversion of a definite culture, e.g., Israelite culture, to which an experimenting world plan is deeply committed. Quite conceivably a complete replacement of Caucasians by Japanese in the Australian population would be an "improvement." But it would definitely not be the same racio-cultural experiment. And it would be an absurd naivety to permit in Britain the replacement of a stock of some repute by uncontrolled immigration—a replacement so bitterly and valiantly resisted in terms of a threat of German invasion only thirty years ago! Quite apart from genetic considerations, the open sluice gate makes the crass political assumption, which the U.S.A. and Britain also followed for a while, and which France has long followed to its present detriment, *that the host culture will infallibly predominate over the immigrant culture.* Basically, as far as we know, the mutual impact of human beings is not immune to Newton's third law that action and reaction are equal and opposite. More complex secondary determiners may come into social situations, but we have no right to assume that, with equal numbers, the immigrant will not pull the values of the host culture as much toward his own, as his own are bent toward those of the host culture. "Cultural assimilation" is a fine phrase, but in the U.S.A. it is obvious that immigrants to an originally English culture have sometimes learnt the English language without, for example, the English sense of law abidingness and restraint of violence, which have long characterized that culture.

[9] In search of some quantification we can turn to sheer economic terms. The cost of a

given cultural advance is typically emphatically less for the community when undertaken by genetic methods. For example, to bring 1000 children of I.Q. 90 to precisely the same level of school achievement as another group of that size, but of I.Q. 110, costs enormously more in teacher effort, the ratio of teachers to pupils, time in school, etc. Genetic selection, could, even in one generation, raise the average child I.Q. and achievement level beyond anything that has proved economically practicable by increasing the intensity of schooling. (For example, if the next generation were bred only from the upper half of the intelligence range the new mean I.Q. should be, according to standard genetic assumptions, about 112.) Instead of repeating generation after generation the costly (and in some extremes, actually impossible) labor of reaching culturally required standards despite poor intelligence, the geneticist would once and for all make learning easier.

[10] This seems to have been the principal issue in a somewhat emotional attack on Sir Julian Huxley by Mirsky in a recent *Scientific American* letter. Speaking of what social workers have long called the "social problem group" (not to be confused with the accidentally disadvantaged or the under-privileged – in simple language the poor) Huxley had said that "improved environmental standards have very little direct effect on the general shiftlessness, the low educational attainments or the size of family of such groups." Mirsky questions this and denies Huxley's eugenic remedy on the grounds that we do not know enough about the mechanics of behavior genetics or have definite aims as to what is genetically desirable. These are the two old chestnuts, which, though roasted by the counter arguments for most of a century, are still brought up. Mirsky is a distinguished molecular biologist but not a social or educational psychologist or a behavior geneticist. The former is shown by his being unaware that Huxley's observation on "low educational attainments" is very widely supported by research reports on the recent very costly intensive programs for the backward. The latter is instanced by his argument that we cannot tell how much is due to heredity until we have evened down environment. If a behavior geneticist could not draw genetic conclusions without this he would be no better than a child in first year physics who cannot think how to tell the pressure of a gas when volume and temperature are both altered.

The answer to the two perennial objections, is as above, however, that perfect information is not a logical prerequisite to reasonable action, else there would be no action. Somehow our benighted ancestors, undeterred by such arguments as Mirsky's, knowing as little about the metallurgical chemistry of tin and copper as we know about, say, the full genetic picture of intelligence today, carried mankind through the bronze age, from the stone age to the historical iron age. If as Mirsky says, progress is impossible because the breeding society lacks "clearly defined aims," the sooner it reasons to them the better.

[11] Because democracies have permitted freedom of discussion it happens that the first ideas of genetic improvement arose largest in the British liberals of the nineteenth century, such as Darwin, Galton, Huxley and Spencer. Unfortunately, there supervened on the one hand, the racist caricatures of natural selection concepts by Nietzsche and Hitler, and on the other, the equally distorted but hostile caricatures of genetic ideas in Russia. Meanwhile the smart journalism of *Brave New World* and *1984*, fell into the shallow view that any high degree of community self-control and foresight can be achieved only by dictatorship. There is nothing "totalitarian" in a society preferring one kind of genetic formula as more suited to its own ideals than the genetic choices of other societies. In any case, it may well happen that genetic experimentations will take place mostly in voluntary, self-directing sub-groups

and enclaves within the total society. As Darlington (1969) points out, this is, in fact, happening, though not with such a degree of deliberation, in inbreeding religious and professional sub-groups in complex societies.

[12] Concerning the real scope and advantage of hybridization, there is still much doubt. Darlington (1969) stresses the best side of it, yet in the broader sense, it is ultimately subject to a biological equivalent of the Second Law of Thermodynamics — the growth of entropy — in the sense that there comes a point when there are no more distinct groups to blend. This is actually also well realized and expressed by Darlington, as a geneticist, when he points out (1969, page 373), "What is important is the preservation of separate, racially distinct groups. So that they can, by very slow hybridization, convert themselves into [new] socially distinct groups, cooperating and competing." Praise of hybridization, however, has helped give an undeserved bad name to inbreeding, which prevents dilution of a pool of good genes, and brings bad recessives to the surface, to be selected out, whereas cross breeding leaves them indefinitely in ambush. Repeatedly in history, it has raised high powers to still higher powers, as (until an optimum is passed) in various aristocracies. And, even when carried to the extent of brother and sister matings, as with the Egyptian dynasties, it actually produced the longest-lived capable aristocracy known to history.

[13] It is one of the oddities of scientific history that Hermann Muller, who received the Nobel Prize for the first demonstration of a scientific possibility of speeding up mutations, was somehow maneuvered into the position (partly by his own Humanistic horror of war) of being a complete Cassandra with respect to the dangers of nuclear radioactivity. If ordinary natural selection alone were allowed to operate, a raised mutation rate would have to be accompanied by a proportionately (some calculations say disproportionately) raised birth rate and greatly increased severity of natural selection, in order to preserve the small number of good mutations and eliminate the much greater number of harmful ones. Without this as Muller, Haldane and others have reliably pointed out, we should face genetic disaster from an overload of deleterious genes. The principle would be the same then as now: that the price of evolution — indeed, of living at all — is the acceptance of a number of unfortunate genetic endowments, because all mutations are experimental. However, if far more research support were given to early detection of defective embryological development, the ratio of healthily adapted to defective individuals might be raised above the present level, by quite early abortion of detected defectives.

Genetic variation, as a prelude to selection, occurs, of course, both through mutation in the true sense, and through "accidental" events in the process of developing the new zygotes, as in crossing over, etc. The changes from new chromosomal and gene combinations are particularly powerful in hybridization.

[14] Effective progress in a human population by the path of hybridization requires three conditions: (1) an initially shrewd choice of the second race, to ensure it has the properties most likely to augment a new pattern, and (2) an unfettered exercise of the right of self-determination in terms of knowing when firmly to put the lid on, and let the melting pot boil, i.e., to begin the second phase of isolation and consolidation, in which the best is to be "brewed" from the mixture, and (3) institution of an effective within-group eugenic selection program to screen out the many defective combinations, positively guided by the culture and genetic knowledge.

[15] Analysis of what happens in this "blind" process is a whole field of social psychology in itself. Even with the most "accidental" initial generation of a particular moral value or

law, we may suppose some dim, preconscious appreciation of its connections with social usefulness. In any case, being in existence, the rule is always rationalized, though for aeons of time the "rationalization" is likely to be nothing but legend, magic and anthropomorphic "explanation," e.g., "It is bad luck to walk under a ladder."

[16] As pointed out earlier, most statements making an antithesis of the "importance" of individual and group are missing the point. Thus when Dobzhansky asserts "We do feel that individuals should not be sacrificed for attainment of social aims" one is compelled to say it is a meaningless statement. (It is on a par with such a statement as "The government can afford to support everyone at a high standard of living" which means "Smith can afford to support Smith, etc.") If the individual can gain full and more permanent self expression only as part of a culture, his "sacrifice" — like that of Christ but on a lesser scale — for social aims is his glory.

CHAPTER 5

The Moral Directives from the Beyondist Goal:
II. Inter-Group Ethics

5.1 THE NATURE OF GROUPS AND THE PRIMARY ROLE OF THEIR COMPETITION

Many philosophers, from Plato through Spinoza to Bentham and Comte, have sought to put ethics on a logical basis. Virtually without exception they have thought in terms of a universalistic ethic — the man to man morality viewed in the preceding chapter — and it has been left to writers like Machiavelli and Treitschke (viewed askance by the philosophers) to raise the possibility that the ethical laws governing the interaction of nations should be different. These writers have only said what every practical statesman knows, but in saying that inter-group behavior from the beginning of history may be not only realistic but right, they have become scoundrels to the purveyors of universal ethics.

Yet according to Thucydides, no less a statesman-philosopher than Pericles himself, explicitly denied that compassion and even honest expressions of intention are proper to state governments in dealing with other countries. However, it is only after Darwin, e.g., in the writings of Spencer, that one finds a respectable philosopher recognizing the natural appearance of a "code of amity" among citizens and a "code of enmity" among groups, though as a description more than as an endorsement. Parenthetically, a product of the misunderstanding of these specialized developments of ethics is often cynicism about the idealism of politicians. Thus, in the case of Pericles cited above, Aristophanes, like a cartoonist, made such satire out of it that he probably contributed to a decay of morale through individuals being led to believe that they should behave like the state.

From the natural tendency of the human mind to "reason by analogy," as well as from the imperialism of universalist religions most men have, indeed, taken it for granted that the moral directives would be identical. However, it must be at once obvious from the basic position adopted here that we must put aside traditional thinking and recognize that the primary state between groups must be one of competition. The ethics of reduction of strife and selfishness among individuals is derived from this more primary truth as a special situational inference. The issue remains wide open as to whether any special considerations require any abatement of out and out competition among groups. To anticipate eventual conclusions let it be said that competition *a l'outrance*, and especially war, are not what one is led to by a ruthless pursuit of the logic of evolution. Indeed, the Beyondist ethics in this area turn out to be more subtle and complex than those among individuals, since they must embrace not only the dealings among nations but those of individuals belonging to different nations and in a variety of roles.

The issues that need to be raised here concern the ways in which both existing and possible modes of group interaction—economic competition and cooperation, cultural emulation, migration, invasion, propaganda, political alliance and war—contribute or do not contribute to an evolutionary goal. By that touchstone an approach to the ethical values desirable among groups can be reached. But first we have to decide what a group is and for what types of groups our generalizations are intended to hold. So far the implication has been that they can be nations, and perhaps religious communities, but could they also be social classes, or political parties, or regional cultures as in the Union of Soviet Republics, and so on? Psychologically there are several kinds of groups, and definition by "a common purpose" (frequently in textbooks) is less satisfactory than "a set of people in which the behavior of all is necessary to satisfy the needs of each." This dynamic definition makes the group an instrument by which the individual satisfies his needs, and shows whence the group derives its energies (its synergy).

A psychological taxonomy of naturally-forming groups would be a long story. From the standpoint of aiding evolution in the senses we have discussed most of the types would be irrelevant or ineffective. Two major conditions are required for efficiency of evolutionary action: (a) some coordination in each group of its genetic and cultural life, i.e., they must go on long enough together—some centuries—for their symbiotic value to be tested, and (b) a sufficient integration within the group, and independence without, for the consequences of the group's own behavior to be largely born by the group. Neither condition holds true, for example, of social classes, from which people constantly pass in and out, or of members of religious groups, whose economic success or failure has little to do with the religious affiliation. Nations are the natural, sufficiently self-contained biological and cultural units, though some religions, e.g., the Jewish religion, have sufficed in the absence of nationhood, to provide a bio-cultural unity with some degree of that economic, breeding, mutual defense, and cultural integration which makes an organic group.

Groups overlap in their membership and competition for the primary loyalty of individuals has existed through history among national, religious and class groups. This has resulted in a dynamic, shifting equilibrium, but nationhood has increasingly prevailed[1]. Within his nation the individual communicates, learns his values, marries, organizes his defense, shares his taxes, and rears his family — bound in a common language and culture. His standard of living and reproduction rate are tied up with its economic success, and it is for its existence that he is called upon to die in military defense. The nation is thus the most self-contained group which, in our technical terms, takes the largest portion of the individual's synergy, i.e., is the means of meeting the largest fraction of his psychological needs. If anyone doubts this, a perusal of Barker's work (1948) of nationalism through history should give perspective.

The three chief units that rivalled the national group are the family or tribe, the universal religious group and the social class, but the first is subsumed in the nation and only the latter cuts across. The imperialism, as we have called it, of the universalistic religions in extending, to a substantial degree in a false sense, the within-group ethics across all groups has at times successfully competed with national loyalty. In medieval Christian Europe, and for a time in the Mohammedan countries it created a political unity out of a religious unity, with which the growth of national cultures at the Renaissance produced dramatic struggles. In spite of attempts to make the second type — the social class — more organic than the nation it came into its own only once, in the extreme conditions of 1918 in Russia[2]. It quickly, as in the rivalry of Communist China and Russia, recrystallized into a predominance of the national type of group. The essential situation, then, is that we are examining the interaction of men and groups where *men belong to organically developed groups so that their fortunes are bound to those of the culture and genetics of the group to which they belong.*

A mistake which seems to need constant correction in discussions here is the assumption that competition among groups is simply a matter of direct struggle and strife among governments. Natural selection among groups works partly through (a) relative success in handling the physical and general environment, and (b) relative success in struggle with other groups (both of which are partly decided by the level of internal morality). Although we do not know for certain what their relative importance may be, yet by analogy with the animal kingdom it seems likely that the less dramatic "economic" competition in (a) decides more in the long run. Inter-group morality has largely had to do with (b), but the rise of interest in ecology and pollution problems now extends ethics over the rights of a group to exploit its material environment, too.

Now the arresting conclusion from evolutionary law — and one difficult for many to digest — is that natural selection should be allowed and encouraged to act freely among groups[3]. This is the *primary* law, and any later modification of it that we may discuss derives from secondary and lesser considerations. Defective internal morality, failure to control birth rate, unwillingness to sacrifice luxuries to education, adherence to superstitions, and many other deficiencies may cause a group to fail either in the struggle with another group or in the economic tussle with nature. At that point external "charitable" support from other groups, or even their failure to expand as the defective group retracts, are immoral acts militating against evolution. They are to be avoided in the interests of the highest inter-group morality. For, by the basic laws of learning, such rewards merely reinforce the strength of the faulty community habit systems. Or, if the defect is genetic, they postpone the reduction of genetic defect.

By inter-individual standards, where altruism and mutual assistance are ethically enjoined by the goal of group survival, this is likely to seem a ruthless conclusion. But in as much as a group — even a last remaining human group — can survive entirely on its own, where a single individual cannot, the rules become different. In any case, by laws which no choice or interference by man can alter, the struggle in the first area, that against nature, is completely ruthless. Earthquakes, disease germs and ice ages give no quarter. In a sense that will be clarified later, in which struggle with other societies is seen as a nursery preparation for the struggle against nature, and a reminder not to be off guard in periods when nature happens not to present challenges, there is an argument for inter-group competition being conducted by the same rules as are fixed in competition with nature.

Outright competition — unless modified by strategic alliance — is thus the primary required condition in inter-group relations. This is not incompatible with the term cooperative in "cooperative competition." The fact that patient and surgeon are cooperative does not deny the necessity for the surgeon to cut deep. Nevertheless, as one scrutinizes this competition more closely in relation to its purposes certain secondary principles appear which make progressive evolutionary competition something more subtle in values than the law of the jungle.

5.2 BY WHAT SECONDARY RULES CAN MAN AID COMPETITIVE GROUP EVOLUTION?

The area of international dealings is one beset by certain standard misperceptions – largely springing from the real discrepancy of inter-individual and inter-group behavior laws. Whole classes and generations of men are shielded by their governments, state departments and foreign offices from the stern realities of an international game in which no mistakes are forgiven. They are shielded as effectively as stall-fed domesticated animals are shielded from the worries of the farmer. A natural consequence of a sheltered population thus believing that inter-individual ethics exist among groups is that the practising statesman is accused of cynicism.

In Europe he has been accused over centuries of paying lip service to Christian principles while failing to support these values in "practice." In America a great deal is made of the disinterested benevolence which America shows in giving "foreign aid" to "underprivileged" countries. Admittedly it is decidedly more than most countries sacrifice for this purpose, but it turns out to be less than one percent of the gross national product and it is skillfully directed to making cultural alliances likely to be useful in any future struggle for survival [4]. If there is any hypocrisy in this it is not in the mind of the realist statesman, who unashamedly dubs it welfare imperialism. The hypocrisy is fostered by the requirement of the man in the street that his within-group values not be jolted in crossing frontiers, and that his image of himself as doing all for Christian charity not be upset. However, many a politician seems not to be much clearer than the man in the street on the different principles involved: he simply acts more realistically, and has not been as conscientious as he should in discovering the true ethical bases for inter-group action.

Nevertheless, it may be that secondary restrictive rules on international, and inter-religious community competition are required by evolutionary morality. Although the full character of these requirements will not be discussable until we have surveyed in this chapter the actual nature of the present competitive acts among groups, some of the broader modifying principles can be introduced here and now in this section.

The aim of Beyondism is to ensure that natural selection operates clearly and effectively among groups, but since this is in the service of a further goal—human progress—certain boundary conditions seem required:

(1) That struggle shall not reach an end point of domination by a single power, creating a world cultural or racial monopoly, thus interrupting variation and selection.

(2) That in a period when legalistic international dealings have been explicitly accepted, one group does not annihilate another by the sudden pounce of war. There are too many accidental circumstances in war to make it desirable to rest the decision of survival of a group on one war.

(3) That all groups do not totally and simultaneously destroy one another. This is put in for logical completeness, but except in science fiction and those laboring under a nuclear nightmare, it is to be viewed, realistically, as an extremely remote chance.

(4) That, at the opposite extreme of competitive activity, groups do not abandon competition, and coalesce into a single homogeneous stagnant mass.

(5) That parasitic relations are not set up such that either of the parties becomes unable to survive should it be set on its own.

(6) That—and here we turn to the more ideal construction needed for the future—a comity of nations sets up a federated research organization to keep records and exchange analyses in order that readjustments in faulty directions of progress may be made early from comparative, competitive performance.

It will be noted that these conditions do *not* include the unmitigated imposition of rules by a totalitarian, supra-national power and police force, such as a government provides among individuals. There is a great temptation, into which some progressives have fallen, to suggest instituting formal "rules of a game" to take the rigor out of inter-group competition. Proviso (a) above comes very close to this, justifying some debate on the issue. It is primarily in regard to outlawing war that the issue arises, though on closer scrutiny one might also find certain other forms of competition, e.g., severe economic strife, and, particularly, the propagandist deceptions of psychological warfare, equally repellent.

The problem is surely that any fractional restriction of the possible range of competitive behavior makes selection, by that fraction, less comprehensive and sound. Boxing gloves and the Marquis of Queensbery Rules, for example, make differences in intelligence between the combatants less important than if all means of combat were left open, and, similarly, the rules of chess make the physical fitness of the contestants almost irrelevant. If the competition among groups is to be effectively on a par with, and integrated with, the purpose of competition against nature, one must look on restrictive rules, even though hugely desirable from other "humanistic" standpoints, with suspicion. Nature pulls no punches, and since communities are in training for the battle with nature, direct group competition could be a poor training if made too artificial. Thus condition (2) above, indeed, needs to be critically examined in further research.

Realistically, there is something to be said in this area even for not eliminating fear of sudden attack. The biological innovation which produced a shift from the reptilian to the Pantotheres mammalian order in the late Mesozoic age demanded substantial increase in physiological complication (notably in the birth of young). This was a costly and precarious innovation but among the circumstances that made it worthwhile was the fact that the warm blooded mammal could both make and avoid "the sudden pounce" far more quickly than the reptile drugged by cold weather. Is it not then part of the test of the evolutionary advancement of a nation that it can organize better to survive the sudden unheralded military attack? Probably "yes." But, it will become increasingly apparent that an important task of a federation of nations is to reduce the element of fatal *chance* in evolutionary evaluation, and one potent way of doing that is to extend the time over which tests are made. As war possesses ever more powerful weapons, the chances of decision by a single event increase. The premature or single event judgment can occur in other fields too: the first small mammals just discussed could have fallen victims to some chance parasite, and perhaps Cromagnon man (a noble creature by archaeological records, as Graham (1970) points out) was wiped out by a germ to which he happened not to be able to develop immunity quickly enough. The story of evolution is full of tragedies and absurdities—mostly set right later, however. One great gain of man's perceiving and embracing its purpose lies in his becoming an intelligent mid-wife to its creations. He needs to find out how by demanding the right rules, he can best aid in the natural selection of groups.

5.3   THE MODE OF OPERATION, AND ETHICAL STATUS OF CULTURAL AND RACIAL TRANSPLANTATION

In all discussions on groups as on individuals it behooves us to keep an eye on the fact already focussed that human group evolution is both cultural and genetic. Movement in both follows the basic rules of variation and natural selection, but after that they differ in a number of ways not yet fully understood. That cultural movement can be much faster both in progress and retreat, than racial movement has been pointed out. So also has the fact that cultures can move independently of races, though that independence is incomplete because in historical record, cultures and races commonly migrate together. And there is evidence that a different race usually subtly modifies the culture to fit its temperament when there is direct borrowing across races. Furthermore, cultural transplants are usually more complete and radical than racial change, e.g., after the initial expansion of the Inca culture it was very rapidly destroyed by the Spanish Mediterranean culture, but the Spanish racial infusion was comparatively slight.

In the cultural progress domain itself there is an old debate between those who explain cultural advance by a parallelism of spontaneous mutations, supposing that similar ideas appear in many places, and others who stress culture borrowing, arguing that in general the mutation appears in one place only, and spreads (*see* Ardrey, 1961; Benedict, 1934; Malinowski, 1937; Kroeber, 1958; Sorokin, 1937). Darlington, for example, maintains not only that the bow and arrow appeared at only one point, in Asia, but that diffusion thereafter was not so much a "culture borrowing" as a *racial* spread of the people who possessed the bow. The analogy of the "independent origin" and the "diffusion" schools to the relative emphases on mutation and hybridization in genetics is rather striking. Conditions in modern times of rapid communication and explicit imitation are so different from the past that it may be difficult to imagine ages in which very different mechanisms operated and the same invention was made in several places, as has happened in different genera in biological invention, e.g., of color vision.

The basic laws of cultural and racial transplantation which we would like to summarize in this section in relation to the evolution of intergroup ethics have never, as yet been sufficiently studied. Different aspects of such study are scattered, with insufficient integration, in different academic fields. In history, for example, one encounters the diverse evaluations of Childe (1950), Darlington (1969), Gibbon (1910), and

Toynbee (1947). In anthropology and sociology there are the interesting discussions of Mead (1955), Kroeber (1958), and Merton (1957). In physical anthropology the valuable insights of Coon (1962a,b) and Hooton (1946), and in archaeology the broad sweep of evidence summarized by Cottrell (1957).

But many necessary ideas can also be obtained from disciplines commonly less considered, such as international law, the business administration of large corporations (including the now quite intensive studies of monopolies); the social psychologists' studies of small interacting experimental groups (Borgatta, Cottrell, and Meyer, 1956; Borgatta and Meyer, 1956; Cattell and Stice, 1969; Singer, 1965), and, last but not least, the recent multivariate analytical studies on international behavior (Rummel, 1963; Jonassen, 1961; Cattell, Breul and Hartmann, 1952; and others).

Diverse as these sources are, one thing is clear from them: that an enormous variety of cultural and racial mutations occur, and that relatively few win through to a comparatively permanent place in history. And on the cultural side, one is impressed by the almost absurd abundance of "pointless" inventions which blossom and vanish.

The social and political "liberal" or "intellectual," educated as he typically was in the last generation, more in literature, philosophy and the arts than in science, would clearly like to believe that natural selection acting on relatively blind social trial and error plays a miniscule role compared to dignified "rational progressive" innovation. "Progress" he sees as proceeding in orderly fashion by education within the group and missionary endeavor abroad. On the contrary, our conclusion is that in the *creative* process itself *a priori* logic alone, i.e., rational as distinct from scientific construction, can play only a small part, and even that sometimes, a misleading one. But admittedly, when a society is reasonably sure of the goodness of its product, an educational and missionary enthusiasm can efficiently spread it. If the bestower is more active we may call it "propaganda"; if the recipient, then "cultural borrowing" or imitation. Normally, there is both, as in the rapid shift from the American intention of exerting propaganda, e.g., in Commander Perry's visit to Japan in 1854 to the deliberate Japanese intention to imitate effectively.

Propaganda was once more prominent in religious and universalistic movements than national ones, though of late Britain, America, France and Russia have worked hard to convert others to their national cultural values, and most are now jealous of the prestige value of their public "image." From the early efforts of the Hebrew, Christian and Moham-medan religions to modern Communism, and to Democracy (propagated in Germany and Italy after World War II) this propaganda has quite fre-quently been somewhat aided by the sword. On the other hand, the graft-ing of a new cultural element is often accompanied negligibly by force in any propagandist "push," but is aided by the admiring imitator, as in the adoption just mentioned of industrial culture by Japan; of the old English public school values (including sports and athletics) by various European nineteenth century educational systems; or of French cooking by the restaurants of the world. Because of the assimilative adjustments which follow, the end result is rarely a simple replication.

The mode of operation of *deliberate* propaganda, and its ethical status as judged by its consequences are considered in more detail in Section 5.8 below. Here it is important to look at the more generic process of *transfer* assuming that pull, i.e., willing imitation, is as much or more involved than push. And it is enlightening, here as elsewhere, to consider transfer of culture *and* of racial type under the same concept, though with recogni-tion of distinctions.

Now since imitation will be deliberately directed to that which seems to be "successful," the transfer process might seem to *guarantee* evolution-ary progress. Like a team of climbers on a mountain, one member finds a way to the next level, all are hauled up, and a fresh dispersion and reconnaissance follows to find the next breakthrough. However, some inefficiency, at the least, is inherent in this process, because of the dif-ference between what is actually successful and what acquires prestige because by popular acclaim it is considered successful. Doubtless Sodom, Gomorrah and Sybaris in their glorious sybaritic heyday excited much imitation except in such crusty sceptics as Lot.

If the imitation is of that which survives, it automatically has the stamp of authentic progress, but if cultural borrowing is a mere magpie collection of things that glitter it could be unrelated to real progress. Imitation, as a source of evolutionary progress in fact faces two difficulties (a) confusing the superficially attractive with real survival potency, (b) being uncertain what element to imitate because social science cannot yet tell *how* some desirable visible end result is in fact brought about. For example, the crime rate in Britain is decidedly lower than in America, but social scien-tists show little agreement in pointing to the features in the British style of life that are responsible for this difference.

Thus the group interaction concerned with propaganda and imitation requires intelligent social science evaluations if it is to be useful, and may go to excess or stop short of the optimal degree. As to the latter a much neglected psychological truth is that learning tends to occur after frustrations. Failure in war and economic failure however, have often had to go far before old habits are rejected and imitation is seriously practiced.

The greatest problem in regard to the evolutionary value of culture borrowing is that it reduces the desirable diversity among groups. Once cultures become sufficiently diverse, however, a single element from one pattern fits poorly into another. Moreover, organic and vigorous culture patterns develop like diverse individuals and species, by an immunological, rejective reaction to one another's values. The obvious usefulness of borrowing is high in mechanical gadgets, etc., and as anthropologists have recognized, it is more likely to occur for these than for moral and other values. Thus in major matters the popular view of the importance and desirability of culture borrowing, as contrasted with independent creation, may be excessive, and its desirability more questionable than we yet realize.

Turning now to the second form of transfer—the genetic equivalent of cultural movement—in which there is actual inter-group invasion or migration, we recognize a process that has had enormous play in history. Here the "push" is far more prevalent than the "pull" equivalent to imitation, except in rare cases of invited immigration. This last has often been determined by economic and other circumstances having nothing to do with the genetic desirability of the infusion (Heape, 1931); but instances of invited immigration believed to be on quality exist, such as the clever Flemish weavers brought into Britain, the acknowledged value of the Swiss as dependable guards and soldiers in Italy and much of Europe, the lengthy period over which the U.S.A. has given larger quotas to, say, Northwestern Europeans than to Asiatics, and the habit of Britain, Russia and other countries of taking their aristocracies from Germany.

Whether by invasion or invitation the tendency for the observed or experienced advantage to be sufficiently genetic in origin for the infusion to bring genetic gain has probably generally been low. But the successive waves of improvement in prehistoric man, e.g., in cranial capacity, suggest that in the long run this replacement has worked effectively. However, even in recent historical times this has worked far more blindly and clumsily than an explicit acceptance of the principle and careful evaluation of the average genetic type of a people would finally permit. For example, the opposition in some countries to Jewish immigration has overlooked the fact that in many groups they have significantly higher mean I.Q. than the populations they would join. And Britain, after resisting to the death invasion by gifted peoples like the French and the Germans, opens its doors casually to Jamaicans whose intelligence test and school performance are significantly below the English average. It even makes no distinction between the Hindu, whose intellectual performance is probably actually higher, and other peoples and racial mixtures in the Commonwealth whose cultural contributions are, and probably will remain, significantly lower. In the day of reckoning this could matter. The lack of clear policy is partly due to our ignorance of the real genetic differences of peoples around the world, of hybridization effects, of our fatuous belief that we know what culture is worth defending but not what our aim is genetically, but especially of emotional confusions between traditional universalistic ethics and a clear, logically derivable evolutionary ethics.

The expansion of peoples by military invasion, as in the replacement by European racial stock of the American Indians in North America; of the Mediterranean-Celtic by Nordics over much of England; of the Ainu in Northern Japan by the Japanese, and in innumerable earlier and vaster movements have usually required some group superiority of organization to bring them about. Modern man's quite appropriate sensitivity over avoiding war, is apt to blind him to the main fact that it is through this constant surge of conflict and invasion that the tide of evolution was able to lift itself from Neanderthal to modern man[5]. With the outlawing of war and the disappearance of available empty spaces for foreign migration, by reason of native population growth, is genetic evolution going to come to a halt? As far as this relative replacement process is concerned, it probably will, and genetic advance will come to depend more and more on within-group selection or chromosomal borrowing.

Any fear that a real arrest of genetic transfer and change is bound to occur as a result of these conditions fails to reckon, however, with a new possibility. That possibility is the promise, through biological education, that more sophisticated political and popular opinion will develop. As will be discussed later, though it may seem absurdly idealistic at this moment, increased awareness of these biological principles may well lead to one group deliberately inviting immigration of specially chosen groups, and spreading within itself, by artificial insemination or genetic engineering, new mutations which have been sufficiently demonstrated in the social life of another group to be valuable. That is to say, cultural and genetic borrowing may eventually proceed by still more deliberate and parallel procedures.

In concluding this brief overview of culture transfer and genetic transfer in inter-group interaction one must ask how they mutually relate. Primarily, is there any *mutual induction* between cultural and genetic changes? How far does the transfer of a culture from group *A* to group *B* tend to produce, by within-group selection, some shift of *A* toward the genetic pattern of *B*? In a racial mixture a culture might, over many generations, favor the survival of one genetic ingredient over another, to the extent of producing very significant changes. In what we called above *culture-genetic positive conditions* this would certainly produce genetic movement toward that of the group from which the culture was borrowed. For example, the adoption by group *A* from a mechanically gifted group *B* of a mechanically complex culture would favor (through jobs and in other ways) a genetic increase of mechanically gifted people in group *A*. Thus the equivalent of transfer of a racial type has taken place without invasion or migration [6]. Conversely, genetic changes may favor particular cultural changes; for there can be little doubt that some races find certain cultural developments more congenial than others. (We have argued that this is particularly true of religions, and that the flow of such universalistic religions as Buddhism, Catholicism, Protestantism, though equally determined by economic and historical forces has shown a distinct tendency to halt at racial boundaries (*see also* the value data of Morris, 1956).)

A form of culturo-genetic interaction which may, however, have sinister possibilities resides in a special form of parasitism. Imagine, for example, that racial variant *A*, which is, say, highly creative but commercially or militarily less efficient, has all its inventions copied by an independent group *B*, which is genetically rather lacking in the plasticity required for creation, but efficient in ant-like aggression. The numbers and political or military power of *B* could increase to the point where group *A* has a negligible command of the future. Some historians – and we may include H. G. Wells – have considered the history of Greece and Rome respectively to offer an example of this relationship of imaginative creativity defeated by brute efficiency – after borrowing. In short, due to ease of culture borrowing, the rewards of genetic gifts in cultures become increasingly dissociated from the genetic causes, so that valuable genetic variants with high creativity do not get multiplied. Thus through culture borrowing natural selection between groups becomes confused.

In sum, despite certain inefficiencies and possibilities of error in cultural and genetic transfer, one must conclude that this type of inter-group action probably ends up, after selection gives its verdict, in real progress.

That is to say, it is favorable to development of groups with higher survival potential. Consequently, cultural and racial transplantation constitute an ethically desirable form of inter-group behavior unless countermanded by other principles to be considered.

5.4   POLITICAL STRUGGLE AND THE ETHICAL MEANING OF IMPERIALISM

Whereas the last section dealt with group interactions which result in transfer of culture or spread of genetic types the present deals with those aspects of competition in which one group exercises its effect on another by political power. Some overlap of the two is, however, part of the nature of the things. For political power sometimes means cultural or racial invasion. Meanwhile political power also has connections with the economic and military strength commonly backing political power, as taken up in sections below.

It is wise in general to take up complex questions at a time when they are not vexed by public emotionality, and when terms are not being twisted to local meanings — but in the case of political imperialism no such time exists. At no time has the imperialism of other countries been viewed as other than tyranny and oppression, and in the United Nations' debates today, imperialism is a dirty word. Yet expansion and retraction are as inherent in life as growth and decline. The only guarantee of an end to expansion and subsidence among nations and individuals is a rigidity of death. What we can aspire to is the introduction of dignity, justice and understanding into the continuing process of adjustment: we cannot hope to abolish growth. It can and should be pruned, but not eradicated. To make the above change of valuation clear perhaps it would help to use another word such as "expansionism" or "power adjustment" or simply "group growth," instead of "imperialism." And yet perhaps it is the most honest term.

The baleful aspects of political power seeking among nations which immediately come to mind are the risk of war, the possibility of imperialism as an oppressive rather than an educative influence on backward peoples, and the obvious fact of history that political power alliances are often made with little regard to cultural and racial similarity and so confound the simpler evolutionary mechanisms. (As witness Britain and Russia aligning against Germany, or America and China against Japan and now Russia.) Alliances are made largely according to threats to frontiers, i.e., are determined by propinquity, and magnitude of power threat and such other influences as Rummel has skillfully analyzed quantitatively (1963). But despite this tendency of the maneuvering to be locally tied, as disputes over sitting room are likely to occur between only the adjoining people on a bench, coercion by power offers some degree of *general* adjustive action. In power struggles, there is the risk of war, but when sensitivity to power is learnt, nations handle political power without war.

As to the better, more defensible aspects a well-known biologist has made the perspective-giving observation that every living organism is an imperialism. For it is always seeking to convert more of the external world, living or dead, into the pattern of itself. To condemn a process so inherent in the guts of life, is to condemn life itself. Moreover, where a country like Russia, after annexing to its sphere of influence half a dozen countries in as many years, condemns "capitalist imperialism" it is time for even the most short-sighted and doctrinaire liberal to recognize the debating game and give the word a new value. And similarly when an American president asserts his aim to be "to ensure the peaceful development of nations free from coercion" and "the creation of conditions in which we and other nations will be able to work out a way of life free from coercion" (Truman, Congress, March 12, 1947) shortly after ordering Japan to be "coerced" with an atom bomb, we realize that between the truth-seeking language of science and the popularly preferred language of political hypocrisy there is an unbridgable chasm.

The standard history of cultures and peoples is one of fluctuant growth and recession. In short, immediate success is no token of final success. However, over, say, a thousand year interval in the case of culture (or tens of thousands in the case of race) one sees clear instances of a trend toward total gain or recession, as in the Roman and English-speaking "empires" (and now the Russian) in which distinctive cultures have increasingly prevailed over a great number of other now forgotten cultural experiments.

In most countries the schoolboy is today taught that empires and imperialism are wicked, and that the proper aim of the United Nations Organization is to put a stop to all future changes of boundaries, but he lives in a country whose statesmen are daily wrestling with all the strength at their command to gain political advantage. Meanwhile, a native of Britain or France has only to look back to the Roman Empire to realize how much of his valuable culture came from it, or an American, an Australian, a Jamaican or a Hindu to realize how much his existence as a country depended on the shield of the British Empire.

With some prejudices to thinking hopefully thus set aside let us ask in what manner today political power might be wielded ethically, i.e., with advantage to racio-cultural evolution. First we must absorb some recent findings of social psychology. The correlations across some eighty nations show, in the "cultural pressure" factor in Table 4.3, page 136 above that most indices of political assertiveness—treaties concluded, pressures brought to bear, and increase in area controlled—are expressions of the same cultural pressure influence as affects degree of urbanization, scientific creativity, technological advance in industry, and creativity in music and the arts. This unquestionable association among nations, of high aggressiveness with high cultural creativity, is one which less realistic "liberals" among intellectual people would like to believe does not exist.

Although we are only at the beginning in psychology of a replacement of psychoanalysis—which served its turn in the progress of psychology—by what has been called the *dynamic calculus* we can begin to get some idea of the transmutations in emphasis which occur between the spectrum of "instinctual" needs which the individual puts into the group and the spectrum of the group synergy which comes out. For example, individuals experience hunger, sexual, security, gregarious and assertive needs, but groups interacting with other groups have lost such ergic components as sex and gregariousness and manifest an emphasis in the ergic spectrum only on such goals as power (self-assertion), parental protectiveness of members, and need for security (fear).

The blood-curdling analysis by journalistic amateur psychoanalysts of sadistic, etc., roots in imperialism are wide of the mark. Experimental dynamic calculus analysis (Horn, 1966; Delhees, 1968) show that self-assertion is a pure desire for mastery, such as motivates much scientific work and is no more connected with pugnacity or sadism than is any other drive. (The popular term "aggressiveness" is a bastard term confounding assertiveness and pugnacity, but pugnacity and hostility in fact arise quite as readily from frustrating any drive, e.g., sex or parental protectiveness.) In inter-individual morality, sex, pugnacity and probably self-assertion, show up as the ergs most frequently banned and involved in repressive conflict. And here as elsewhere, there is the usual tendency to extend to group interaction the same value judgments as among individuals [7]. The sublimations of sex and other drives which have been discussed as the last stage in the mechanisms which develop cultural pressure (page 137) end by leaving self-assertiveness and parental protectiveness as two of the major resultant expressions of group synergy.

Recognizing these needs as features of the normal dynamic psychological make-up of the group, and rejecting in the inter-group situation any simple carry-over of the inter-individual moral values which condemn assertiveness, we have now only to ask whether such assertiveness is functional—in a moral sense—in inter-group natural selection. The answer is surely that it is, for the capacity to exert political power is favorable to the expansion both of the group and its culture. Moreover, the correlations on which Table 4.3 (page 136) is based, show that the same powers as are involved in pure cultural creativity and in the impersonal struggle with nature are as effective in political imperialism. For example, the countries with the greatest military strength today are also those most active in astronavigation and astronomical research.

Four weaknesses—from a Beyondist standpoint—nevertheless remain, which point to the need for special ethical control of political expansiveness: (a) That without new machinery (discussed in Section 5.6 below) for reapportioning resources peacefully, expansiveness may bring war. (b) That expansiveness implies (as do some other interactions) the possible danger of an ultimate monopoly, in one world conqueror. (c) The possibility that the expansionistic displacement of a seemingly moribund or degenerate society may be premature, since the latter may have unrecognized germs of development, and (d) that power is a function of so many characteristics besides level of internal morality, and the soundness of the bio-cultural experiment—notably sheer size—that relative survival on this basis would contain much error. (Are the U.S.S.R., China or the U.S.A. more advanced cultures than Switzerland, Holland or New Zealand?) The experimental evaluation by power is optimal only with equality of size.

5.5   THE FUNCTIONALITY AND MORAL VALUE OF ECONOMIC AND POPULATION GROWTH COMPETITION

Readjustments of territory, population and cultural and political control have been a major means of cultural and genetic advance over the last half-million years of covering historical and pre-historical human group development. But, the drama of history balances a tale of triumph and pride of accomplishment with one of bitter complaints of brutality and "injustice." The process we are now about to survey—that of better relative survival by purely economic and population growth successes—has not been so roundly condemned, even by those who see only the "justice" of the *status quo*. And indeed, it may truly offer one of the more acceptable means for making effective and peaceful readjustments today with positively no change of boundaries.

Assuming all territories held inviolate, and war outlawed, how far could the greater survival capacity of any culturo-genetic group express itself? Success would now be capacity to maintain a larger population, at an equal or better standard of living, and, as we shall see, the capacity to program more varied cultural and genetic experiment. The limitation to existing areas is, of course, a severe one: an excellently run Luxembourg could not support the population of a poorly organized Africa—but no one method of natural selection is expected to be perfect alone.

For the sake of compression we are considering economic and population growth together here, and it is justified by the consideration that population growth without economic growth to maintain the standard of living is far from being the expansion of a culture. Indeed, not only the esthete and the ivory towered intellectual but even some harder-headed social planners, e.g., the MIT, Club of Rome, *Limits to Growth* (Meadows, *et al.*, 1972), question booming industry as the true index of vitality of a community. Nevertheless, the fact remains as Table 4.2 (page 135) shows that though the luck in possession of natural resources may temporarily give middle class incomes to working class minds, there is in general, a substantial correlation of average real income of a society with its educational level, especially in technology. The Affluence-Education dimension (page 137) we are considering (Cattell, Breul and Hartmann, 1952) certainly involves general educational level and educational organization too, though not necessarily the cream of intellectual culture as such. Let us consider also that part of economic success which has to do with wisdom of expenditure rather than technological production. Here a good economic level is a sign of freedom from unnecessary debts, such as follow from unemployment, crime and vicious luxury expenditures. In any case we must admit that the library, the symphony orchestra, and the laboratory rest on the physical, economic substrate of life being adequate: "affluence" is a *necessary*, but not a *sufficient* condition.

In asking of what features of a competing group economic well being may be considered a consequence and a cause, serious consideration must be given to the possibility that differences of average intelligence as well as of the average education produced by the group may contribute. Research psychologists such as Horn who are in process of developing more refined forms of culture fair intelligence tests, may yet live to see proof that the Affluence-Education dimension is a function of both. If this should prove true, it would account for the seeming relatively minor role of geographical resource and land size in this factor, for "resources" become what an intelligent people cares to make into resources. One people will stand with resources unused under its feet, whereas other (the German chemists who drew nitrates from the air; the Dutch tamers of the Ijsselmeer; the U.S. oil prospectors in the Gulf of Mexico; the Japanese sea farmers, and so on cited earlier, are cases in point) will draw wealth from what others look upon as a barren waste. Incidentally, this does not deny that countries such as the U.S.A., Russia and Bahrein may be living at a better level through "luck." And, regarding the magnitude of the role that native intelligence will play in the "enlightened affluence" factor let us note that the more education succeeds in leveling up countries educationally, by cultural borrowing, *the more important will become genetic intelligence differences in determining the ultimate differences of level of nations on average wealth.*

It is easy to see that better intelligence and other genetic qualities, as well as higher altruism and internal group morality in general, will contribute to a given area maintaining a larger population, and at a high standard of living. In speaking of population growth, in what follows, we shall assume that living standards are held constant (or improved). The size to which a population can develop depends primarily on such parameters as available sources of energy (which includes food), skill in disposal of waste (avoidance of pollution) and psychological capacity to adjust to crowding. Granted that these conditions for a larger population are met, groups with larger populations have the following competitive advantages:

(1) A greater resilience in sustaining catastrophic blows from the physical world. This victory from numbers the lord of the earth — man — has had flung in his face by the bacteria and the insects, who, in spite of his devastating chemical and other attacks upon them, survive with ease.

(2) A probability of being used as a source of population supply in relatively empty countries. (After the population increase by better farming in Scandinavia in the ninth century, in Europe by the industrial revolution, and in China by community habits making dense populations viable, these countries became sources of population for many other neighboring and remote regions.)

(3) A greater formidableness in the event of war. (*Vide* Mao: "China can accept casualties three times as great as the enemy and be the surviving power.")

(4) Possession of greater resources to sustain cultural propaganda.

(5) A greater likelihood of successful genetic hybridizations.

(6) If we may deal simultaneously with population and prosperity increase then we should add the fact that an increased standard of living and leisure means reduction of irksome restrictions and increased capacity for that "play" which is necessary to find new cultural mutations.

Of these advantages from the biblical injunction "Be fruitful and multiply" only the fifth may need some explanation. Since mutations are rare, and advantageous mutations still more rare, particular advantageous mutations might not appear *at all* in small populations. Secondly, many instances are now known where some mutations are perhaps mildly disadvantageous in themselves, but where a *conjunction* of such mutations is strongly advantageous. (An example lies in the mutations called "purple" and "arc" in drosophila, which, jointly, powerfully favor longevity.) The chances of *simultaneously* obtaining these mutations and of creating a sufficient pool of intermarrying persons with the new, advantageous combination of genes, are quite remote in a small population. (Haldane once calculated that fruit flies would have to make a ball bigger than the earth for certain combinations to appear!) A mixed cultural and genetic aspect of this population size advantage occurs also where the genes would operate in *different* individuals, i.e., where it is not a question of a particular combination of genes in one person, but of appearance of a sufficiency of certain genes in different people who will cooperate in the culture. (As an extreme instance in a population of three, there is a twenty-five percent chance that reproduction will be impossible, i.e., that all three will be of the same sex.)

The notion propounded here, that improvement of standard of living is a contributor to cultural survival is likely to be overcast by thoughts of the loss of moral discipline and spiritual direction which historically have sometimes occurred with such changes "where wealth accumulates and men decay." Today men dig for the luxurious city of Sybaris, wonder of its age, in a desolate marsh. Obviously, whether an access of wealth is to be the advantage it *could* be, depends critically on further factors. It would be wise to expend most of any increase in wealth in increase of population and in scientific research, rather than in cluttering life with distractions and self-indulgences [8]. Incidentally, critics of such "useless" expenditures as the U.S. and U.S.S.R. space programs should reflect that apart from the ever-valuable scientific knowledge gained, these countries are building a reserve of skilled manpower of survival value both in war and in meeting natural catastrophes. This is no "irrelevant" expenditure; but an enormous amount spent on recreation is.

A different kind of abuse by which the economy-population level survival gains can be dissipated is the accumulation of what we shall later define as the culturo-genetic load or handicap. This consists of an accumulation, beyond the inevitable minimum in a morally well-organized society, of a hard core of unemployable, borderline psychologically defective, systematically criminal, drug-addicted and similar groups which, whatever else they may be, and whatever their cause, are definitely a "welfare load" on the rest of the community. All such non-positive individuals, i.e., members whose contribution to effective group synergy deviates negatively, necessarily compete with more positive enterprises — medical research, higher education, interplanetary exploration, a better birth rate among the conscientious and capable — for the available community resources. Success or failure in inter-group competition for survival via economic level will be decided partly by the intelligence, and the level of social-scientific knowledge, but also, and, especially, by the moral strength and clarity with which this problem of the load or "handicap" is handled.

Economic and population gains thus have their dangers and in any case are easily dissipated. One other fact deserving comment is the finding by factor analytic culture pattern research (Cattell, Breul and Hartmann, 1952) that countries with large populations systematically suffer from difficulties in cultural integration (a tendency to excessive bureaucracy is another, and minor threat). Also we may well reiterate that not so much from the group's standpoint as that of the *world experiment* in cooperative competition, there would be advantages in having many smaller instead of fewer and larger groups (*see* pages 93, 218). The solution is obviously for any large group to decentralize, except in defense, and deliberately divide itself into sub-experiments within its own general form of culture, as the U.S.A. has done to a limited extent in its states and the U.S.S.R. in its separate regional republics.

It is in examining natural selection among groups along this dimension of economic success and population increase that we encounter in its most easily discussable form the issue of whether cooperative competition does or does not require "succorance" between one society and another. Often, in history the decay of morals and morale in one group leads to its being such an obviously sick society (as, say Turkey was "the sick man of Europe" before Kemal Ataturk) that other nations are exhorted to follow the inter-individual morality expressed in "Bear ye one another's burdens." Is this "compassion" sound inter-group morality?

The primary answer from evolutionary principles is surely, logically,

"No." The internal morality-developing function of inter-group competition is wrecked if any substantial reward is diverted to a group living on standard gains of another group consisting of conscientious and mutually faulty principles of inter-individual morality, by taking from the living altruistic citizens. Only if such a gift is responded to by a re-evaluation of cultural values toward those of the donor group would such action be ethically defensible. This condition on charity though unquestionably part of foreign aid as practiced by far-sighted statesmen, is nevertheless so contrary to our traditionally taught charitable habits (which popular thought tends automatically to transfer to inter-group morality), that it may require more discussion. However, the issue will recur systematically here (notably in Chapter 7) sufficiently for the reader's intuition or habit to be duly tested against new reasons and facts.

The central fact is surely the very simple one that *learning is a product of reward and punishment.* If we are confused by this false extension of inter-individual charity, at least the Communists are not, for as Khrushchev says with characteristic honesty in his autobiography (1970): "The policy of giving 'house presents' to other countries must be pursued intelligently so that our generosity will always repay us economically and politically" [9]. This seems to be an echo of Xenophon, that "The only way to conquer a country is through generosity." Putting aside for the moment this special contribution to the survival of the giver, the wider inter-group morality requires that a society which shows itself obstinately unable or unwilling to change vicious habits that are killing it should be phased out, by some generations of a geometrical rate of population reduction, and die in peace.

It must be repeated that, for learning to have any chance, habits should as rapidly as possible encounter the discipline of natural consequence. Every psychologist knows from learning theory that any haphazard dissociation of reward, obscuring any relation to contributory behavior is fatal to progress. If the rewards of individually more well disposed and morally clear behavior in the better group are to be reduced or washed out by a severe tax to aid the morally retrograde and obstinately perverse group such a form of group interaction becomes an unethical transaction. Some progress results from it only if the latter group solves its moral confusion by faithfully adopting values of groups of the first type, that have, by the evidence, found better ways of living. Certainly the trend in recent popular political argument that the "have" nations must give to the spawning "have-nots" because the latter will otherwise revolt and cut their throats is neither a moral argument, nor, in view of the technology of modern warfare, a realistic threat.

In arguing as here for "charity only with strings attached," we are illustrating the evolutionary action of economic and population gain through process (4) above (page 193). To those who find their feelings too tender to accept this view we can only reply that in the light of history they are guilty of a gigantic hypocrisy of reasoning. Christianity, from Pomerania to Peru, was imposed with the sword of cultural power in one hand and the Bible in the other. Missionary activities have succeeded partly because of the systematic tendency to imitate the more powerful. Besides, the soundness of a "charitable" act is to be weighed by the redemption accomplished.

All this, of course, applies only to differences of culturo-genetic fortune that are due to some systematic inherent causes. But *since there are vicissitudes and accidents* as well as *systematic consequences of the true character of a culturo-genetic experiment*, charity comes in two distinct varieties. It is as sound a principle in inter-group as in inter-individual morality that one should distinguish between the succor of the momentarily unlucky, and that of the systematically defective or vicious in moral habits. This matter is systematically discussed below (page 226) in connection with inter-individual behavior where a distinction is drawn between Type A and Type B misfortune, and where the popular belief that compassion is invariably a virtue is severely questioned. Intelligent and morally enlightened compassion reacts differently according to circumstances, whereas, like any instinctive behavior, blind compassion can harm the very individual it seeks to help — or harm other deserving souls more than it helps the recalcitrant.

In the realm of inter-group behavior the further argument can be made that Type A or "accidental" misfortune is less common than among individuals. The reason for this is, first, a statistical one — that a group mean averages out accidents to individuals — and, secondly, that a group's control of its environment is greater than that of individuals. Just as the physicist can make only a poor prediction of what the speed of an individual molecule will be a second later, but a very good prediction of what the pressure of the whole collection of gaseous molecules will be, so "accident" is smaller in determining what happens to large groups. Thus with certain resources partialled out, the effective behavior of a group is a very accurate measure of — indeed the definition of — the average morality of its citizens. Consequently, there is always a certain amount of injustice among individuals when the price of belonging to a given society has to be paid; but there is comparatively little injustice in the relative rewards received by groups.

5.6 SOME EMOTIONAL ASTIGMATISMS THWARTING ATTEMPTS TO REDUCE WAR

Since war is a possible consequence of the existence of independent groups, and since, as Andreski (1954, 1964) points out, in 4000 years of recorded history, only one year in fifteen has been free of war, i.e., without a major battle somewhere or other, its role in the cooperative competition of groups needs to be objectively studied. Let us be quite clear that in a Beyondist framework, war has *no necessary role*, since the aims of cooperative competition could be ensured without it. But let us freely recognize that it is unfortunately a common degeneration of competition which Clausewitz (1943) perhaps too nonchalantly accepted as "the continuation of policy by other means."

It can surely be said that nine out of ten human beings, i.e., excluding some psychopaths, idiots and theatrical types, profoundly desire to put an end to the waste and horror of war. For that reason, and because war is a social disease as complex as cancer is as a physical disease, parades and pious protests[10] against governments caught in wars are a waste of time compared to contributing to social research. How can people expect their governments to show the superhuman qualities necessary to save them from wars, begotten in the end from the selfish unrealism of the people themselves? For most of those same people cannot exercise the self-control, say, to give up the deadly habit of smoking, or to stop murdering more people per annum by drunken driving than are killed in the world's wars?

For good reasons, a rational man can admire the response of the pacifist and the *ad hoc* conscientious objector. Inherent in their position, however, there is the conclusion that "If war has even *a chance* of occurring as a result of inter-group cooperative competition, then group competition (and therefore distinctive groups) must go." Incidentally, this has enough counterparts in other realms of belief to be recognized as a mode of thought which has been a systematic weakness in human reasoning, e.g., in certain Buddhist sects which believe that since all life is accompanied by pain and frustration the ultimate aim is the negation of life. (Compare our folk saying: "It's your wants that hurt you.") To eliminate war by eliminating competitive life presents itself, to anyone with faith in evolution, as the supreme example of throwing out the baby with the bath water. Besides, it will not work. Nothing happens simultaneously in all groups, and the group which forces its government to lay down arms, regardless of the issue, at once receives a new government from the invading group, which is not pacifistic. To create a power vacuum in the hope of controlling power is automatically self-defeating. The solution lies in the opposite direction, in an as yet undesigned machinery to produce a controlling equilibrium of positive forces.

Even in well-designed machinery for a federation of nations — machinery which permits war-avoiding adjustive expansions and contractions as discussed below, and which sets up emotional idealism more appealing than war — criminals and madmen will arise. A police force will then be necessary, and, as two intransigently pacifist intellectuals — Russell and Einstein — belatedly came to admit to their followers — a "police force war" is appropriate. Thus some well-known conscientious objectors ceased to be so in World War II, when the particular views of the enemy became apparent to them. (The slogan appeared: "War is Hell. Hitler is Worse. Stop Hitler Now!") Einstein in 1931 urged all young men to refuse all military duties. But in 1941, he was writing to President Roosevelt urging construction of the atom bomb. And even Bertrand Russell suggested a deterrent war against Russia, in 1948, believing that if the nuclear weapons race should shift in favor of the Communists, the world would have a less benevolent dictator than the democratic United States. He unrepentantly adds in his *Autobiography* (1968, Vol. 2, page 8), "Nor do my critics appear to have considered the evils that have developed as a result of the continued Cold War itself, and that might have been avoided . . . had my advice . . . been taken in 1948." (The same argument for preventive war by the U.S.A. was made by the leading physicist and mathematician von Neumann.)

To a student of moral values it is worthy of serious thought that the leading universalistic religions, though they have sought to reduce war, have never categorically denied its morality in all situations [11]. It is not helpful to this issue to say that "Christianity (turning the other cheek) has never been tried," for even if theologians agreed that this is the interpretation of Christianity in the given situation it could not, for the above reason, work out. It is true that there is universal emotional repugnance, which has disproportionate political value, against seeing passive resisters (for any cause, good, bad or indifferent) calmly massacred for their obstructionism. But, even within this century, the massacres of the Armenians, of the Jews, and of the Kulaks of Russia has nevertheless proceeded. It only requires a Ghengis Khan, a Stalin or a Hitler with the courage of his convictions (however much we may dislike his convictions) to "call the bluff" of "passive resistance," on no matter what scale. "Civil disobedience" was blown up into a saintly cult by Ghandi, and masqueraded as a vague new religion (especially among the literati) only because the tradition of the English Christian gentleman made massive executions impossible. When the English withdrew, slaughter on a grand scale, as an inevitable consequence of the inherent insincerity of the mutual impositions and aggressions hidden in "passive resistance," followed almost immediately.

Central in the vortex of emotion which denies to our nauseated senses some objectivity of perception of war is, of course, its horrifying cruelty and loss of life. Without denying these last by one iota, let us recognize that one may negotiate the edge of a precipice better by not letting the horror overwhelm one's judgment. Since we are all to die, in fashions that we do not care to anticipate, the question could be raised whether a life is not more fulfilled and significant by dying for one's cultural principles than for nothing. A perhaps shorter life lived for the spiritual values of one's group is better than a longer one lived merely as a digestive tube. Men die from over-eating, drunkenness, dangerous sports and much else about which no "conscientious objection" protests are made. For example, in the last generation deaths from lung cancer, largely associated with a trivial luxury of smoking, exceeded in this country all deaths from World Wars I and II by a large margin. By automobile accidents in the U.S.A. alone, 48,000 were killed on the roads *in one typical year* (1969) and the manner of their death was no less horrible and cruel than that of the 43,000 war casualties in nine years in Vietnam. (Analyses report that 28,000 of these deaths were through drunken driving and 60% of all accidents occurred on "recreational joy riding.")

When people think some activity is worthwhile, they are apparently willing to risk death or agony for it. The agony of surgical operations before anesthetics did not convince people (Mr. Pepys of the famous diary, for example) that operations should not be performed.

Among the horror writings that have not added to perspective are the lurid predictions (not unlike those made to Londoners before World War II) that the new weapons—and now one refers to nuclear explosives—would wipe out 95% of the world population and essentially end the human race. Kahn's (1960) passionately attacked dispassionate analysis of these dire predictions is more realistic than most, and, in accordance with what is known throughout history about the balance that grows between offense and defense, he reaches very different conclusions about the "end of civilization." Even if, as an imaginary exercise, we suppose this 95% dying in war instead of in peace, it is horror-mongering to call this the end of the human race. Estimates have been made that ten to fifty thousand men, from the usual assortment of occupations, possessed of one good surviving university library, could reconstruct our present culture. (Perhaps we may suppose the remote island of Tasmania, well away from "rocket alley," to survive; or at least, the city of Hobart, sheltered under Mount Wellington, and this alone would supply more than our minimum.) Indeed, if we are being realistic, there is even to be mentioned on the credit side that a re-peopling of areas from a fresh start would permit a planned increase of cultural and genetic quality from the beginning, much as the re-peopling toward the end of the last ice age marked the rather dramatic change from the Neanderthal to the Cromagnon and modern races of man.

In seeking to place some of these disturbing emotional situations in perspective let us debunk the pro-war illusions equally. There are certain philosophers and writers (not to mention poets and song-writers)— Nietzsche, Fichte, Gumplowicz, Treitschke (1916) and Andreski among them—who have argued that war is a boon, bringing an indispensable bracing experience to the human spirit. General Patton (1947) as one might expect, asserts "War is the supreme test of man, in which he rises to heights never approached in any other activity." But one might well listen more sympathetically to Eisenhower (Ottawa, January 10, 1946): "I hate war as only a soldier who has lived it can . . . . Yet there is one thing to say on the credit side—victory required a mighty manifestation of the most ennobling of the virtues of man—faith, courage, fortitude, sacrifice." So broad and perceptive a psychologist as William James (1962) considered this true, but was then moved to search for "a moral equivalent to war." Much literary reaction to war in the past—at least up to Tolstoy, Sassoon and Owen—was as falsely jingoistic and theatrical as some modern writing has been misguidedly cynical and disparaging of military heroism. It remains a grim but indisputable fact that facing the dirt, the waste, the sordidness, the weariness, cruelty and death of war, for loyalties beyond the comfortable interests of life has tempered and tested the morale of men and women to limits to which they would normally never find themselves challenged.

One is reminded of the mountaineer Hornbein (1965), on climbing Mount Everest: "Everest is a harsh and hostile immensity. Whoever challenges it declares war. He must mount his assault with the skill and ruthlessness of a military operation." Perhaps the distinction between survival as against nature and against other groups in conflict is in the last resort not a basic one.

Nevertheless, when some balance of comparative emotional sanity has been reached regarding this sanity-testing topic the fact remains that there are better ways—less costly, cruel, sordid and brutish ways—of settling inter-group clashes without aborting the primary purposes of evolution. If there is any single cause of war it is low emotional frustration tolerance, and one trembles to recognize that if there is any characteristic of youth over the whole world today distinguishing it from the past, it is low frustration tolerance, begotten of ease and a parental unwillingness to teach restraint. Every psychological index as of 1970 points this way, and trouble may be expected when this group gets into the saddle.

In terms of the effects of war upon culture, e.g., in increasing willingness to search for new knowledge, in chastening arrogance, and in breaking the grip of obsolete habits, there are good as well as bad effects. It is in regard to effects on human genetic resources that war—at least in the last century or two—has to be regarded with deep suspicion. There is much circumstantial evidence that war kills off the best. It is amazing that social scientists have so neglected to follow up some pointers on this issue (Ardrey, 1966; Darwin, 1859; Bogart, 1919; Cattell, 1933a; McDougall, 1925), so that today we know nothing with certainty about any genetic effect of war. We know with fair certainty that in physical fitness and mental stability drafted men are superior to the general population. The rejections in the draft for low intelligence, neuroticism and psychotic tendencies are a substantial fraction of all rejections. On the other hand, by occupational status selection higher managerial and professional levels are less exposed to attrition. However, among the young, who are the bulk of the inducted, these have not yet found their status and stand to be lost — as the poets Owen, Brooke and Seeger, and the budding scientists Lodge and Moseley (the discoverer of atomic numbers) were lost — as much as any other types. Among those actually selected for the military forces, the differential casualty rates show little selection on gross indices, but there can be little doubt that war is an area where, contrary to civil life, *the reward of enterprise and altruism is death*, and that of apathy and selfishness in life, (the evidence of Meeland, Egbert, *et al.* (1954) on the psychology of those who take higher risks in war is relevant here). It is little wonder that the cultures of countries very frequently involved in war (as Athens and Sparta in classical times and Spain before the Renaissance) are apt to end in mediocrity and stagnation.

5.7 THE FUNCTIONS OF WAR AND THE DEVELOPMENT OF A FUNCTIONAL SUBSTITUTE

Any discussion such as the above seeking to gain perspective on the causes and effects of war is likely to become more involved with navigating through the baffling currents of popular emotion than with the relevant data themselves. But here let us turn to some more objectively analyzable issues in "What to do with war." Except for the presumed dysgenic effect—in the combatant groups themselves—which only a substantial replacement of population of the failing by the succeeding group might rectify—one could argue that the outcome of war on the whole is likely to favor the more competent culture and thus effect a general advance. For centuries, and particularly in this century, war has become so much more complex and costly that it has shifted the basis of natural selection relatively from dependence on tactical accidents to systematic performance in economic, cultural, and scientific competition. For it is competence in these which makes it possible for a nation to maintain adequate armaments. It is noteworthy, for example, that such advanced countries in science and general organization as Sweden and Switzerland have alone developed such technically adequate nuclear weapon shelters as could save a large fraction of their national populations.

The social scientist naturally wonders whether it would not be possible to stop a war by showing that one can reliably predict before a war actually erupts, what the outcome will be. Then, as happens in areas of the law where professionals can predict accurately, the contestants would be more inclined to "reach a settlement out of court." This has not been seriously considered as yet, because of the immaturity of the social sciences, but countless writers have enjoyed the idea of some limited substitute "game," a David and Goliath duel, for example, or a joust between the members of the opposing governments. That evolution has developed something not unlike this "token war" in *battles between members of the same species* shows that the idea is not entirely nonsensical. Ethologists tell us that among animals, *particularly the males in the same species* (so that the species does not lose its males), a hierarchy of "political" power is established by ritual conflicts. These contests by a ritual of threat, visual evaluation of the enemy and advance or retreat (Ardrey, 1966; Kropotkin, 1902; Carr-Saunders, 1936; Lorenz, 1966) stop short of destruction. Unfortunately, the degree of "unrealism" in evolution and in the acceptance of a token surrender also permit animals stupidly to accept quite damaging aggressions (the cuckoo's stealth) if performed without heraldry of battle and by another species.

One may doubt whether with humans one could find a game so complex, and involving so much of the spirit and resources of the group, that it would be accepted as a true "equivalent" in outcome[12]. Moreover, would it also be an acceptable equivalent in terms of discharging the causative frustrated emotionalities? Fortunately, there is a sense in which a kind of mock-war is already available, with the necessary realistic equivalences, in the sheer capacity of nations to maintain deterrent defenses against being overcome in war. These defenses include much of culture, for modern war is no low-browed performance. It requires not only high technological skill but a high general cultural level in all areas for effective defenses to be maintained. And we have already recognized a substantial correlation between cultural advance and success in war itself—if war is forced upon the country. Finally, let us not forget that the level of morale is an ingredient that is not only very intangible but also most important. A recent historical instance of the last is seen in the collapse of France in World War II, despite its having greater resources than Germany, through what Shirer (1969) documents as a widespread decay of morale and public morality.

Now the burden of maintenance of military insurance is in every country in competition with the demands of other "burdens"—the social degeneration or welfare burden; the cost of crime; unemployment; idle rentiers; civil corruption and poverty; the unnecessary part of the educational burden (through maintaining slow learners and defective teaching methods); the excessive recreational extravagance burden for those who are not happy in their work in civilized life; and so on. In President Eisenhower's farewell address (January 17, 1961) he pointed out: "We annually spend on military security alone more than the net income of all United States corporations." So long as there is no international police force, this is the cost of insurance against being overwhelmed by sudden attack.

So long as the magnitude of these defenses makes the cost to an attacking country an obviously prohibitive one, their maintenance succeeds in indefinitely extending a period of peace. Beyond their immediate pay-off these costs have the valuable function of providing an inter-group selection akin to that in war; for there is constant need to improve morale and mode of life if the group is not to be dangerously weakened by this burden. Furthermore, the challenge of preparedness demands good levels in that "potentiality for survival over a wider span of conditions than actually exist" which has been discussed above (page 88) as a hallmark of a more evolved culture.

While this argument is no mere attempt to make a virtue out of necessity, and recognizes that there may be better substitutes, yet we should be grateful that this necessity functions tolerably well. For one thing, its pressure makes society more alert to reducing *other* equally undesirable and more avoidable burdens. For example, the inauguration of radically new and effective eugenic steps to reduce the burden of the occupationally incompetent and the mentally inadequate in school learning, which should have been undertaken directly as an idealistic goal by a sensitive public conscience, is at least more likely to come about when nations are severely squeezed by armament insurance and the realities of competition from other nations. The state of "war without war" or what has been called the "war of nerves" thus in fact is capable of performing much of the function of war. If a nation appreciates that a war is lost before it begins, should it break down under its burden of armaments, a badly organized country may seek to rectify its ways. When all is said, the cost of being prepared for war is a more humane source of natural selection than war itself[13]. It was, incidentally, in this sense that the poet Robert Frost, in his visit to Russian intellectuals, gave explicit thanks to this mutual national competition. And anyone familiar with the reactions of American science and education realizes that they too owe much to Sputnik.

The stimulus that lives in the residual threat of war has just been discussed largely in economic terms, but that "war without war" which, with some self-pity, has been dubbed the "war of nerves" has more direct psychological effects. What does it do to culture in the intellectual and technical sense? Just as cultural pressure correlates over countries with frequency of being in a war, so over time there is a tendency for war to give at least a temporary fillip to cultural activity, even if it drains from it later. It can be noted, for example, that Dutch painting and Scottish philosophy reached peaks at the times when the countries concerned were in their direst straits from stress of war and the same might be said of English poetry (though not in World War II). A psychologist is inclined to see the cause of this in the contagion of superego demands from the aroused and temporarily dedicated general population to the leisured intelligentsia, as well as to the loosening of accepted traditions and the increased readiness to transcend individual goals ("Now, God be thanked Who has matched us with His hour") absent in the turgid luxury of peace.

That greater progress occurs in *science* in wartime is well known,

However, this calculation may rely on artificial academic standards, and there will sometimes be countries that will justly claim a right to revert from the judgment of the court of "social science" to war[14]. More often it will be mere low rationality that will provoke such rejection of the court verdict. Then all we can say is that war is functioning to bring the residual insanity in man into full leaf, that it may be pruned. In this respect, war is a guarantor of evolution toward higher sanity ensuring that when societies are not advanced enough to maintain steadily the indispensable state of cooperative competition by civilized and effective means, competition will infallibly be maintained by a fall to a lower level. Thus it has been pointed out in history that World War I, not to mention many others, was partly due to an intolerance of the strain of commercial competition. Someday the rebound from this nadir of rationality will generate such self-discipline in men that they will never again need to fall back into war.

5.8   THE NATURAL SELECTION VALUE OF INTELLECTUAL CULTURE AND PSYCHOLOGICAL WARFARE

Beyond the comparatively concrete and widely recognized – if not yet widely understood – effects of competition in territorial, political, economic, population, and military senses surveyed so far in this chapter, there extends a more subtle and tenuous, but not necessarily less powerful domain of cultural group interaction and competition. In the space available here we shall attempt to explore it in terms of drives to cultural supremacy, educational exchange and imitation, missionary activity and conscious and unconscious propaganda. Cultural anthropologists have unfortunately failed to give our language two words – one for culture in the generic anthropological sense, as we have mainly so far used it, and another for culture in the specific sense of intellectual, technological and artistic activity, as it is understood in salons and universities. (There is an even narrower sense, in which it is made synonymous with literature and the arts, or even that subdivision which was called correctly "humanism" before the recent use of the latter for an ethical system.) But what we mean by the first topic in the present section is a more rounded concept of intellectual culture as the range from mathematics and science to literature, art and music – roughly the "liberal arts" as taught in universities.

In regard to the process of transferring culture in the broader sense of Section 5.3, we decided that although in its main action it obviously aided the process of evolutionary advance, there are backward eddies, notably from interference with organic patterns and from creative groups being hurt by their "patents" being stolen and used against them. Much the same issues must be asked here about imitative borrowing and missionary propaganda in the narrower sense of intellectual culture. Indeed, first one should ask what good a large investment in intellectual culture can do for the culturo-racial group concerned and secondly, what mechanisms of interaction among groups in this commodity are favorable for general evolutionary advance.

Some differences must be noted at once between the arts and the sciences. As we have seen, investment in the latter translates itself so quickly into survival value in the technical gains of industry, in economics, war potential, etc., that its relation to inter-group competition and status needs no further comment. But invention in literature, the arts, in political structures and social values, on the other hand, is susceptible of no such direct evaluation. It is a shibboleth of education as a craft to assert that all expenditure on education in the arts and literature is good. Indeed, as C. P. Snow (1959) points out, among the literati the world of art, languages, literature, drama and the classics *is* culture, and the world of science is by them regarded as a possibly unfortunate emanation from the activities of blue-collared mechanics. In the eighteenth century the latter was to some degree true — but this is not the eighteenth century and the intellectual challenge and beauty of science has now generated as great an emotional-aesthetic quality, for those with eyes to see, as any of the decorative and entertainment arts constructed by man. Nevertheless, among the well-to-do and leisured, who often find the initial discipline of science too much, art, music and the performing arts *are* "culture."

As an avenue to truth, rather than as a means of emotionally enriching truth, we have already in Chapter 1 had definitely to reject the claims of the arts. That is not at issue here. The issue is whether the arts are, from the standpoint of group survival, a worthwhile investment, an irrelevance, or a dangerous drug. Briefly, the main functions the psychologist can see in art are:

(1) Catharsis, which means a discharge of more primitive emotion, remaining unsatisfied as a complex society elaborates past the average genetic fitness level. Most aesthetic catharsis takes relatively harmless forms. Therein the arts charm the frustrated primitive emotion from unnecessary revolt, or prevent the overloading of social issues with irrelevant personal emotionality. This usage is particularly obvious in the function of drama in discharging the perennially excessive drives of sex and pugnacity (hostility).

(2) Education of the emotions and shaping of the personality to the cultural norm. Astonishingly little education of the emotions is planned as an integral part of formal academic schooling (except in schools modelled on the old public schools, in England and abroad). Most of it takes place through vicarious experience in drama and novels, where the individual learns to handle triumph and tragedy, and to interact better in complex emotional relationships. Especially we see that literature can produce effects by inducing empathy on the part of the reader into admired characters. "Cultured" writers (and unfortunately other writers) are increasingly left as the guardians of spiritual education, for churches and sermons have appreciably gone out of fashion.

(3) The arts explore spiritual values and help to integrate the culture, making it "livable." There are social scientists who claim that to get the fullest loyalty from its members, in times of storm and stress, a society must "pamper" their luxury "needs." It must go beyond the usual compassion for the helpless young and old, and provide maximum welfare and entertainment everywhere. The decoration of the dwelling of the society by the arts is part of this. The theory has a cogency, though it is not complete, for men of societies such as Sparta or the Carthusian monasteries have shown exemplary loyalty ("Tell them in Sparta that we die at Thermopylae for Sparta") despite a bleak interior. Perhaps the value of the art and literature of a society is not so much that it entertains as that it integrates [15].

(4) A more subtle function, related to but different from the "safety valve" function of (1), is what we shall later call the *condenser function*. Here it is implied that emotional energy is brought by art to more refined form and in some sense stored for modes of discharge in real life suitably subtle to meet the demands of new developments in a socially complex community.

In viewing these functions of the arts, however, the possibility must not be overlooked that the whole intellectual investment may indeed go awry and head in the direction of unrealistic and maladjustive mutations. For in "art for art's sake" there are absolutely no dependable adaptive checks, as in science. As Chesterton implied of Baudelaire, and as intelligent people have complained today of the widespread publication of de Sade, Frank Harris, *et al.*, art may also have its poisoned gifts, its pathology and its degeneration. Perhaps for this there is no check like that on a false science and it may be ended only by the end of the culture which bears it.

As to borrowing of their intellectual cultures, nations have rarely had any objection, and have viewed imitation by others as a flattering invitation to cultural imperialism. In science the buyer is safe, but except for some probability of safety when it comes from an obviously viable culture, the buyer of the values of literature and the arts should respect the Latin tag *caveat emptor* (let the buyer beware). Possibly the musical beat from the jungle, or even the mood of the literature of Dostoyevsky, introduce incompatible elements in, say, Anglo-Saxon culture; and much that was brought by Europeans into Polynesian culture proved destructive, without any guarantee at the time of being superior[16].

As if by some healthy sixth sense, cultures have in general been far more conservative and wary about absorbing artistic and religious emotional values from abroad than in borrowing scientific and technological elements, as anthropologists noted fifty years ago. The natives met by Captain Cook were quick to admire steel knives or to borrow the trick of constructing houses with nails, but seemed indifferent about the white man's views on interest on money, Sunday schools or the theological correctness of the Thirty-nine Articles of the Church of England. By similar utility-borrowing the American Indians quickly took to horses, rifles and drink.

The generalization can safely be made, however, that utility-borrowing seldom stops there. After the deliberate, insightful choice of what is useful there commonly follows a conscious or unconscious acceptance of what is not so well understood. There is a tendency to buy the rest of the pattern — or, at least, a great deal of it. The Japanese set out to build steel battleships and finished by becoming essentially a Western culture. Nevertheless the Japanese steadfastly retained many artistic and religious values of their own, which, perhaps to the surprise of those who believe art has to be organically connected with the rest of the culture, seem to fit industrial and scientific culture no worse than the art and religion of the West did.

Spread of intellectual, verbal and artistic culture is now extremely rapid because of high mutual cultural awareness and rapidity of communication and travel. It is limited only by the obvious utility of the trait concerned, its prestige, and the willingness of the second culture to readjust and absorb. Nevertheless, the fact remains that the soundness of the borrowing — in terms of "progressiveness" — is always suspect and the increased speed of transmission of fads and fashions today is only an increased possibility of indigestion. The chief value of emotional, artistic (as distinct from scientific-technical) transmission probably falls to the transmitter in so far as he obtains disciples and prestige in a cultural alliance which is likely then to extend to politico-military alliances. A classical instance is the capture of Roman culture by Greek culture after the military defeat of Greece.

Although prestige from other sources may determine (along with political closeness) the direction of flow of cultural values, some direct appreciation of cultural level in turn determines prestige and indirectly affects other borrowings, along with political and military alliances. American troops stood alongside British troops on the shores of Normandy partly because educated Americans read Shakespeare, Milton, Dickens and Shaw. A recent survey in some Eastern countries, however, shows Germany ranked in cultural prestige above the U.S.A. and Britain, and apparently this ranking had its roots partly in a high regard for German universities (which America also imitated) and German cultural contributions in the last hundred and fifty years. One may surmise that the treatment of the Japanese, at surrender, as something other than ruthless barbarians (in view of their treatment of prisoners), was prepared for by the contributions of Japanese art and science to world culture. Cultural standing could be shown by many such illustrations to have important survival value through political alliances, etc.

The limitations to the help which group evolution can expect to get from culture borrowing reside primarily in (a) the lack of any guarantee that what people choose to borrow is *actually* better than what they have, (b) the danger that the cultural advances made by a genetically more gifted group will be very quickly turned against them by other groups, and (c) the danger of reducing group variety by a widespread borrowing of common features. As to the first, one can perhaps reasonably conclude that the probability of insightful borrowing being actually of something superior in contribution to survival is greater than chance. Thus in terms of effective evolutionary advance the borrowing of a whole pattern of culture is a most direct and powerful impetus to raising the world average, granted that it meets this definition of being truly a superior variant.

As to the second, the same argument holds for liberal arts culture as for general culture, including those technological inventions which, as pointed out earlier, might be used by a more regimented culture against the creative culture which produced them. For it is the argument of this section that "practically useless" liberal arts creations bring more subtle survival contributions through "cathartic," education of the emotions, and other functions. Of cultural elements that survive in this way it may be said that, except for the lack of reward to the genetic origin, just mentioned, evolution is aided by survival of the cultural idea *as such*. It proves itself *through* a group, but there is no need for it to remain *with* a particular group. Group interaction inducing transmission of cultural elements that work belongs therefore among the most ethical of forms of inter-group behavior.

The thought will suggest to the reader several historical examples which come close to this. Thus the Jewish bible raised the moral cohesion and social vitality of, at least, most European nations while the Jews were allowed to go homeless, the French in 1870 claimed (speech of St. Claire Deville, French Academy of Science, 1971) that they were defeated through the discoveries of their scientists over a brilliant century being turned against them; in 1940–1945 it is certain that the same happened to Germany (the discovery of atabrine alone saved thousands of soldiers in Africa to kill the sons of the inventors); and now we see China, which contributed virtually nothing to the imaginative discoveries in physics from 1500 to 1900 readying nuclear weapons to which the countrymen of Galileo and Planck at present have no defense. (The issue is clearer in science, but it must be repeated that it is not confined to science.) Intellectual, cultural and implicit emotional value gains are sometimes reached through internal conflict and bloodshed in one group, but used at no cost by others. For example, Britain in the end benefited from the experiment of the French Revolution.

In this situation scientists find themselves in a quandary for which their traditional values ill prepare them. Every schoolboy is taught the glorious ideal of the international citizenship of science. If he is an American he may be reminded also of the sentiments of Jefferson, who disliked patents and copyrights, saying "He who receives an idea from me receives instruction himself without lessening mine." Apparently inconsistent with this is the realism of statesmanship and business, the former maintaining "top secrets" as long as possible in the name of survival, and the latter holding patents as long as possible because the research costs necessary to discovery and invention must somehow be sustained. To the human being who enjoys the gains of science and medicine the logical conclusion is not the pious repetition of "the blessings of science are for all mankind," but that he should contribute personally toward directing the blessings of support more powerfully to the culturo-racial groups which consistently make the real gains. (For the differentials of contribution, per million of members of different groups, are of the order of ten to one.) If more scientific progress is what the world wants, then reward should be directed where an increment of support gives the largest increment of output. And if capacity to gain scientific knowledge is believed to be a desirable quality to evolve, a relatively greater feedback of the benefits of science to the groups which manifest and support scientific capacity is indicated not only for the advance of science but for the evolutionary advance of man. But this conclusion is not digestible along with the simple "international world of science" belief, and there are complexities[17] which we must examine further in Chapter 9, in implications concerning international organization.

Finally, in the intellectual cultural interaction of groups we have to consider that sinister caricature of missionary endeavor and facilitation of culture borrowing which is being called psychological warfare. The intention here goes beyond propaganda and aims either (a) to ruin the morale and cultural loyalties of members of a hostile culture or (b) to win friends in neutrals.

Even if a scientist is ready to admit that, living being a precondition of pursuing science, national secrecy in science may be at times desirable, and that science thus cannot be innocently or naively international, he may yet feel revolted by the deliberate intention to distort and obscure truth in "psychological warfare." To the morally sensitive it is even more repulsive than physical warfare. Indeed, anyone who esteems truth is likely to feel violent opposition to *deliberately* confusing people's minds on matters of logic, scientific truth, philosophy and religion for ulterior motives. (Incidentally, *Life* editorialized recently (1971) that a "Department of Political Warfare" has been suggested in the C.I.A. (Central Intelligence Agency) but that this "shocks the traditional morality of the nation.") Yet, we should not be so shocked for, in fact, the partial statement, the deliberately emotionally biased word, and the cunning omission, are already the stock in trade of ordinary internal politics, of much advertising and of personal argument wherever it may occur. The journalist-psychologist who, like a Goebbels, becomes a specialist in psychological warfare, merely tries to brew a quintessence of this widespread poisonous skill.

The strong and probably ultimately justified aversion of the scientist to this "tact" and "persuasiveness" is that his own art is directed to conveying the maximum of unbiased truth in the minimum of words. The arts of the politician are directed to precisely the opposite: a seeming maximum of warm communication activity accompanied by as little precision of commitment as possible, and a maximum appeal to emotionality of decision. Furthermore, it must be admitted that the best we have yet been able to create in law — trial by jury — uses just the same procedures. Nevertheless one suspects that both in politics and in law social science could invent radically better designs for the social procedures concerned *when the aim coincides simply with seeking truth.* They would be designs restricting certain practices, such as rhetoric, and substituting calculation and the computer for emotional harangue. But the possibility of achieving logical, mathematical and scientific forms to heighten the morality of within-group civic debate does not help us in treating psychological warfare if our principles of competition above should say that absolutely no restriction ought to be placed on between-group competition.

Psychological warfare has two weaknesses when considered as a genuine aid to the goals of natural selection. The first it shares with culture borrowing, namely, that it succeeds in spreading what fallible human beings think — or can be made to think — is desirable, instead of that which actually works. The second it shares with all warfare — the distraction of community effort from the battle with nature and the development of ill-balanced overspecialization in methods of direct combat. The part that is aesthetically revolting to a scientist — the Machiavellian persuasion and deceit — must surely in the end be accepted on the grounds that groups lacking in intelligence and critical sense will more easily succumb to it than groups which equip themselves with better education — as occurs in high pressure advertising. Incidentally it is democracies that are most open — if not most vulnerable — to psychological warfare, since the attacker has only to outwit the average man, whereas in organizations like Communist Russia or the Catholic Church a magisterium of trained intellects shields the less intelligent individual from subtleties against which he cannot personally defend himself. Something more professional, in a scientific, psychological sense, than a skirmishing line of journalists is needed if democracies are to guard themselves — with their openness to ideas — from the propaganda of totalitarian societies, aimed with great skill at adolescents and adults of low mental age.

In this connection some psychological solution is needed to the problem of the young — and older persons in a "trusting" situation — who, being accustomed to benign teacher-friend relations, often do not realize when a complete switch over is needed to critical, sceptical defenses. The problem is as old as history, and the "con-man" is as unavoidable in everyday life as the stair-step that sometimes is not there. It is no answer, where the child is concerned, to breed a distrust of everyone, for a sufficiently rapid education in an extensive culture rests extensively upon docility in absorbing information from a trusted source. Every society we know of has believed that children should be protected against exploitation by "psychological warfare" and has maintained what is a quite definite censorship of unsuitable literature, suggestive art, drugs, etc. The only matter in dispute seems to be the age – actually the mental age – at which this protection by the cultural magisterium can be withdrawn – without risking the onset of a paralyzing cynicism and suspicion. This issue of the role of censorship in an age of psychological warfare is taken up again in Chapter 9, Section 9.5. Meanwhile, one may conclude that the interaction of groups by psychological warfare, even if it *could* be eliminated, probably should not be. Elimination is in any case extremely difficult because the line between unconscious deceit through defense mechanisms and deliberate deceit is very thin. In any case, as some small consolation, this is an instance where the more intellectually capable groups enslave the less, not vice versa.

5.9   SUMMARY

(1) The aim of inter-group cooperative competition is to produce by variation and natural selection group genetic and cultural patterns with the highest survival potential in a changing and indifferent universe. Our purpose in this chapter has been to ask what the value of the various possible forms of group interaction is in relation to this goal and what regulations, if any, need to be developed by a world federation of nations to facilitate the best kinds of interactions.

(2) It is a popular error, notably in universalistic ethics or the common misunderstandings thereof, to assume that between-group ethical rules will prescribe the same behaviors as the within-group ethics do for individuals. The injunctions for (a) inter-individual behavior among citizens, (b) inter-group behavior, and (c) behavior among men at large, derive as an interlocking set of moral laws from the same basic goal, but distinctly differently, as a result of difference of situation. The primary law among groups is outright competition, which is secondarily modified by the groups in the world also being a community and requiring boundary conditions to be set up for the most effective outcome from competition.

(3) An instructive instance of the difference is that whereas mutual charity is the major law among individuals, outright transfer of gains from one group to another frustrates and confuses the feedback of proper reward to good cultural habits and genetic inventions. It thus constitutes not an equivalent of "charity" between individuals but a pernicious and evil interruption of group evolution. The highest inter-group morality calls for goodwill and fair play among groups in a plan of adventurous separate group experiment. These two systems of lawful behavior –

individual and group—have to be matched by the development of two patterns of moral injunctions in the individual's thinking, ultimately consistent in goal, but *adjusted* respectively to individual and group behavior. Moreover, they must be integrated under a single individual conscience.

(4) Nations are in the process of generating new cultures and new races. The two are organically interconnected, in that all cultures may not fit naturally and without modification on all races. Nevertheless, there are good arguments for a group's initiating new developments independently in both fields—genetic and cultural. For cultural innovations may lead to genetic innovations otherwise not directly conceivable, and vice versa. The latter—experiment primarily at the level of planning genetic hybridization and inducing mutations—is especially important. For without independent genetic experiment there is some danger of the genetic development merely conforming in a final equilibrium to the "cocoon" of its own culture. Although the disturbing outer challenges from other groups may prevent an arrest of evolution in such a virtual homeostatic equilibrium of cultural and genetic oscillations, the stimulus from introducing independent genetic innovations may be quite important.

(5) The ways in which the primary principles of competition and the secondary rules which adjust it to best action are to be built into an inter-group morality depends on the type of group involved. Groups need to be specifically defined as here, psychologically. Social classes, religious congregations, professional "craft" groups, etc., do not meet the optimum requirements for units in a group evolutionary scheme as well as the self-conscious nations that have developed (to the number of more than a hundred) since the Renaissance.

The model of variation, natural selection and fresh variation which sustains racio-cultural evolution in such groups takes up to a point the same form in the genetic and in the cultural fields. However, the laws of cultural mutation, borrowing, hybridization and differential survival are at the moment not nearly so well understood as those in the genetic field.

One important requirement which the model indicates in both domains is the need for a cycle of hybridization or culture borrowing, followed by isolation and inbreeding. The latter is needed to realize the potentialities of harmonious pattern formation from certain ingredients acquired by hybridization and mutation. Any evolutionary scheme has to cope with the inherent fact that most mutations are "bad." But technical means may soon be developed for speeding up mutation, genetically and culturally, while avoiding a high percentage of adult failures by pre-natal detection, in the genetic case and rapid objective social evaluation in the case of cultural experiments.

(6) This chapter proceeds on the above basis to examine the effectiveness — and therefore the morality — of the common historical forms of group interaction, in producing maximum evolutionary advance. These can be brought into a classification of five major processes: (1) Transplantation from one group to another of (a) cultural practices, and (b) genetic strains, by a movement of people. The last means either invasion or permitted migration; (2) increase of political power and control ("imperialism"); (3) growth of wealth and population; (4) warfare, and (5) the development of intellectual-cultural ascendancies, and the exercise of psychological warfare.

Some *seven primal conditions for evolution*, as we shall henceforth call them, must be maintained throughout any of these interactions, as follows:

(a) *The avoidance of a world monopoly of power and culture.* In business, among competing corporations it has been reliably generalized that complete *laissez-faire* tends to result in time in a reduction of the number of competitors, and, at least in the U.S.A., precautions have accordingly been taken to draft laws preventing a complete monopoly. The number of automobile manufacturers, for example, fell from a one-time high of forty to the present figure of close to a dozen. However, it is not certain that reduction normally proceeds beyond a certain point, or that conditions could not be easily set up that would allow the defunct to be replaced by new-born entrants to the competition.

The monopoly danger is particularly great through the mode of interaction we call war. Force, has tended to obliterate small countries no matter what their efficiency, if they make one small mistake in alliances. (Finland came near to being so obliterated by Russia, along with Lithuania, Estonia, etc. which were. The disappearance of Venice, Savoy, Naples, etc., as kingdoms through Mazzini was viewed as a liberation from Austria, and thus worthy of liberal enthusiasm, and in this case a common language and closely related culture may justify the absorption.) But history has far more examples where war by an imperial power has gobbled up valuably different smaller experiments.

Fear of war, as well as war itself, favor coalescence in larger groups (often, however, federal). But this seems to be reversible, with return to separate cultures, as occurred to different degrees at the end of the Roman and British Empires.

(b) *The maintenance of the spirit of cooperative competition and diversity*. Two dangers beset competition: (a) that implied in the old saying that competition can act like alcohol—initially stimulating but finally bringing a brutish sameness. In short, it must not become a narrow, single track race, self-consciously aimed at one specific goal, but embrace a generous diversity of goals. For example, the fact that the human race has remained one interbreeding species makes this danger of "one track" a serious one in the eyes of a biologist. Muller, who gave way to no man in his concern that there should be "no place left for *biases* against races or social classes" nevertheless recognized that in regard to the future of human evolution, in relation to the hundreds of thousands of animal and insect species, "It has been intrinsically dangerous for him [man] to have so long existed as just one species" (1966). Rather than move toward coalescence, it is important for man culturally and genetically, to become increasingly divergent in his varieties, to the point where formation of distinct species occurs.

(c) *That cooperative planning and recording is desirable*. It has been suggested in introducing the factorial design of the "grand experiment" that the blundering, humanly-expensive "experiments" of history could be improved upon by deliberate planning. There is no doubt that by keeping groups at an optimum and more equal size, by planned diversification, and by other ways, results could be delivered more rapidly, without any interference with the rules of competition as such, as discussed in the planning of research institutions in Chapter 9.

In this area the notion of *correlative competition* is important. By this we mean essentially that group A, exploring in one direction, may find a valuable new cultural or genetic mutation x, while group B, in a diverse direction, finds a mutation y. These are advantageous enough to put A and B ahead of other groups, but they remain mutually equal in viability by different excellences. It is conceivable next that, unless total pattern effects forbid it, x and y could now be mutually borrowed. Enriched A and B would then start off at a still more advanced level in new divergencies. This we call "correlative" because it supposes a special correlation of effort in cooperative competitive. Its limit is set by the distance or unrelatedness of the two groups.

(d) *The avoidance of total genocide*. In the interest of perspective one must insist that the danger of mutual annihilation in competition, or of total extinction of some genetic strain, has been much exaggerated in some current emotional discussions. However, in the frustration-pugnacity spiral discussed under (e) below, the danger of such attitudes developing as "Delenda est Carthago," or Hitler's "final solution of the Jewish problem," or Roosevelt's "total surrender or total destruction" is psychologically very real.

Now total disappearance of cultures and races at long intervals *is* something that a rational ethics has to face as part of nature. Speaking of one of the periods of highest evolutionary effectiveness Haldane (1928) reminds us that "innumerable species, genera and families disappeared from the earth" and Darwin (1917), "The greater number of [past] species in each genus, and all the species in many genera, have left no descendants, but have become utterly extinct." (Biologists count that ninety-eight percent of the one hundred million species ever on earth are now extinct.) Tennyson (1908) after ruminating that the death of the individual does not mean the death of the species, reflected yet again that to nature even the species is not immortal:

> "So careful of the type?" but no,
>   From scarfed cliff and quarried stone
>   She cries "A thousand types are gone.
> I care for nothing, all shall go."

If the earth is not to be choked with the more primitive forerunners a condition of birth of the new is the disappearance of the old. However, it is part of that *cooperativeness* of competition that an emotional harmony with the total purpose should eliminate the barbarities and emotional misunderstandings which have constituted the brutality of expansion and contraction in past history. Newer and more humane methods must prevail. For the tragedy of the death of the individual is magnified in the death of a culture and a people.

Unfortunately, wherever a question of relative reduction of a population is concerned the word "genocide" is today being bandied about as a propaganda term. Nature constantly commits both homicide and genocide, and there is no question that both individuals and races are born to die. But at what point voluntary euthanasia by individuals or genthanasia by groups becomes appropriate is a difficult question. As regards animal species, we are today inclined, for aesthetic and scientific purposes, to make sanctuaries and reservations for species obviously heading for extinction, and still more extreme and scrupulous consideration is indicated before allowing a breed of humans — however maladapted — to become extinct. But it is realistically questionable in both cases how much space the more vital species will continue to allow for museum "storage." The maintenance of the *status quo* cannot extend to making ninety-nine hundredths of the earth a living museum. Clarity of discussion on these solemn issues of rise and fall in culturo-racial groups would be aided if *genocide* were reserved for a literal killing off of all living members of a people, as in several instances in the Old Testament, and *genthanasia* for what has above been called "phasing out," in which a moribund culture is ended, by educational and birth control measures, without a single member dying before his time.

(e) *That degeneration of competition into pugnacity must be avoided.* In inter-group competition, as in any, if aggression is permitted, the instinctive tendency is to respond with counter pugnacity, and one then often sees a *frustration-pugnacity* spiral. In some periods of history, notably in that from which Christ sought the escape of turning the other cheek, this can reach impossible levels of hatred, cruelty and destructiveness. Any competitive situation whatever has to be monitored, like a chemical reaction which has positive feedback upon itself, against this degeneration into pugnacity. The modern trend to reduce (socialistically) competition among individuals is likely, incidentally, to increase competition among groups. Not decrease of competition, but avoidance of degeneration into pugnacity, should be the aim of social research.

(f) *That a humanity-dominated group environment must be avoided.* The condition needing attentive monitoring in (e) is actually a special case of certain more general problems that arise when groups begin to constitute too large a part of each other's environment, obscuring at the same time the basic importance of the competition of each group with nature. In these circumstances of a radically altered ecology it has been recognized since the time of Kropotkin (1902), that various cooperative and parasitic relations can develop among members such that they are no longer subject to selection on a "fair" basis as regards independent competence *vis-à-vis* the environment. As Hardin (1964) has well said, "The coexistence of species cannot find its explanation in their competitive equality." Indeed, it has obviously been a weak point in our "operational test" of fitness by the criterion of survival that certain lowly types have survived for an enormous time unchanged, e.g., the oyster, the brachiopod, the opossum and the antique New Zealand lizard (Sphenodon). That is why we did not accept this criterion alone. However, as one looks over the broad spectrum of zoological species, from man to amoeba, it is perfectly obvious that they are "equal" in surviving only as long as each has the advantage of its particular ecological niche or geographical isolation or adapted source of nutrition. Put them out in the open arena, and demand that quality of "adaptation to a broader environment" which has been our additional touchstone of evolutionary advance (page 88), and one will find decided inequalities in survival. There is nothing wrong with evaluating evolutionary level by survival and "efficiency," provided the test is applied in a sufficiently broad environment.

(g) Doubts in trusting competition arise when direct inter-group competition begins to account for most "points" in survival. A more subtle difficulty than any faced in the first six processes and conditions above then arises in that a group of groups, in a restricted environment such as the earth, may provide an inadequate testing ground. Any misunderstanding of the purpose and conditions of inter-group competition by this community of nations could set up conditions of inter-group dependency, and striving for artificial, community-approved goals ("best fitted for this company") that would either distort or invalidate the whole experiment.

In summary, there appear some seven limits within the boundaries of which inter-group competition needs to operate if the best conditions for evolution on earth are to be maintained. Before any regulation by a federated group of nations can be safely put into effect, however, intensive research is needed on the full effect of these and other boundary conditions.

(7) An examination has next been made of the specific virtues and possible malfunction of each of the five principal modes of group competitive interaction that have long been in operation. *Cultural and genetic transfer* from group to group (borrowing, propaganda, migration) probably contribute *effectively* to evolution only under certain conditions. When groups, for example, borrow simply what is admired, it may turn out to be actually deleterious. On the other hand, *relative population gains* (with economic level maintained or increased) lead to group interactions favorable to a higher survival potential for the group concerned, e.g., through political influence, stimulation of mutations, survival of major catastrophes.

(8) Although imperialism has slyly become a term of reprimand, the political and territorial expansion of groups has in the past probably been the most rapid and unadulterated of aids to the supersession of culturally and genetically inadequate groups, and to a general evolutionary forward movement. No sound morality of group interaction can take as its goal the maintenance of the *status quo*; but new conditions and rules for a justice of expansion and contraction remain to be worked out by social science.

(9) Any discussion of the complex role which war has played in the

evolution of cultures is difficult because of our intense emotional reactions to it. The central problem in connection with avoiding war is that of keeping competition at a higher level without relapse into what is a degenerative pugnacity, and also in finding alternative expressions for emotional frustration. Contrary to the simple minded stereotype which contrasts the level of "civilization"-(vs-"barbarism") of a country with its proneness to war there is a positive correlation of cultural pressure with frequency of involvement in war, and of level of advance in education and technology with military success. Innovations are suggested which, by performing certain functions which war now has, would have the best chance of eliminating it. War has the dysfunction of diverting development from the conquest of nature, but at the same time it may keep communities braced and adaptive through periods when nature offers no inexorable challenge. However, it counts size above quality and has the constant danger of leading to a monopoly, by conquest. A sudden and complete cessation of the social habits ending in war, which persisted over millennia is unlikely, but development of federated groups is likely to phase war out.

(10) The survival value of intellectual culture ("civilization") as such is hard to see, and it is often disputed that it has any "practical" biological value for group life. The practical survival value of the scientific half of culture is unquestionable; but psychological analyses also suggest real survival value, both *vis-à-vis* nature and in rivalry with other groups, from intellectual development in the arts—in literature, music, art and drama. This resides in their assistance in emotional adjustment to a complex culture, in developing loyalties expressly to the culture, in guiding emotional learning, and in creating attractiveness and status for the group relative to other groups.

(11) Psychological warfare, as a special, deliberate and self-seeking form of cultural interference in another group, is repugnant to the scientist in that it aims to go beyond logical argument and use all manner of deception and emotional persuasions. In this, however, it is no different from advocacy at law and from partisan politics. As between groups at least, it makes for the relative success of the more intelligent and disciplined in thinking, though it may deceive *both* aggressor *and* victim in time. The design we call a democracy is here at a disadvantage, for in a dictatorship or oligarchic elite, persuasion is examined critically by a magisterium of trained minds, and the average man is shielded from anything but the consistent picture they give him, whereas in an open democracy the average man is more vulnerable to the Machiavellian arts of psychological warfare. However, this has the advantage of producing

higher levels of intelligence and emotional education in the general population—*in democracies that survive*. A model that should not be overlooked in obtaining better understanding of the evolution-generating action of psychological warfare is that of the bacteriophages, which, as Delbrück showed, are parasitic viruses that invade the host and supplant its genetic code with their own, thus using its initial energy resources for a totally different culture. Definite analogies to this exist in business competition and in attempts by rationalist intellectuals directly to capture, for extreme, doctrinaire positions, groups of positive viability derived from experience, e.g., the Eisner take-over of the Bavarian government in 1919.

(12) The above processes of group interaction as they have existed since historical records began seem to our examination to be on the whole positively functional for the goal of evolving higher types of group. However, organization to bring scientific mid-wifery to bear on the process is now called for because (a) the "factorial" design could be better applied, and extensive record keeping and analysis would reduce trial and error waste in discovering the promising directions of progress; (b) not all existing procedures—especially those of culture borrowing and migration—are as substantially effective as they could be. Finally, most have degenerative forms needing to be avoided—notably in warfare and some forms of psychological warfare—which threaten to deny the required *primal* conditions (avoidance of world monopoly, etc.) necessary for evolution; (c) with the approaching crowding of the earth there is risk of conditions arising in which man constitutes too much of his own immediate environment for him to react to the realities of nature.

Unless new steps are taken this will tend to result (i) in directions of advance more concerned with adjusting to other group pressures than conquering nature, and (ii) in excessive imitation, uniformity, and failure to branch out into more divergent cultural and genetic types. Here we note that so long as man remains a single species he is vulnerable to any single noxious influence that might destroy a species; (d) also a risk arises in these circumstances that self-conscious man will deliberately, for his own ease, seek to arrest the natural selection process. This issue is scrutinized in the next chapter. Meanwhile, since the group natural selection process is one on which all else hinges—notably the derivation of within-group, inter-individual morality—it needs a volume of study that would make this chapter seem a mere trickling head stream to a Niagara of new knowledge.

5.10   NOTES FOR CHAPTER 5

[1] In history up to the present the primary survival-determining groups have been first, family — in the primitive, wandering, hunting and food gathering which held for hundreds of thousands of years; then the tribal and larger family; and then city-state and nation. The last — for Egyptians, the descendants of Sumerians and a few other peoples — has now been going on for five thousand years. Since at least medieval times we may take the nation as the primary group to which in practice we refer.

Christ — and other religious leaders — have bid the individual give up family and other groups for the sake of the religious group and such overt competing and bidding among groups for interest and loyalty is common. A central theme in the drama of Western history has been the battle for human allegiance between the "sovereign" nations and the universalist religious groups — Christianity and Mohammedanism. Most possibilities of sociologically interesting variants have been realized at one time or another — Japanese Shintoism where nation and religion are joined; Hebrewism where the unique is claimed to be the universal; the Holy Roman Empire and the Communist block, where (if Communism may be generically considered a religion), nationalisms are like eggs in the basket of a common religion — and so on. But virtually all are now variants on the main theme of (a) nations, and (b) universalist religious congregations.

[2] That conjunction was of (a) a stage in the early spread of universal and free education in which there were unusually gross disparities in class educational values; (b) a phase in industrialization which had produced enormous economic disparities; (c) a greater awareness in a half-educated and untravelled working class of the between class, than the between nation differences of values; (d) an anarchic breakdown of loyalty to the nation in the extreme stress of war and defeat. The "class war" was a brief episode, which has now reverted to the stronger polarization, as Communist Russia opposes Communist China.

[3] A well-known instance of an explicit assumption that inter-group rules ought to be a simple copy of inter-individual rules is in the speech by Woodrow Wilson which triggered the entry of the U.S.A. into World War I. "We are at the beginning of an age in which it will be insisted that the same standards of conduct and responsibility for wrong done shall be observed among nations and their governments that are observed among the individual citizens of civilized states. . . ." (Congressional Record for April 2, 1917.)

[4] Perhaps by taking too close and concrete an example at an early stage of the argument we run a risk of confusing basic principles with local values. Nevertheless, the above instance of "foreign aid," as practiced by many countries besides the U.S.A. and U.S.S.R., is worth an illustrative pause. When countries flounder in poverty because they take no decisive steps to control their birth rates does it help to send famine relief, or does it not rather, as most neo-Malthusians have concluded (see Hardin, 1964), merely reward and make still more resistant to change the cultural habits which systematically produce famine? In short, relieving famine, in some cultures, increases the likelihood of greater famine. Or, in more subtle fields, let us ask what happens to cultural evolution when countries which accept all kinds of benefits of organized science, developed at the research cost of other countries, fail to alter their values one iota from superstition toward science or to make the least contribution to mankind's need for scientific knowledge? When we simply give, with no strings attached, to corrupt communities, devoid of a sense of social duty, do we aid movement toward civic responsibility or help develop in the individual the habit of a well-organized daily life? More likely he is encouraged to persist "all along of dirtiness, all

along of mess, all along of doing things rather more or less" in a spirit of cheerful irresponsibility.

A much respected world commentator on international and interracial dealings, Gunnar Myrdal (1968), renowned (or suspected) for the liberality of his viewpoints, has recently confessed that he sees the causes of Asian poverty and other Asian endemic problems in something far more culturally pervasive than he first thought could be cured by legislation and loans of technical task forces. From a wide sampling of observations he believes the roots of underdevelopment lie in: "Low levels of work discipline, punctuality and orderliness; supersitious beliefs and irrational outlooks; lack of alertness, adaptability, ambition, and general readiness for change and experiment; submissiveness to authority and exploitation; low aptitude for cooperation" and so on. (Let it be noted that these conclusions are well in accord with psychological measurements which, though still on too small a sampling, consistently show in the populations of certain countries (Cattell and Scheier, 1961; Lynn, 1971) higher scores on personality traits of a neurotic (unrealistic, unstable, emotional) kind, such as have been shown in small experimental groups (Cattell and Stice, 1969) seriously to damage most kinds of group performance.) Whatever, the correct quantitative statement may be about the relative standing of particular cultures in these respects (and America with its hippies and Britain with a few incorrigibly featherbedding trade unions are not spotless of permitting cultural sabotage in the interests of emotional-self-indulgence), it is obvious that countries vary enormously along some of the dimensions we looked at in Chapter 4. The technical decision may to some seem premature, but the writer considers that we can say with substantial probability that the status of any country on certain of these cultural dimensions will have predictive value for ultimate survival.

[5] Shifts in people alone do not guarantee shift of culture. Instances of invasion with seemingly no lasting cultural effect are numerous. The invasions of Egypt by "barbarians" for a time seemed to leave the culture unaltered; the flooding by the hordes of Ghengis Khan, just cited, seems to have led to the Tartars leaving remarkably little cultural or even much racial residue in the areas dominated, and much nearer home one may cite the case of the English in Ireland. On the other hand, as with the Russians in Czechoslovakia, the Americans in the Phillipines, Japan in Korea, or the Romans in Gaul, Spain and Britain, the cultural effects are not to be underestimated.

By contrast, those expansions which have been accompanied by substantial supplanting of the native peoples, as in the migration of the classical Greeks into Greece, of the English into England, of the Americans into the Indian lands of North America, of the Japanese into Hokkaido, and of the English into New Zealand and Australia have shown quite permanent cultural and (as far as available measures go) genetic gains (and the incursion of the Israelites into Arab territory may show the same). A generalization of this kind, if sustained with more detailed correlations (such as Darlington (1969) begins to supply), is an argument for the superior effectiveness of *joint* racial *and* cultural transplantation, such as occurs in what we are designating "expansion." It may rest in part on the argument above that there is some inherently greater viability of a particular culture when developed in a particular, congenial racial group, whence culture and race survive better when transplanted together.

[6] As indicated in the last chapter we properly use the word "racial" for a distinctive genetic complex — a *pattern* — of gene endowment — embodied in a population. Yet the future is actually likely to be little concerned with the haphazard, existing races. Far more important in the future will be the races yet to be formed in the crucibles of culture and scientific design. The formation of existing races and peoples (conglomerates of races) has rested, as far as the half dozen primary races are concerned, (Coon, 1962a; Hooton, 1946)

on independent mutations occurring in some degree of climatic and geographical isolation of the group. Within-group selection genetic drift, and other forces then operated to produce distinctive groups (Dobzhansky, 1960; Keith, 1949) as now studied with mathematical sensitivity by the geneticist. These seem to have accounted for the three or four major and many more minor races now recognized by physical anthropologists. The monumental work of Coon (1962a, b) and others presents the fascinating story of this creation of races over the last fifty thousand years. Deliberate cultural selection and the engineering of the molecular geneticist rather than the climate or the physical environment are likely now to become predominant. That is to say, the demands of the particular cultural environment that has been distinctively created by each group and the isolation of an interbreeding set of people within the culture is likely rapidly (*in terms of archeological time*) to mold as many new biological sub-races as there are cultures. Furthermore, we must not forget that, as suggested above, a people will consciously conceive and attempt to engineer its own genetic destiny in some independence from the cultural molding.

Thus modern nations and language groups – the genetic make-up of each of which usually consists originally of a "people" – i.e., a hybridized *mixture* of the older races rather than any pure race – are in process through their action as inbreeding pools, of creating new races. Let us leave to the physical anthropologist and population geneticist the question of how many generations may be needed for these re-structurings of the old races to produce the statistically recognizable homogeneity of new "cultural races." The world traveller can already glimpse some partly culturally produced races, especially where marriages within a culture have been long continued, as say, in the Jewish religion, the Scandinavian language group, or the Japanese nation. The triumphs of cultural advance are obvious; those of genetic advance are more abstracted from the surface behavior and likely to be evident only to the precise and analytical methods of the scientist. And in pace they march with the slow, deliberate tread of centuries. This is perceived also by the deeply probing poet, as by Thomas Hardy in *The Dynasts* (1904) where he speaks of that which can:

> In curve and voice and eye
> Despise the human span
> Of durance
> That is I.

In evaluating mutual induction effects by culture and race, one of the main problems with which the social scientists of the future will have to work concerns the decidedly slower responsiveness of race to historical influences. (Though, for that matter, culture also has more gyroscopic momentum from generation to generation than the molding school teacher likes to believe.) Thus what Communism molds in Russian culture may yet prove to be superficial compared to deeper attitudes which it does not touch. In culture pattern statistics comparing Czarist to Communist periods it is surprising that the bulk of the factors show considerable persistence. Probably genetic make-up has a still more persistent and steadying (or retarding according to viewpoint) flywheel influence on the cultural experiments that are embedded in it. In this connection behavioral geneticists have still to ask such questions as "How far does cultural borrowing among groups tend to restrict the natural and desirable divergence that would otherwise occur among the genetic types of different groups?"

[7] The type of mentality which for lack of a new term we nowadays still call "liberal" (and which is common, for example, among younger students and the literati), though it lacks the

independence and realism of thought of the nineteenth century liberal, is often prepared to be highly inconsistent on the subject of pugnacity. A person of this type cheerfully condones violence in the anti-social individual while being allergic to disciplined force in the group. He is captive to superstitious stereotypes about force, while claiming to be emancipated from any "stereotypes" on sex. Regardless of whether his "rationality" about free sex expression is as logical as he believes, the fact remains that no early Victorian experienced greater difficulty in looking sex in the eye than this type of literati experiences in trying to look objectively at natural assertiveness. To judge by modern journalistic emphases, the great preoccupation today is in controlling "aggressive" impulses as it was sexual impulses to the Victorian. It is true (see Chapter 6) that experimental personality study shows the sexual and pugnacious ergs as most involved in suppression and the growth of cultural pressure, Table 4.3, suggests that the latter would become more salient in this generation. However, there is a difference between recognizing these facts and entering upon the panic-stricken repression processes which make intellectual acceptance of "imperialism" as a normal need (sometimes inappropriate or perverted) as difficult as Victorians found the acceptance of sex (sometimes inappropriate or perverted) as a normal need. This is not a condonement of war: it is a recognition that by treating expansive forces realistically and without immediate condemnation we could often avoid war.

As to political hypocrisy over the term "imperialism" let us note that Russia, for example, during the period in which it spoke most scathingly against American and British imperialism, took over Lithuania, Latvia, part of Finland, part of Poland, Western Mongolia, and several other areas. Britain, in the same period, adjustive to realities of political power, and without conflict, withdrew its political control from countless thousands of square miles and millions of native peoples scattered about the earth. Germany contracted; Japan enormously expanded and then contracted. Poland (to spell such changes, for one example, more specifically) lost through World War II a noticeable percentage of its area, 22% of its people and 40% of its national wealth, and suffered the migration elsewhere of Lithuanian, Ukranian and German minorities.

All these changes of people and political area occurred in the middle of a century which likes to assert that expansion and contraction of frontiers are gone for good.

The fact is surely that successful expansion, as the case of the Chinese into early Japan, the English into North America and the Spanish into South America, is more often than not accompanied by qualities which by any historical test, represent superior culturo-racial capacities. The sterile expansion of a Ghengis Khan type is an exception, and is likely to become still more exceptional as military action is increasingly eliminated from international adjustments and military capacity itself becomes more closely tied to educational and technical levels of the cultures concerned.

[8] Indulgences may be too specific a term for the often merely "irrelevant" or superfluous expenditures — irrelevant to any group goal or cultural stimulation — which are prone to occur in a society with a rising living standard. The "dissipations" which have weakened societies have by no means always been patently luxuries and perversions. They have ranged from accumulating what can only be called "cultural bric-a-brac," through somewhat excessive comforts and sensuous modes of self-expression to another area where the activities can superficially be called Humanitarian or intellectual in character. In forms ranging from beatnik festivals to self-acclaimed *avant garde* art, our society at the moment is obviously heading into such a climate. Parenthetically, the chief bastion against the flood of sybaritism in phases of plenty has been, in Western cultures so far, in what since Max

Weber's time has been briefly designated "the Protestant ethic" (without prejudice to Catholicism also having strict encyclicals on the subject of waste.) The Protestant ethic as so defined by Weber and others is centrally an ascetic assertion that good works and simplicity of living should prevail. As such this would lead to excess wealth being converted into a larger population operating at acceptable living standards and higher educational levels. This is one more instance of a Beyondist derivation leading to the same values as a "revealed," ethic – in this case the revealed ethic of Quaker Christianity.

Of these puritan bearers of the new ethic Weber says: "Over against the glitter and ostentation of feudal magnificence which, resting on an unsound economic basis, prefers a sordid elegance to a sober simplicity, they set the clean and solid comfort of the middle class home as an ideal" (Weber, page 171, 1904). Defining it in more specific goals, he says: "To be avoided are (1) Worldly vanity; thus all ostentation, frivolity, and use of things having no practical purpose . . . (2) Any unconscientious use of wealth, such as excessive expenditure for not very urgent needs, above providing for the real needs of life and the future." (Paraphrased from R. Bavelay, *Apology for the True Christian Divinity*, London, 1701, 4th Edition.)

[9] A recent popular article asks "How long can Western culture go on being an island of plenty in a sea of misery?" It is one of many rhetorical appeals in perhaps half of which the suggestion is that if a country reaches and maintains a good living standard the envious have-nots will gang up and destroy it. The other half of the apologists does not threaten, but takes the alternative attack that Western culture should feel very guilty about its relative social success. In either case the "solution by charity" so long as many Asian countries retain their fantastic birth rate, corruption, and the habits Myrdal has so well described (page 226), would be as rational as trying to dry up this "sea of misery" by throwing some lone island piece by piece into the ocean. Japan, one may note, once looked as if it were part of that sea, but is now likely to become the world's most prosperous nation. The control of population in Japan began not as a result of charity, nor through acceptance of propaganda for Christian values (which it rejected), but as a sequel to defeat in World War II. There is no evidence that even if charity is used as a means of teaching it is any more effective – or even as effective – as a realistic and perhaps punishing challenge from environment.

Actually the real extent of charitable aid among nations is far less than the man in the street might imagine from popular discussion about the aid programs of his own nation. It extends from two or three down to less than one percent of the national product in the "giving" nations, and is, of course, zero over most of the world's countries. However, this same amount if bequeathed to social scientific and medical research would seem an enormous gift. The unfortunate deception in calling it charity rather than an investment in propaganda or political influence is its worst aspect. A typical newspaper editorial, when the U.S. responded to Russian infiltration through aid, by raising its own, ran as follows: "President X is a believer in foreign aid. He accepts the proposition that developed nations should give aid to underdeveloped ones – that rich nations should help poor nations." Thus are illusions fostered and maintained. No more complete instance of the fallacious confusion of what morality indicates for inter-individual and inter-group morality could be found however, than in the following, apparently sincere argument by a senator:

"The obligation of the rich to help the poor is recognized, so far as I know, by every major religion, by every formal system of ethics, and by individuals who claim no moral code beyond a simple sense of human decency. Unless national borders are regarded as the limits of human loyalty and compassion as well as of political authority, the obligation of the rich to the poor clearly encompasses an obligation on the part of rich nations to poor nations. Indeed, it is no more than common sense to recognize that, among nations as within them, *the security of the rich is best assured by providing hope and opportunity for the poor.*

"Neither we nor any other nation, however, have yet accepted an obligation to the poor nations in any way analogous to that which we accept toward the individual poor and the poorer states and regions within our own country." (Fulbright, 1967). I have italicized the sentence in which truth breaks through rhetoric.

However, the U.S.A. is at present giving $100,000,000 aid a year to Egypt; the manners and morals in which are very different from our own (Cattell, Breul and Hartmann, 1952). In the context of within-group "charity" such gifts would, of course, be a legitimate part of support of within-group common function. In between-group dealings, short of saving life in mortal peril, these "gifts" have other functions. It is unfortunately part of the common assumption among politicians that rhetoric requires cant which causes them to speak of this as "charity" instead of honestly defining it.

[10] Among these cobwebby delusions that a quick broom must necessarily remove, are the still repeated clichés that wars are due to armament manufacturers, capitalists, world Jewry, professional military men, patriotic citizens, and so on through a list of scapegoats. Not far removed from this is the idea that journalistic "incitement" has much to do with war, but this mental excitement is trivial compared to the fundamental frustrations that might be operative. What can one say when the supposed work of mature thinkers, in Article 26 of the U.N. Declaration of Rights, proposes penalties for "the advocacy of hatred and hostility between men"? With the characteristic superficial, indeed, journalistic, omission of regard for fundamentals which has unfortunately characterized the early place seekers in U.N.E.S.C.O., this condemns Christ's hostility to the pharisees and the money changers in the temple. It ignores the psychological reality that the height of a hatred simply reflects the profundity of a prior love. For instance, no "hatred and hostility" equals that of the unselfish love of a mother when frustrated by a threat to the life of her child. In 1953 the U.S. Government declined, with admirable perception of cant, to sign the above clause. And it has declined again in June 1970, to sign the so-called genocide Convention, which actually appears merely designed for general international interference. To eliminate war we do not have to contemplate the elimination of life, i.e., of love-hatred or the diversification of living communities.

Yet another form of tragic misdirection of the young or immature is the presentation to them of the proposition that world peace is immediately achievable by a simple decision. (*See* Congressional Record Speeches, e.g., April 28, 1970, or most any issue of political speeches here and abroad.) No social psychologist would be ready to predict that a social habit definitely known to be 5000 years old and in less organized form two million years, is going to be dropped by all in a decade. Purely technical innovations, such as the eradication of smallpox, can be achieved in a generation. But these depend on the disciplined intellects of trained and coordinated scientists, whereas pressures for war come from the gut of the man in the street and from the ballot box. A Norwegian statistician has shown that in 5560 years of history surveyed there have been 14,551 distinct wars (2.6 per year). Let us bring into conjunction with these figures the fact that just since 1900 A.D., 1,700,000 people have been pointlessly killed in traffic accidents, more than all the military personnel killed in every major war from the American Revolution to Vietnam (Congressional Record, 91st Congress, W. L. Springer, August 10, 1970). We are unlikely to control the former more complex problem if we cannot control within our own borders the latter, simpler but actually more deadly problem. The most optimistic estimate possible, supposing as much advance achieved in social science in this century as took place in medical science in the last century, is that the disease of war may take at least as long to bring under control as multiple sclerosis, arthritis, cancer or schizophrenia.

[11] Christianity, taken as an example of a universal religion which advocates non-aggression in inter-individual behavior, has never in its officially responsible writings claimed that wars are indefensible in inter-group behavior, though masses of people seem to have assumed that it has. Historically, in the Crusades, recently in opposing Nazism and Communism, and in countless other instances the highest theological opinion interpreting Christianity — Protestant, Catholic and Orthodox — has in fact made it clear that conscientious objection to war is not an integral part of Christian belief. "I came not to bring peace but a sword" is considered meant by Christ also to define the Christian's duty to sacrifice his life in battle against earthly powers of evil, as well as in the original abstract sense of spiritual strife. In Mohammedanism the positive duty of war was still more clear in the great Mohammedan expansion. Paradoxically, as some would think, the verdict of history is that Christian nations have been the most active and successful in war of any. Their formidableness is surely partly due to the excellence of the internal morale, as contributed by religion. Conscientious objection still needs a base in some nonsubjective ethical system.

However, the present argument rests on the fact that the principle of "turning the other cheek" where groups are concerned, vanishes as a cultural value along with the culture which espouses it.

[12] In 555 B.C. Sparta and Argos sought to settle a war by combat of 300 champions a side. One Spartan and two Argives were left alive. Dissatisfaction led to a real war in which the Spartans were decisively victorious.

[13] It should be noted that no role or credence whatever has been credited here to the supposed psychological need for war as an expression of innate "aggression." (*See* Carthy and Ebling, 1964.) The self-assertive drive which is undoubtedly one of the human ergs, does not include pugnacity, as the bastard concept "aggressive" traps many into believing. As pointed out earlier the pugnacity erg, which is distinct from self-assertion (need for eminence and mastery of environment) draws its energy with equal facility from thwarted hunger, love or curiosity, as from frustrated self-assertion. Consequently, the only sense in which human nature has a "need for war" lies in the effect of the complex frustrations of a complicated culture. Reversion to hitting people on the head may be a momentary relief, and in this sense we can speak of a "need for war," but it is not a specific "need of aggression" but derives from the situational frustration of any and all life ergs.

The position taken here is thus precisely the opposite to that taken by Russell above (page 199), when he views a "preventive war" as being not too great a price to pay for the cessation of the Cold War and the cost of armaments. It is precisely by such emotional intolerance of the tax of ongoing competition that nations rush into war. It is a specious as well as a tempting short cut, for it invariably leads to a new, post-war "cold war." Happy acceptance of a healthy and continuous inter-group competition by all concerned is probably the most important prerequisite for avoiding war.

[14] Great though the temptation may be to "civilize" all rules of inter-group competition, more extended examination of the proposition is likely to show that this can be done in no simple "Sunday School" fashion. Civilizing steps may almost certainly safely begin with the outlawing of war, and the control of unfair competitive practices. But should they proceed to restrictions on certain forms of competition that seem, to many, inconvenient or disturbing? Such good intentions could nevertheless be the beginning of the end of man as an

evolving creature. A stagnation of this kind is more likely to happen, of course, through the action of some world organization which has failed from the beginning to espouse Beyondist ethics (and which like U.N.O. tolerates too many false assumptions, as in the absurdity of one vote for one country regardless of its size or culturo-political standing!). But it is a *possible* degenerative disease even with a world federation which aims to be Beyondist in conception from the beginning.

It is this possibility which brings us face to face with the difficult question explicitly raised but not fully answered earlier, namely: "Must the forms of competition be free to take *any* shape, however, severe or rigorous ("cruel"), or can the same quality of natural selection be achieved less wastefully and painfully by artificially "domesticating" this "game" of competition, by rules that a world council could devise?" The most constantly recurring demand here, as far as outlawing some one form of competition is concerned, is unquestionably for rejecting all resort to war. The above beginning of an objective examination may suggest that at the moment a clear answer cannot be given. Here one may hope that no intelligent reader has made the mistake of misperceiving our respect for realism in seeking the truest expression of Beyondist principles, as any mere harshness or callousness. The aim of organizing social science for human betterment is expressly the humane one of finding radically less wasteful and cruel, *but still effective* measures. A typical instance is the Beyondist's emphasis on eugenic measures, which can give us continuing natural selection without all the individual suffering of allowing the congenitally potentially diseased to be brought into the world to die or can allow the phasing out of a culture without its destruction by war. However, in all such "once removed" steps (as we may call them, to prevent our forgetting their relation to real, immediate demands), human judgment may go wrong. Fortunately, inter-group selection always has the "fail-safe" device that in the end we are recalled to the judgment of nature.

Thus the first move is to inaugurate birth control instead of famine; economic competition instead of genocide; a social science calculation instead of a war, and so on. But in such moves one must distinguish, on the one hand, between genuine humane substitutes for the harsher forms of natural selection, effectively operating at one remove, and, on the other, purely artificial, non-functional substitutes that are humane but useless. "Idealists" must be reminded that decision between nations by the Olympic Games is no functional substitute for the verdict of more "total" forms of competition. Although the battle of Waterloo may in a sense have been "won upon the playing fields of Eton," the full economic resources, manufacturing skill, international diplomacy, popular morale, and military skills of Britain and France at that time could never have been contained or evaluated in a game beside the Eton wall. Countless reasonably efficient substitutions (as, among individuals, of a classroom examination for a demonstrated professional life performance) can function tolerably for a long time. But these convenient predictors always fall short of full predictive validity. When real doubts accumulate *there can be no final substitution of a game with rules for competition unbounded by rules.* Whether the freedom implied here for groups to break into entirely new domains of competition is likely to be understood and condoned by a too staid world organization remains to be seen. Maybe history will have to await periods of temporary breakdown of federated control and overthrow of the world organization in order to change the rules and scope of action from time to time.

In short, the creation of a substitute for something with such diverse and complex functions as war is not easy. Certainly it will not be effected by any doctrinaire "bureaucratic" decision, of the type now offered by U.N.O., in which maintaining the *status quo* is the sole consideration and which is remote from the realities of organic group life. In the growth of empire, watching the inadequacy of the bureaucratic pedant, Kipling (1940) asked:

Ah! What avails the classic bent
And what the cultured word.
Against the undoctored incident
That actually occurred!

Douglas MacArthur (1964) said it again with equally admirable terseness: "In war there is no substitute for victory." (Or, to transpose MacArthur's values into science, no scholarly dissertation can be a substitute for discovery.) So, in whatever adjustive procedures we substitute for war, it must be an evolutionary realism, not a set of doctrinaire statements about equality or ancestral rights, that prevails.

[15] We have questioned, at several points earlier, the common assumption that a group which is more indulgent, protective, compassionate, "humane" and protective to its members receives better service from them in all respects. At least, we have argued that a sceptical examination should be made of the proposition that a justification can be made for an unbounded degree of group cosseting of individuals on the grounds of such a functional effect. However, psychological research yet to be done *may* prove that certain kinds of single-minded devotion to creativity, research and community service appear more frequently where the members of the group are indeed shielded to a high degree from the gross demands of war, crime, exploitation, and harsher economic conditions. Correspondingly, in the functioning of the individual, clearer thoughts may appear, as to Archimedes, in the relaxation and seclusion of a comfortable bath.

The issue of creativity and protection is no trivial one. But as far as the "comfort" of the arts is concerned, the alternative should be considered that their good effect comes from a contribution to emotional integration rather than mere consolation. To take a familiar culture for the sake of illustration, it could be asserted that the literature of England has done more for the "morale," enjoyment and integration of its mode of life than the literature of some other countries of lesser creativity has done for them. From Shakespeare's "precious stone set in a silver sea" and "happy breed of men" through Kipling's (1940):

Trackway and Camp and City lost,
Salt marsh where now is corn;
Old Wars, old Peace, old Arts that cease
And so was England born!

and so to Brooke's (1943):

Her sights and sounds; dreams happy as her day;
And laughter, learnt of friends; and gentleness,
In hearts at peace, under an English heaven.

all tell of a culture fortunate in having resolved ma.y of the inconsistencies which wrack other cultures. And that integration and perception was apparently achieved by "humanistic" culture. It built, however, upon a general culture of sound "material" and political foundations (which latter are seen incidentally, by dialectical materialists, as the *whole* story). The contribution to group survival of the spiritual discoveries in the national culture, when hurricanes strike is surely as great, as the history of this and other countries may show, as that from grosser material factors. And in extreme cases, as that of Israel, the possession of a spiritual, written culture, may even be able to reconstitute a nation whose physical existence and political structure have been completely scattered by the tornadoes of history.

[16] A psychological study of the mechanisms and the fate of various types of culture borrowings is greatly needed. One suspects that decidedly less than half of the transplants attempted are successful and of those that are viable, another half may turn out in the end to be definitely noxious. It is instructive to think of the main importation—other than venereal disease—of Elizabethan culture from the Carribean Indians—tobacco smoking! (Though further afield, it is true that the voyagers found the potato.) Borrowing by the criterion of present prestige may also often turn out to be ill-timed. Doubtless there were cities that borrowed more from effete and luxurious Egypt in 1000 B.C. than from the far more promising but rough early Greeks. Because of political and military interaction Britain has borrowed more from France than Nordic Europe since early times and certainly since 1900. One's doubts would concern mainly the probability that much has been missed by this preoccupation with one direction of borrowing and the conservative influence of keeping to a classical, Humanistic, conservative Mediterranean emphasis in education rather than opening up to the more modern scientific and socially exploratory movements of Germany, Sweden, America, and lately, Russia. (To take a small but revealing concrete instance, Britain hesitates almost neurotically between accepting practicing psychologists as constituting a new profession, as in America or Holland, and proceeding conservatively to recognize only the medically qualified "psychologist" as in Italy, Spain, and most of France, accepting clinical psychology as only a branch of the ancient medical profession.) Furthermore, in this overview, let us recognize again that the borrowing of "good" elements can sometimes be as destructive as bad, if they are from so remote a culture as to be indigestible in the functional pattern. For example, the Japanese influence in the East Indies, in the form of modern medicine, created a problem by cutting down the death rate but not the birth rate.

[17] It must not be overlooked that the triumphs of intellectual culture are achieved at environmental and genetic costs. As to the former, it could be argued that if Germany had deflected more talent from science to politics in the nineteenth century it would not have entered the twentieth century with alliances defective relative to those of the diplomatically more experienced Britain and France. Since restriction of birth rate is characteristic of the talented (geniuses as parents, seem never even to *replace* their numbers), and of the educated middle class, the intellectual contributions of, say, France, may be said to have been achieved by a genetic expenditure. And today the U.S.A. and U.S.S.R. are contributing (by taxation or its equivalent) five to ten times more per head of population to scientific and medical research for the world at large than most other countries.

Incidentally, the fact that some scientists themselves tend to ignore the need for or to protest against (Oppenheimer, 1955) any national restriction of discoveries should by no means be considered evident of nobility of ideals or of concern for all mankind. Any scientist familiar with scientific personalities and conferences knows that a child-like enthusiasm for what he is doing will render the average scientist blind to all kinds of consequences and moral obligations, much as a group of fox-hunters will (unless watched) ignore damage done to farmers' lands. And in this state of excitement he feels nearer to a fellow scientist from some alien culture than he does to his fellow citizens who are not scientists. If this is wisdom and idealism so was Nero's enjoying chamber music with his fellow exquisites while Rome burnt.

Part II   The Impact of Beyondist Principles
and the Institutions Required by them
in the Modern World

CHAPTER 6

Psychological Problems in Human Adjustment to the New Ethics

6.1 THE CLASH OF MORAL CULTURE AND HUMAN NATURE: ORIGINAL SIN

The whole concern of Part I above has been to state the bare and absolute essentials of the logical and scientific origins of Beyondism. No question has been raised about either (a) how it may fit the emotional life of the individual, or (b) what the nature of its popular reception may be. As to (a) the approach here has been diametrically opposite to that of most searchings for religio-moral values. In these latter the searcher has characteristically looked inward for man's deeper feelings and inarticulate aspirations. Then when a religious state of mind has been defined, he asks, secondarily about the forms of moral behavior to which it will lead. Here, by contrast, the moral directions have first been deduced from looking at the universe in scientific terms; and the whole question of the education of the emotions to sustain that position has not yet been touched.

Accordingly, to examine how the harmony of belief and emotion can be reached and to ask how Beyondism should express itself in the political, educational, social, and family life of the citizen today, is the object of this second part of the book. Even in this development, however, we shall not start with introspection on the feelings, convictions, and perceptions that might express this harmony. Instead, consistent with the canons of scientific behavioral psychology we shall treat the human nature-vs-moral rules problem as one for psychological dynamics and biological adjustment. Only when the possible adjustments of human nature to the ethical requirements have been explored can we finally ask what this means in terms of perceptions and introspected feelings.

Since most—but not all—of the adjustment issues that arise in connection with Beyondism arise also for any other major moral system, most of this chapter—though not of the later chapters, which ask about special social action appropriate for Beyondism—are concerned with human nature in relation to morality generally. Indeed, it cannot be claimed that the specific injunctions of a Beyondist ethics have yet been sufficiently worked out by research to permit us to see what the problems of some of its more specific impacts would be.

The general impact of moral systems upon man, denying him this and requiring him to do that has always been precisely what one would psychologically expect—aversion, grudging acceptance, boredom, and the psychological defenses which all the world's wits could produce. Oscar Wilde, whose writings (until the *Ballad of Reading Gaol*) showed a conspicuous rejection of moral values, quipped that he usually found interest in moral questions in men who were hypocrites or women who were plain. The psychological forces on which ethical interests can depend are indeed few—in terms of the heredity from the jungle which man still substantially possesses. They consist of the slender forces of man's unique gift—the evolving superego—and such situational incursions as fear and anger at others' wrong doing, self-assertion in reaching a desired self-image, and compassion. With most people these have to suffice to keep oneself, the neighbor, and the children in order. Particularly, for most of the young, morality has meant a nagging restriction, from birth, upon what they feel—perhaps correctly—to be their "true natures." With the dawn of Weltschmerz and altruistic feeling at adolescence, and the issues of "rights" which arise in growing up socially, young people normally then become relatively strongly interested in justice and moral values. But, psychological observations suggest that there is a greater variability of mankind in moral maturity and therefore, ethical interest than in almost any trait, intelligence not excepted. Incidentally, this is characteristic of evolutionary developments "on the way in," like superego formation—and also on the way out, like appendices and canine teeth.

Although for most average men ethical discussions thus have little more interest than household rules for the dog, or a prisoner's concern for the clanking of his chains, by a minority they have with sudden insight been perceived as the most important thing in life, enough to justify a Franciscan life of menial service or a Savonarolan death at the stake. And, indeed, no one perceives the role of moral rules as the mainsprings of civilization—which they truly are—and the key to progress to the greater society of tomorrow, without realizing that as morality becomes a psycho-biological science, it will indeed be the king of sciences.

Nevertheless, the fact is written broadly in police forces and prisons (and mental hospitals according to some psychiatrists) that moral rules and human nature are initially antithetical—regardless of whether the morality comes from a revealed religious or a scientific, Beyondist origin. Yet morality is in this respect not different from the rest of the environment—the winter that is too cold, the summer that is too hot. It is merely part of the broader antithesis—the "I want" properties of living matter, on the one hand, and, on the other, the "It is" properties of the world to which living matter so slowly achieves some degree of intelligent adaptation. At the heart of Darwin's discovery of evolution is this meeting between what we may call Life and Labyrinth. The labyrinth is the waiting path of complexities in the external world into which life, as it grows and ventures must extend itself. Living matter impinging on the rocky faces of the labyrinth, is engaged in a journey, in which, by more or less blind trial and error, it gradually grows out into an increasingly extensive replica or an internalized image of the pattern of reality. In this respect it is like an ill-fitting key which has to erode itself to fit a lock. Gaining this "fit" is education, as Aeschylus realized when he wrote, "God whose law it is that he who learns must suffer."

The moral demands are (except for the seer) not the demands presented by the whole outer world, or even by the whole culture (most of which ministers in friendly-wise to the individual), but by that part we may call the *ethical cultural demands*. It is a sterner part than much else in the mother culture. Culture, in general, is a middleman between the desire of genetic human nature to live, and the inexorable demands of the physical world. Much of culture represents hard-earned habits which men have learnt and passed on, e.g., in habits of food-getting, protection from the weather, treatment of disease, and generally coping, as a society, with the demands of external reality. But, as a middleman between the innate man and his cosmic environment this body of transmitted physical and moral learning, *is necessarily more "advanced" than the natural genetic make-up of man. It tends to lead the way, dragging the genetic endowment all unwillingly to school, with a resultant conflict between human nature and culture.* (An interesting development of aspects of this truth appears in Donald Michael's *The Unprepared Society*, (1968).)

Nevertheless, the advance of culture on genetic make-up is relative, for culture is itself a pupil of cosmic realities. There exists, thus, a two-stage learning process for man: to learn his culture and to find how to advance it further. But since culture may itself learn its lessons poorly it remains true that some fraction of every culture is "wrong" and misguided and that human nature is right in not being willing to adapt to all culture. However, let us recognize that its unwillingness is mostly a conservative dislike of adapting to anything (what the psychologist calls "disposition rigidity"). Thus when human nature rejects culture because of its difficulty, it may yet end up by having to adapt to another culture still more difficult. But in his revolt man may find some elements that are better.

The statement that the properties of the cosmic reality — "environment" — on the one hand, and the life demands of a given species — "the will to live" — on the other, are *independent* needs qualifying, if one wishes to be exact. For the make-up of the animal already represents appreciable adaptation to its particular corner of the cosmos. An unqualified statement would be that every *new* demand by environment for adaptation, and every direction of "spontaneous" new mutation in living matter *are* independent of the creature's pre-existing make-up. Culture itself is, we have recognized, a "middleman," an ambassador from life to the cosmos, who has already brought back a message about some things that have to be learnt, and some hints of how the germ plasm can, therefore, maximally adapt to the new requirements. The result is that there is already a discrepancy and a conflict between the pathfinder, human culture, and human nature, since the latter is born each generation afresh in the old pattern. Out of this conflict, much of the drama of human experience and much of the torments of "sin" and the consolations of religion are born. Present research unquestionably already reveals dimensions of temperament in which maladjustment to our society is largely in one direction (Cattell and Tatro, 1966), i.e., in which one direction of endowment can be said to be "desirable" and another "maladjustive." Although experimental work has not yet reached the motivational, dynamic traits we are discussing there are undoubtedly some drives besides sex and pugnacity that may be genetically excessive for modern life.

Sentimentalists in education — and this, alas, includes many so-called "liberal" writers — like to adopt the axiom that all education should be and can be pleasant. As pointed out below (page 375), this runs counter to the fundamental principle that no animal will alter his way of life — say, in getting food — unless the old way fails. *Modification of behavior begins in frustration.* New, long-circuited, but more reliable, ways of reaching a goal are not acquired unless the direct way is blocked. It was through the Ice Ages that the early ancestors of man learnt to control fire. (In Africa that cultural step came far later.) Since a child cannot appreciate what frustrations he will eventually avoid in later life by the act of learning the multiplication table, we give him temporary motivation by praise and affection for his efforts. But, implicitly (though this is not readily admitted), we frustrate him by restraining him from doing what he finds more easy or interesting and by withholding praise and status if he refused to try to learn[1].

In discussing the nature of the cultural pressure factor (page 137) we were careful to bring out that the experience of frustration actually ends in a net instinctual satisfaction gain (with long-circuiting of means, however). So here, the newer cultural patterns adopted may ultimately be found more rewarding, and certainly have survival value. But the fact remains that the child has to be forced or cajoled into them initially. Going to the dentist is such a case, and the most convinced rationalist in progressive education will usually have to admit that reasoning is not enough, but that need for assistance from bribes or threats is almost the rule with children of average intelligence or less. The experience of accepting more remote, instead of ultimately less rewarding immediate, satisfaction, we will call *deflection strain* (the abiding adjustment problem in cultural pressure); and the partly inherited capacity to tolerate it, *frustration tolerance* (or, in man, ego strength). There are also differences in the amount of deflection strain according to the age at which the learning is brought about, "imprinting" usually being easier at an earlier age (Hess, 1959).

Experiments show a continual rise in frustration tolerance capacity from lower animals to man (measured in simple cases by number of seconds the animal will restrain movement, after a signal, in order to get a reward), and among men there is evidence (Cattell, Blewett and Beloff, 1955) of substantial hereditary components in differences in the restraint we call ego strength. That the educational acquisition process in complex, modern culture stretches the genetic endowment in frustration tolerance of present day man close to its limits is shown by the temper tantrums and tears of childhood; the disorders and mental anguish of adolescence; the normal adults' need for appreciable time out for primitive recreations; and the alienation, mental disorder, drug addiction, and rising delinquency rate in early adult life. The connection is further shown by the tendency of these disorders to increase as cultural pressure (used in the technical, measurable sense of page 137 above) increases (though demoralization factors are independently at work).

And, consistently with the distinction between *adjustment* (to frustration) and *adaptation* just made we see that as complication increases, so also, and primarily as a result of the technical growth, does the length of life, the goodness of food supply, the length of education, time available for recreation, and the real standard of living. In short, the complex society satisfies substantially better the *ultimate* human adaptation in terms of life, health, food, and leisure; but it does so at the cost of this psychological burden of complex "long-circuiting." That is to say, there is such deflection of the natural "instincts" as to strain the genetically given intelligence and frustration tolerance of the average growing citizen to their limits.

From this discussion two concepts arise important for further steps. First, that of the *cultural lead* and *genetic lag*, i.e., the notion that cultural growth itself and the learning it institutes, move (at least in part) in the right direction for better group adjustment to reality, but that the genetic endowment moves more slowly, and "lags." And, secondly, as a consequence of this, that one can, in principle, measure a *culturo-genetic adjustment gap* (or "CAG," for briefer reference) which is the magnitude of the discrepancy at any given time between cultural demands and the genetic endowment of the race concerned. That part of the CAG which has to do with the moral demands of society, we may call the *moral adjustment gap*. Social psychologists are at present most impressed by the CAG (in some term or another) with respect to intelligence level, because the substantial degree of inheritance of intelligence is now realized and problems caused by the increasing demands of education on intelligence are embarrassingly documented. Yet, to a minority familiar with recent behavior-genetic research, the indications of substantial inheritance also of ego strength (*C* factor) and even, less strongly, of superego strength (*G* factor) are becoming impressive, and the implications of this for the CAG are being realized.

Although we know next to nothing about what the inherited elements are in the human conscience — the superego — it seems unlikely that they appear as emotional revulsions against things specifically tabooed in the culture, e.g., cannibalism, incest, murder. Yet there is some sense in which certain individuals are genetically better adapted than others to the restraints or demands for action which the morality of a culture makes. The indications are that higher endowments in these areas are measurable in the psychological capacity to control impulse and accept deflection strain (*C* factor) and in a greater sensitivity to guilt and altruistic feelings (*G* factor) in general. From measurements now being made on *G* factor (Horn, 1965; Gorsuch, 1965; Cattell, 1957; Dielman and Krug, *In press*), it begins to appear as pointed out above, that it has the wide developmental scatter one would expect in a trait undergoing rapid evolution. Thus an even greater range in this capacity for moral sensitivity, seriousness, and altruism may exist than has already surprised psychologists (*see* Burt, 1917) in regard to intelligence — which is also a trait in rapid evolution.

The intent of the above digressions into psychology has been to give a little concreteness to this paramount, yet not widely recognized fact that the genetic make-up may be said to lag behind the cultural demands, which in turn lag behind the present or potential cosmic demands. It has just been proposed that the first form of lag be conceptualized, ready for measurement in social psychology, as the CAG (the cultural-genetic adaptation gap) within which a special subset of measures may define the MAL (moral-genetic adjustment lag). It may seem a long way from these social psychological research proposals to the notion of "original sin" — which has cropped up in various forms in various religions (Aquinas insisted on original sin), but most clearly in Calvinism.

Indeed, to some modern rationalists "original sin" is a rather amusing museum piece. But the social scientist of the future may find himself substantially concerned with the CAG and MAL concepts. As behavior genetics advances and measurement improves some individuals and, therefore, some groups of peoples are likely to be found innately more difficult to educate than are others to a given level of morality of behavior as defined in a given culture. Indeed, already the grosser observations of the clinical psychiatrist have accepted the reality of the *psychopathic temperament* or sociopathic personality as that of a person "born short" of a normally expected potential for altruistic tendencies and self-control.

Of course, the standards of any given society are themselves partly erroneous reflections of what evolution (as "cosmic demand") requires; so the "gap" with respect to any *particular* culture cannot be given a universal currency value. But probably there are certain demands common to all cultures against which a more widely acceptable evaluation of genetic lag could be made.

6.2 ADJUSTMENT TO MORALITY IN THE LIGHT OF GENERAL PRINCIPLES OF PSYCHOLOGICAL ADJUSTMENT

To understand what the possibilities and dangers are in society's adjustment to a moral system — any moral system, though Beyondism is our more specific concern — it is necessary first to be more specific about the nature of the cultural-genetic adjustment gap — the CAG. Secondly we need to apply what principles psychology yet has available to understand and predict the nature of the expected adjustments or maladjustments that will ensue.

As to the first, while we are not necessarily committed to the biblical view that human nature is "desperately wicked," we must unequivocally recognize, in the CAG concept, that *no matter what direction cultural evolution may take, genetic human nature will initially be unadapted to it.* And if we grant — as it seems we must — that culture has skill in following the scent of what the cosmic realities are demanding, then a maladjustment will typically exist between the genetic and the cultural pattern. In regard to this maladjustment the greater part (but not all) of the required adjustment will ultimately take place by the genetic nature shaping itself toward the cultural pattern. (An exception might be when genetic mutations have produced a higher intelligence than the culture, as perhaps in the Dorian Greeks, when the rustic culture had to grow to become intelligent enough for the people.)

If we ask now for more substance and illustration concerning the genetic lag, the data and principles might be expected to be found in social psychology, criminology, and psychiatry. They *are* found there, but as yet not organized — except for a few original slants like Lombroso's concept of "atavistic reversion" (1911) and Freud's *Civilization and Its Discontents* (1930). The quantitative, systematized, psychologically and physiologically penetrating, picture of man's basic deficiency and his epic attempts to adjust, has still to be written (*see also* Cattell, 1950c, 1965, 1971; Cartwright and Cartwright, 1971; Epstein, *In press*).

Necessarily grasping at straws of evidence at this stage let us estimate that as far as man's ergic ("instinctual") make-up is concerned the biggest discrepancies between endowment and need are in the fields of pugnacity ("aggression") and sex. The evidence lies both in the disproportionate number of crimes motivated by these, and on the psychiatrist's finding that the maximum number of inner conflicts and repressions are connected with them. The objective, multivariate experimental analyses of the modern dynamic calculus show evidence for some ten distinct functionally unitary drives in man — such as hunger, sex, fear, gregariousness, self-assertion, curiosity, pugnacity, etc. Both the researches of the ethologists (Tinbergen, 1951) and the experimental researches of Harlow (1949) essentially agree with the drives in the primates — and even most higher mammals — being basically the same as those found by the multivariate analyses of the dynamic calculus in humans. But among all of these some serious maladjustment seems to be reported mainly in the two mentioned.

The maladaptation in sex may be comparatively recent, if we consider the Pleistocene period as man's history. With such small fractions of the earth's surface as were occupied by man, and such high death rates as characterized, for example, the Middle Ages, an incessant attention to procreation was desirable for survival. Nowadays, sexual activity of a few days in a lifetime would biologically suffice[2]. The problem that arises centrally in regard to a Beyondist evaluation of sexual morals, i.e., concerning the group survival value of various injunctions, stems from the rather widespread assumption that if a practice does no positive harm, it cannot be deleterious to survival. This overlooks the widespread evidence in evolution that nature is opposed to waste, and if any function is useless — though "harmless" — it is rapidly bred out.

At present the theory is well entertained among psychologists that all ergs in some way draw upon a common reservoir of energy, though the dynamic calculus, as yet, gives no evidence thereon. But, regardless of its validity in specific "dynamic" *psychological* energy terms, it is obvious that the individual has only a certain sum total of *physiological* energy and of time, and that the organism which wastes this on activities which do not contribute to biological adaptation will fail compared to one that has a more functional relation of energy expenditure to activities useful to survival. The reduction of the vermiform appendix and of body hair in man may have nothing to do with any real disadvantages of these in relation, respectively, to our eating and acquired dressing habits, but only to the deflection of physiological processes to maintaining a superfluous expenditure. Nature, when examined closely, is invariably a superb economist and accountant.

The gravamen of the charge against sex is both that its excess interferes and conflicts with the life of the community, and that it is an "economic" loss in terms of energies that might otherwise appear in the other ergs. There is, of course, no lack of writers with ingenious explanations of how sex created civilization, as in Morris's best seller *The Naked Ape* (1967), to which one can only say that it could equally be considered an account of how man became civilized in spite of sex. To the question of what might be done about the "lag" in a progressive society we return in Chapter 8, Section 8.6, since our concern here is first with the survey of maladaptations as such.

Regarding pugnacity, a fuller discussion is undertaken in connection with the *total* adjustment process analysis below. But it is worthy of notice that as cultural pressure — urbanization, complication, educational demand, heightening of military destructiveness, and general deflection strain — over the three or four thousand years of increasing urbanization, from Sumer and Akkadia to the time of Christ, a crescendo of pugnacity and cruelty must have occurred. We have referred to this elsewhere as an explanation of why Christ took the extreme way out of "turning the other cheek." Despite its being so stated in uncompromising terms as a philosophical and moral principle, perhaps it was only intended — human nature being poor at compromises — to be a corrective, cooling the fire of competition rather than putting it out. At least, if this is not the case there is ultimately an escapist unrealism in the Christian doctrine. Since aggression begets aggression it is obvious that a simple "positive feedback" system of this kind can be stopped in extreme circumstances only by reversing the principle. But social scientists should before long be able to invent a control system to dampen the extreme oscillations which would follow from crudely reversing the principle.

Meanwhile, many other societies than those following Christianity have been acutely aware of the pugnacity problem and have instituted strong moral inhibitions against it. It is a prominent element in the CAG. At the present moment in history, however, one cannot but smile that the Copenhagen liberals still regard "violence" with abhorrence, but, throwing away the insights of revealed religion, welcome all excesses of sexual expression.

A third genetic endowment in the dynamic field that deserves scrutiny as a possible major source of cultural-genetic adjustment lag is what psychoanalysis calls narcissism (in technical psychological discussion and here shortened to "narcism" [Cattell, Radcliffe and Sweney, 1963]). The existence of a natural pattern from some single drive which desires sensuous, self-directed sexual, and general self-indulgent satisfactions has been verified by multivariate experiment and named the *narcistic erg*. Its purpose would seem to be the physical preservation of the individual, by careful attention to comfort, avoidance of stress, etc., and it appears to have the auto-erotic overtones which Freud assigned to it. In any case, it comes nearest in natural drives to what would popularly be called selfish overconcern with the individual's physical and emotional satisfactions. It is not surprising that a substantial negative correlation exists between strength of narcism and strength of superego (Cattell, 1957; Gorsuch, 1965; Horn, 1966). However, among its myriad forms of defensive and devious expression, some even include an apparent altruism[3].

The social effect of narcism is less likely to be seen in crime or positive offenses against the culture pattern than in draining away energy from those group enterprises which require definite altruism. It is *par excellence*, the instance of a component in behavior which has *individual* survival value but parasitically undermines the survival value of the group. Many a malingerer from the battle front, or escapist before draft boards, has unquestionably survived when spiritually fitter men have not, but civil life has countless more subtle instances. The connection with the sexual drive which Freud noted at the clinical level and Darwin in animal self-display may also indicate that it has some functional tendency to favor higher reproduction rate. (In fact, in women one can see that narcism may have some function even in the heterosexual mating activity. For perhaps the more narcistic woman, spending more hours before her boudoir mirror, may have better chances of reproduction than, say, the "outdoor" girl.) Whole *societies* which turn narcistic, as classically defined in the city of Sybaris, or the French or Russian aristocracies in their luxury periods before the Revolutions, clearly fail; but narcistically highly endowed individuals may obviously parasitically thrive.

Societies have so far found little defense against this dependent tapeworm behavior since narcism is by its nature a skilled and obstinate *passive* defense against within-group moral demands. Thus, no matter how well society's moral code itself may be oriented to the values of between-group survival, it may fail if it possesses too many individuals of this type capable of unobtrusively avoiding external demands. For, as we have seen, this urge to sensuality, self-indulgence, and over-concern with the self tends to be negatively correlated with superego strength. But in as much as it does encounter the *external* equivalent of superego demands it is clearly the source of many defense mechanisms, such as rationalization, projection and reaction formation. It is the expression of these in the intellectual defenses of the social psychological area — in philosophies and social movements — which needs to be increasingly studied by any science handling the impact of a moral system upon social groups.

Among the other suspected lags of genetic development in relation to cultural needs we can surely include also that of (1) constitutional *mental capacity* levels, (2) levels of *frustration* (or deflection strain) *tolerance* as involved in ego strength (C factor) and, of course, (3) superego strength (G factor). The latter have already been briefly discussed, and there is no need here to become more technical. Meanwhile fuller evidence on what has been called the supply-demand dislocation in the intelligence domain has been set out elsewhere (Cattell, 1971, page 444).

It remains to compact into a couple of paragraphs the psychologist's paradigm of the adjustment process in the individual life, in order that we may be equipped to discuss the whole morality-adjustment problem in perspective. The "dynamic crossroads" model, so ably developed by Cartwright and Cartwright in *Psychological Adjustment: Behavior in the Inner World* (1971) shows how an obstacle to instinctual satisfaction is typically dealt with first by pugnacity and, then, if this fails by a series of attempts to handle the drive pressure itself by the suppression and repression which may ultimately lead to control, sublimation, ego defenses or neurosis as shown in Fig. 6.1.

It is in these defenses at the fifth and sixth dynamic crossroads of the adjustment process that social psychology has to understand the action of some powerful social and intellectual forces which create problems in morality. The response at the first crossroads is the story of crime; that at the last is the story of neurosis; but in between are the diverse accumulating defenses which create a thousand group and individual, behavioral and intellectual modes of evasion of that instinctual "obstacle" which is morality.

One general problem which comes up again and again is that the revolt against existing social authority arises from two very distinct types: the advanced, intelligent, thoughtful individual who is ahead of the norm, on the one hand, and the individual of low frustration tolerance, low superego development and poor intellectual capacity to enjoy a complex civilization, on the other. Both wish to "shatter this sorry scheme of things entire," but for very different reasons, though the ruffian is quick to take on the protective coloration of the idealistic reformer. The history of revolutions shows this unholy alliance repeatedly ending, for the idealist who has attempted to use the mob as "horse power," in a cruder society than that which was destroyed. This particular phenomenon will be examined again in Chapter 8 and Chapter 9, Section 9.4, where machinery for intelligent evolution instead of violence is proposed.

Thus, in the adjustment to morality in its restricting aspect—as part of the "realism" of life—one encounters the same sequences and possibilities as in regard to any form of social authority or physical obstruction. First there is pugnacity; then attempted self-control; then anxiety; and so to repression, and a variety of defenses such as rationalization, projection, regression, etc., ending perhaps in neurotic breakdown. These operate not only in proportion to the strength of the maladapted drives, as Freud pointed out, e.g. sex, aggression, etc., but also according to the degree of weakness of ego strength, intelligence, frustration tolerance, and so forth. Actual formulae for the degree to which different personality

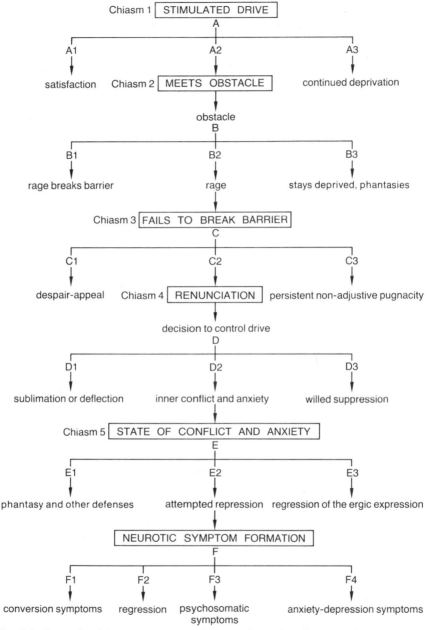

Fig. 6.1 Synopsis of human adjustment to frustrations: the adjustment process analysis chart.

factors are involved in the neurotic process have been worked out by Cattell and Scheier (1961, page 284).

However, amidst these forces is the superego, which is the instrument of salvation in the situation if looked at from one point of view (Mowrer, 1960, 1964), but a source of increased conflict from another point of view (Freud, 1930), and the chief villain of the play according to certain permissive psychiatrists. The next section will focus on the superego and its conflicts.

6.3   THE SUPEREGO AND THE PLEASURE AND REALITY PRINCIPLES

When all is said and done as to the total contribution of police forces, social reward systems, the stabilizing effects of the individual's major sentiments to home, institution, country, etc., the fact remains that adherence to a high standard of morality depends most on the inner structures we call the *superego or conscience* (G) and the *self-sentiment* (Q_3). The latter helps only in the normal person, but would even be counteractive for the individual who has the personal image of being a superb criminal, and we can set it aside in a brief discussion of fundamentals here.

The superego is a structure for which the foundations are partly inherited, and which depends for its growth on an affectionate emotional relation between parent and children. As a result of the bond in early years the child "introjects" the moral standards demanded by the parents, and accepts unquestioningly the restrictions and injunctions conveyed. There is a stern and uncompromising quality about the values that are then built up, and perhaps a combination of love and anger, as in the ancient gods, especially Yahweh, is a normal condition of this growth.

In maintaining a moral system socially there is no effective substitute for the altruism, shunning of evil and desire to do good which are developed in the superego structure. Although one would expect the particular expressions to be highly dependent on the particular culture into which the individual is born, there remains, as we have seen, so much in common to the "maintenance morality" of all cultures that in fact, one does not see such radical differences in superego structure. Psychologists are just beginning to learn a few dependable things about it since Freud's early speculations on its origins and role. Mowrer (1960), for example, has questioned that a strong superego favors neurosis *if other structures are normal*, and the work of Horn (1966), Gorsuch (1965), and the present writer (1957) shows that throughout the normal range (90% of people) *anxiety* (contrary to the psychoanalytic first approach) is actually very significantly *negatively* related to superego strength (a strong superego is tranquility). It also shows some tendency for a stronger ego (integration of drives) normally to develop with a stronger superego — though the correlation is not as large as with the self-sentiment. In modern times there have been some signs of a tendency for reduction of superego in the upper class and "intelligentsia" relative to the lower middle class. Finally, the twin and MAVA (Cattell, Blewett and Beloff, 1955) studies show perhaps as much as 30%–50% of the superego individual difference variance to be due to heredity.

As mentioned in the introduction, the evolution by natural selection of the superego must be peculiarly beset by difficulties. For, whereas a strong superego is advantageous to the group it is obviously disadvantageous to the individual (except in a few highly stable groups, where honesty gets rewarded). This we may call the *principle of counteraction of within- and between-group natural selection.* The half-million years of small group selection — between family and family and tribe and tribe — may have been the chief cause of whatever genetic elevation of superego has taken place. For as soon as large "civilized" groups appeared, parasitic types of relation undoubtedly arose and the tie of survival to superego became inefficient. Men culturally immortal are more often than not biologically eliminated. Men like Nelson, Pasteur, Shakespeare, Galileo, Newton, Haldane, Luther, the Christian saints, and Christ gave greatly to society, but left few or no offspring. Thomas Jefferson, whose life was one long gift to his country and mankind, neglected his own business to the point where he became bankrupt in old age. Everywhere, in genetic terms, we see a decided tendency for societies to undo in their faulty design of inter-individual genetic morality what they have achieved genetically in inter-group selection.

It is very probable that as societies get larger the rise of superego endowment by selection between groups and its subsequent decline in intragroup selection, takes on a longer rhythm. A substantial level of superego development is necessary to form a large society and get it to work. Once formed, however, there is little to prevent a genetic trend, at least, in which the more altruistic and culture-oriented are sacrificed for the idle, the anti-social, the incompetent welfare-dependent type — at any rate in a society, such as most, in which a Beyondist ethic has not extended to control reproduction. Then, as biblical prophets were so often moved to complain, "the wicked flourish." The situation cures

what best fits one's instinctual needs, and it has been extensively measured and analyzed (Krug, 1971; Cattell, Radcliffe, and Sweney, 1963;
Cattell, De Young and Burdsal, *In press*).

Although there are some tendencies for particular defenses to associate
with particular ergs, e.g., phantasy with sex; or with particular personality
qualities, e.g., projection with high dominance (*E* factor) and with pro-
tension (*L* factor), that may be of technical interest to the psychologist
we need to leave all such psychological details here in order to concentrate on the general operation in society of the basic psychological
evasions of morality demands. Our concern will be largely with the
pleasure principle as defined by Freud, with autism, and with the ego
defenses. The broad fact is that human nature sets out to bend the ethical
value system nearer to the gut's desire. Bandits, criminals and psychopaths do so brutally; society as a whole does so surreptitiously, and many
would-be-reformist "intellectuals" do so by elaborate arguments and
systems, which in any philosophical approach must be warily scrutinized.
We make such a scrutiny in terms of specific religious, social and political
movements in Chapters 7 and 8, but in general psychological terms in
the next section.

6.4   EMOTIONAL SOCIAL DEFENSES AGAINST DEMANDS OF EVOLUTIONARY ETHICS

The age-old rebellion against and attempted secession from, ethical
demands is part of an ever-continuing revolt against culture as a whole.
Although this latter *general* revolt is only on the flank of our pursuit, we
should note that the sheer increase in complexity relative to a not-so-
rapidly-changing native intelligence and frustration tolerance is an
appreciable part of the adjustment problem. It is little wonder that the
lower intelligence ranges, experiencing first the bewilderments of school
and then of life, with little opportunity for simple, full-blooded emotional excitements have (until the recent translation of delinquency into a
wider loss of morale, in alienation) contributed disproportionately to
delinquency.

One consequence of these pressures in the general cultural field is an
almost group-organized secession—as in the alienation, anarchical and
hippie movements—but which also shows itself in new fashions within
universities, the arts and even science itself (at least if psychologists in
the group "realization" movements are scientists). The Chancellor of
New York State University, J. B. Gould, recently observed (*World*

News, June 3, 1970), "Some students are more interested in things that are emotional and instinctive than in things that you figure out intellectually." One must add that the intellectual world is full of emotional meaning only for those intelligent and stable enough to study it. Things "emotional and instinctive" we share with the lower animals. From such opposition in the intellectual domain, it is a short step to politically liquidating any leading elite, by an unashamed expression of gross envy and rage, as has, in fact, occurred three or four times in the past two centuries. Incidentally, this danger is seriously considered in the recent book by Graham (1970); while Professor Slater, in a recent address (Gregor, *In press*), has deplored the ominous fact that "elite" has, in this generation, become a "bad word" in certain circles (with educational implications well discussed by Michael, 1968).

Short of succeeding in bringing about some near-breakdown of society, however, the attempt of human nature to attenuate or distort a too demanding cultural or moral value system has to proceed more by what are indirect, though in the end, more important ways. One form of this problem is that there arise whole systems of doctrinaire thought, constituting a crazy patchwork of neurotic, pleasure principle and reality principle thinking. These are not infrequently progressive in intent and of fine intellectual quality—yet showing that combination of progression and instinct-rewarding regression already mentioned here as more saleable than pure progression. In the next chapter we shall have occasion to confront some writings of, for example, Wells and Russell, having this character. Both Wells' Utopias and his *Rights of Man* (1940) are expansive demands for social emancipations of all kinds, sexual freedom and a world state in which competition, war and cultural boundaries are banished forever. The espousal of too many unrealistic, pleasure principle goals inevitably led to his final despair. Incidentally, it was the observant Beatrice Webb (Dickson, 1969) who commented on the influx into the Fabian Society under Wells' propaganda that they seemed to regard "each individual as living in a vacuum with no other obligation than the forming of his or her own character." To this comment we shall return in connection with the secular religion of "self-fulfillment."

In actual political *action* (except where large numbers are temporarily vulnerable to psychological warfare) the unreal soon has to come to terms with the real. In World War I, Trotsky's refusal either to surrender or to go on fighting did not nonplus the German generals, nor, if the logical French had had their way, would Hitler's re-occupation of the Rhineland, when he actually lacked military supremacy, have been possible. Similarly, in, say, the economic field, the claim simultaneously to have full employment and increased purchasing power abroad brings its own answer.

But on paper — since verbal facility and good endowment in the reality principle have never been proven to have any positive correlation — all is possible. And — as will be suggested later regarding the cultural role of drama and stagecraft — reality and purely emotional argument can be divorced even more completely on the stage. How are these poisonous mixtures of unrealism and seeming logic to be set aside? The attack on erroneous systems by logic — brought to a fine art by the scholastics and the Catholic Church — and by experiment, as pursued by modern science, are rightly the main axe and dynamite by which to demolish dangerous intellectual buildings, but they are not enough. Any clinical psychologist, fresh from a session with an intelligent paranoid, will recognize the futility of logic or scientific fact alone — and we are all somewhat paranoid. The hope is that a new branch of quantitative psychology will arise — perhaps a branch of the dynamic calculus — that will prove capable of examining simultaneously the soundness of the logic and the magnitude of the forces of autistic deviation at each step in the argument. The determination of the dynamic connections — the points of application of emotional needs — and of the individual's general liability to autism should suffice to give a psychologist an estimate of the magnitudes of departure from objective reason at various points. Promising calculations of vulnerability in that sense have already been made in regard to answering questionnaires, and with sufficient ingenuity could be developed to aid the straight examination of the logic of argument by logic machines. In a rough way there is no doubt that this approach is used widely: for example, we discount assertions of truth in proportion to the number of cocktails taken by the speaker. And in science, though *ad hominem* arguments are wrong in print we learn to give more weight to the conclusions of a man whose experimental and intellectual integrity we have learnt to respect than to those on an equal sample from another. With all this, it still remains true that false conclusions within a theoretical or philosophical system are vastly more difficult and costly to expose than those evidenced by the fact that an engine or a technological process or a political system does not work.

If the evasions of reality which enter a set of beliefs were only in an individual the problem of rebuttal would be relatively trivial, but when they become organized collectively in social theoretical systems they become formidable. Because evolutionary morality is newly arrived, and by what one trusts is an essentially logico-empirical route, immediate collisions can be predicted with many such systems. In particular the austerity of its message guarantees that it will meet the entrenched emotional opposition of many "leagues for the narcistic defense of this and that"[5]. Actually, the bleak lack of compromise of evolutionary ethics with human nature is no greater than that of some great religious insights, notably the system which Moses brought down from Mt. Sinai. A Beyondist is sympathetically drawn to those moments of emotional misgiving which Moses experienced, when he met the "unenthusiastic" reaction to his laws by his wayward followers. Desiring to be reassured by some glimpse of the glory of God, he was yet denied this much needed reassurance with the answer, "I am what I am" (sometimes translated as "I shall be what I shall be"). Indeed, this "revealed" truth beautifully expresses the essence of what we have here been emphasizing as the core of science: that the nature of the universe, and therefore, of the moral laws which derive from it, are given to us as absolutes. It is our task to discover the nature of the cosmos — the glory of God — and increasingly to understand its message, but not to pretend that it will adjust to *us*.

Although the dogmatic moral systems of revealed religions are more protected against degenerative, "rationalized" compromises — so long as people are "superstitious" — than a man-made ethics, such as Beyondism or Humanism — the fact is that historically they have often bought their popularity, and popular obedience at the price of offering an "emotional sop," given by exchange in some less important area. For example, in the Catholic Church before the Reformation, the pomp of power of an earthly kingdom became increasingly a reward for service to the church, while in Mohammedanism the approval of polygamy was not without its attractions. The specific forms in which these concessions have been or are being offered will be considered more in Chapter 7, where the differences of Beyondism from major existing ethical systems in these respects will be made explicit. The concessions which tend to be made to the pleasure principle also need to be given enriched emotional meaning by being considered in terms of particular psychological drives and tendencies, and that also will be taken up in Chapters 7 and 8. Here our concern is restricted to those fundamental reality-adjustment problems in adopting Beyondist ethics, that are of widest generality.

First among these problems in adopting Beyondism is the apparent assault on the vanity of the individual, his immortality, his right to a loving god, his claim to equality, his notion that God is made in his own image, and his belief that man is the lord of creation, in whom no further fundamental changes will be required. The word "apparent" is used because in being asked to give up some of these it will turn out that he is only being asked to give up rather tawdry toys for the realities a mature person would desire. Thus while science has already given two major blows to human narcism, first when Copernicus took the earth from the center of the universe, and secondly, when Darwin made us the children of early primates, the real power of control, and the sense of wonder it has brought to us in return are gifts we would not give up for the old tinsel.

Probably among the additional austerities the average reader may question in evolutionary ethics are the concept of continuing (though regulated) competition among groups, the suggestion that duties are the truly important aspect of rights, and the denial that human beings (individuals or races) are born equal. In concluding this section, since the two first are more self evident, let us take the bull by the horns regarding the last of these.

By "equal" we cannot mean identical, for that would mean that all of mankind consists of "multiple twins" (a clone, in technical terms) for one individual. But if not identical then "equal" always implies "for some purpose," and individuals are not equally able to survive under defined conditions, equally attractive to their fellow men, and so on. The true meaning of equal in the socio-political context is *initially equal in those rights and opportunities that it is in the power of society to give*, or, *equal before God*. The latter needs further definition, but, as to the more "operational" first concept it is obvious that this birthright does not last for long, since one woman is desired by many and another by few, one boy gets a scholarship and another does not, and one loses his citizenship rights by being sent to prison, while another becomes president of his country. The socio-political equality is a right to be equally examined, not a right to equal reward. Even if society should or could guarantee the latter, as a government, yet the individual citizen would by the equality and freedom argument assert their right to prefer the company or the help of one individual to that of another. And it is questionable whether society *should* aim, after examination by life, at equal reward to those who strive for its goals powerfully and those who do not.

As to "equal in the sight of God" a real meaning can be given to this in the sense that whereas the socio-economic-political equality of opportunity has to be based on, and continually modified according to human experience and evaluation of an individual's contribution, there remain admittedly areas where human evaluation is sharply limited in reliability. From a Beyondist standpoint the basic premise in this extra-social setting is that while it is quite true that some individuals will have greater survival, as types, into the future, it is equally true that no one knows with certainty at present *which* they are. It is clear, therefore, that all men should have equal opportunity and equal legal status *vis-à-vis* society and — which is different — equal spiritual importance ("Importance to God") in regard to an indefinable future. It is no less clear that it is desirable that men and nations *should* be produced with different (and therefore "unequal") genetic and cultural endowments. Indeed, as psychologists like Eysenck (1954), geneticists like Harrison (1961), and physiologists like Roger Williams (1956) have pointed out, the range of human individuality is so great that measurable significant differences are found on almost any variable we like to take: amount of food consumed, length of life, educational level attained, sense of pitch, speed of running a mile, command of language, income, capacity to tolerate alcohol, etc. When Jefferson wrote into the American Constitution that all men are born equal it was rightly understood by anyone with more experience than an infant that this could not refer to innate endowments, but defined the conception of equality of legal and social opportunity as stated above. Nevertheless, in some social proposals and viewpoints, as in the official Communist philosophy (which, for example, recognizes that children may differ in school performance but not in intelligence), this is interpreted as meaning they are *literally* born with the same endowments.

The real problem, of course, is that though we recognize that Jim's excellence in verbal skill is not to be weighed, by any possible index of the "excellence" of the two men, against John's excellence as a long distance runner, yet we fall readily into accepting some superiorities, e.g., in intelligence, geniality, or good looks, as *total* superiorities. But while a comparison of any such "general excellence" is not possible, it is quite possible to reintroduce the terms "superior" and "inferior" *for particular traits as they are valued and needed in particular cultures.*

A culture, as we said in Chapter 4, Section 4.7, has a "right and a duty" to define what it considers desirable, though it does so with full knowledge that it is undertaking an experiment. For example, in our culture, though possibly not in one so simple as that of the Australoids, higher endowment in intelligence is "superior." Consequently a genetic lag can be said to be created when we allow an excessive proportion of the population to be born at a low intelligence level. Possibly even a majority of cultures would today put higher intelligence among their desired excellences, though specific groups will differ on such questions as the desirability of larger or smaller body build, nervous sensitivity, sociability of temperament, etc. And it should be noted that even in one culture it is difficult to assign a "general excellence" value *to an individual as a whole organism* unless one is willing to assign weights to particular components, and to modify them from time to time in relation to the supply and demand in particular institutions within the social body [6].

Thus among the quite different concepts that are confused in discussing "equality" as a social ideal it is important to distinguish, first, *identity*, which must be rejected as a goal since variation is desirable, and population homogeneity or uniformity beyond some calculable functional minimum is undesirable. Secondly, there is *equality of promise*, which, translates itself into socio-political equality of consideration (examination). This equality in the brotherhood of human enterprise can be lost by moral defect, and it never reaches *equality of contribution* to the needs of a defined culture. Thirdly, there is "equality before God," which should give us sympathy with the rebel and the murderer, because we do not know the future.

Thus apart from the fundamental equality of promise, and of consideration by others as participator in the human enterprise to which this gives title, the rest of the possible equalities are actually inequalities. Beyondism explicitly recognizes and adjusts to this. The majority of flattering religious and political systems slur over these awkward truths, if they even recognize them. Yet, like all flattery, this toadying to self-assertion and the pleasure principle has its ultimate penalties. For example, it sometimes wrecks a proper regard in educational systems for the individual treatment of individual differences, and in voting it permits an intelligent and far-sighted minority to be outvoted by the inexperienced and the demanders of immediate satisfactions. One observes that in several cultures and sub-cultures, incidentally, denial of individual differences is particularly passionate and irrational where *innate* differences have to be faced. Indeed, wishful thinking is apt to kick against the whole biological reality of heredity as a "restriction on individual freedom." It is, of course, as pointless as kicking against the law of gravity, both because the latter is a fact, and because it is a useful fact — it enables us to walk about the earth. Similarly, although heredity is a lawful restriction on what direction an individual can best choose to develop himself, without heredity there would simply not be an individual, for it both makes him and ensures his individuality. Its lawful constraint also saves him from the chance of appearing from his mother's womb as an ape or a jellyfish.

When a great biological scientist like Haldane says, "I believe that any satisfactory political and economic system must be based on the recognition of human inequality" (Clark, 1969), the statement deserves especial weight because he had for many years been an enthusiastic Communist writer. At length, appalled by the continuing official party doctrine on genetics, he felt impelled to comment "How surprised they (the Communists) will be when they find they are not born equal." Justifiably extending the biological recognition from individual to group genetic differences (slighter though they may be) one can add "How surprised the environmentalist sociologists and economists will be when some of the 'undeveloped countries' remain relatively 'undeveloped,' and with continuing differences in cultural level and real income after all has been done that foreign aid can do." Incidentally, Haldane comments from his knowledge of the Communist leaders that though the doctrinaire Marxian view is that there are no innate differences, and Soviet psychologists are not allowed to study them, the leaders are under no such naive illusion. And, one may add, that Washington and Jefferson and other makers of the American Constitution, in writing an egalitarian premise, could never for a moment be accused of the uniformity illusion, nor for that matter could Lincoln in his public statements. They all distinguished, as the press and the high school often does not today, among the three very distinct senses of equality.

And as regards genetic differences, one must logically recognize that if there can be behavior genetic differences between two individuals, such difference can also be statistically significant between groups. If people enter the room in pairs, with a genetic difference in intelligence between the pair members, i.e., they are not twins, and a receptionist assigns the more intelligent of each entering pair to the left of the room and the less intelligent each time to the right, it is extremely probable that the left and right groups will soon be very statistically different in their average intelligence. Nature has operated many times in the past as a receptionist, by migratory, economic and other means. It would be very surprising if urban and rural populations, social classes, cities, nations, colonies, etc., did not show statistically significant differences on inheritable characteristics. In short, peoples and "races" can (and one would add from existing measurements *do*) differ in levels on such inheritable components of behavioral and other traits, though no value judgment (except in terms of the values of a particular culture) can be attached to the differences.

The production in Russia, and to a lesser degree in the United States and a still lesser degree in England, of supposedly educated individuals unable to think objectively on the inheritance of individual differences in mental characteristics is a complex disorder. Briefly it may be said to be the result of an historical trauma, a socio-political situation and an unchecked advance of the general pleasure principle in popular thinking habits. The trauma was the crass and unsubstantiated claim of Hitler for German racial superiority, imitating in a new key the Jewish claim to being a Chosen People. It is recognized in psychology that the weakness of mind which makes an individual "suggestible" may also take the alternative form of expressing itself in *contra*-suggestibility, i.e., of asserting exactly the opposite of what it is proposed that he should recognize (as occurs most frequently around two years of age and again in adolescence at periods of ego weakness). Both suggestibility and contra-suggestibility are dangerous weaknesses, to be psychologically contrasted with a truly independent, objective, and untrammelled judgment. In all three of the countries mentioned, especially under the stress of war, it became imperative for some individual to assert exactly the opposite of the Nazi unscientific assertion, even though it might turn out to be equally unscientific. It is an interesting aside on the relative scientific status and development of the physical and the social sciences that the fact that Germany developed the jet engine and the rocket did not in the least prevent allied physical scientists adopting these principles, but in the "social sciences" propaganda enforced a contra-suggestibility-derived error consisting in believing the exact opposite of whatever the "social science" of Fascism had said, sound or unsound.

In accordance with good dictionary practice we may define a racist as *one who asserts the superiority of his own race or people, without perception of the inherent impossibility, in our ignorance, of making such a value assertion.* But both contra-suggestibility and the departures from objectivity due to the pleasure principle have developed a sect equally prejudiced in the *opposite* direction. These bigoted individuals may be called *ignoracists* because in recent years they have totally refused to consider the scientific possibility that races may show statistically significant *differences*. An open and enquiring mind must accept the possibility that observed differences of culturo-racial groups could be as significant in inherited components of, for example, mental capacity and temperament as in the historically acquired cultural features. Both racism and ignoracism are extreme and dangerous fallacies equally unable to lead to happy and realistic solutions of our problems. Beyondism calls for a more mature attitude than exists in either. It demands as a first act of respect to the reality principle that human beings recognize equally the cultural and genetic origins of individual and group differences, and build an ethics of progress on that basis.

6.5   HUMAN RIGHTS IN THE LIGHT OF BEYONDIST MORALITY

In calling for a distinction between what ethics demands in *inter-individual* and *inter-group actions* (and, indeed, as we see in Chapter 9, Section 9.1, in as many as seven distinct value systems), Beyondism also calls for a complete revision of the concept of "rights." And this revision is appropriately studied here because the principal distortions of logic arise from narcistic and autistic intrusions from the nature of man. As throughout this chapter, the central issue being examined is that of the impact of an ethical system upon human nature, though in examining the distortions and compromises which the latter attempts to demand, the opportunity is being taken also to sharpen some of the moral concepts themselves.

In approaching "rights" we encounter ideas which have been dragged out of their true orbits to a considerable degree by that arch enemy of moral demands, the narcistic erg. For it is the essence of narcism to claim the "omnipotence of the individual" (Freud's "infantile omnipotence"). And it is the nature of the social defenses which develop with its aid that they seek to give it the protective coloration of those true refinements of individual development on which the growth of advanced societies does indeed appreciably depend. Thus we have the problem in discussing rights to individuality of distinguishing between the individuality of contribution which is vital to the growth of societies — as in the lonely self-expression of a Bacon, Newton, da Vinci, Marx or Christian saint — and the mere luxuriation demand for narcistic individual "rights" which every parasitic movement claims, and which often steals the noble label of "self-realization." How do we distinguish between the sensuous vanity of a limelighter — all too common in the domain, for example, of the stage — and that claim to the privileges of true "self-realization" which can and must be granted to the genius-innovator?

Actually, the problem of true and false rights to extreme individualism is a very old one in religions, where more saints, seers and prophets abounded at each period than now stand firm in the "time-proven" lists of hagiology. Nowadays a well-developed clinical and social psychology should ultimately have no difficulty in separating the genius and the parasite (e.g., the capable artist from the even more skillfully-bohemian, left-bank, camp "follower," with a large studio). For the genius is odd because his absent-minded originality is so basic that he can (to his regret) not avoid it, whereas the bohemian is so sterile that he must at all costs substitute a striking oddity. The term "self-realization" (once more in fashionable discussion) has been well examined by experienced philosophers and moral leaders who have made the important point that self-fulfillment can have meaning generally only by integrating with something greater than oneself, and in any case only by accepting natural laws [7]. In view of culture having gained this insight there is little excuse for its being forgotten by some modern writers such as Russell (1955), Maslow (1960), and possibly even Huxley (1957). In these, and still more in lesser writers like Ellis (1970), and Leary (1966) or psychopathic ones like Wilde, the perception of the difference between self-fulfillment in an historically meaningful way, and narcistic self-elaboration, has been lost, and one more chapter in wishful thinking in social philosophy has been written.

Self-realization is only one "right" among many that an individual may claim, but it behooves us to study this difficult, subtle concept — and thus recognize its bogus forms — before reaching the more obvious issues of social and political rights. A popular, but not necessarily counterfeit use of self-fulfillment as an ideal value is made by Huxley when he writes "As the over-riding aim of evolving man, it (evolutionary Humanism) is driven to reject power, or mere numbers of people, or efficiency or material exploitation, and to envisage greater fulfillment and fuller achievement as his true goal." The emphasis on the quality of the human spirit makes evolutionary Humanism seem the same as Beyondism, but, as the discussion on Humanism below in Chapter 7, Sections 7.4 and 7.5 shows, there are great gulfs between them [8]. That gulf shows itself in Huxley's failure to make anything but a subjective definition of "fulfillment." Before reacting more constructively to this problem, let us consider another view, in a moving passage by Russell (1968, Vol. 1, page 145). "It is the individual human soul that I love — in its loneliness, its hopes and fears, its quick impulses and sudden devotions. It is such a long journey from this to armies and states and officials; and yet it is only by making this long journey that one can avoid a useless sentimentalism." The point is well made: if one considers only the feelings, one ends only in visceral physiology. Every mental hospital is full of those who have wrapt fantasies around the demands of their feelings and every den of drug addicts of those who live only in intoxication; and both have completely given up on Russell's "long journey." Clearly self-fulfillment cannot be defined as *the most satisfying expression*, for the latter is at its highest when the innate nature is directly expressed, and the innate nature lags behind the evolutionary movement. Even if considered as life-long total satisfaction by inadequate ideals, rather than mere innate expression paying the usual price of impulsiveness, it is still only the "fulfillment" of the healthy animal.

Self-realization, as the "frustrated" saint begins to realize, is definable only in terms of significance of participation in the group adventure. (Not primarily the "armies, states and officials" unfortunately emphasized in Russell's thought, but the triumphs and sacrifices of scientific work, of social reform, of farming the bread for mankind, and so on.) This much needs to be said, in a glutted generation, which uses the term self-realization for the life of the playboy, though the best minds have known the distinction over thousands of years. And this need for self-fulfillment as defined objectively, in relation to the drama of the human conquest of knowledge, must look askance even at the Existentialist's focus of concern with subjective human emotional experience, perceiving therein many narcistic qualities.

Related to this "pleasure principle" thinking which confuses self-realization in a deeper sense with the right to boundless self-satisfaction in an individualistic sense, are the socio-political emphases on "rights," including the above discussed right to equality in all circumstances. The desirability of equality of opportunity, in the interests of evolutionary progress, has been discussed above; but there exist in most societies inequalities of *particular* rights associated with age, qualifications, sex, etc. Certain modern (actually recurrent) cults (*see*, for example, Dreikurs, 1960) show insufficient subtlety in distinguishing between the third *right* above (to be considered as equally potentially important to society, and worthy of esteem) on the one hand, and the right to equality of influence, rewards or specific status on the other. The tendency of some educators, psychologists and psychiatrists, e.g., Dreikurs, to forbid the adult to act in an authoritarian way to the child confuses these; since the greater experience and intelligence of the adult compared to, say, a four year old, requires the former to exercise authority, and no myth about the requirement of "reasoning" the child into doing this or that is either intellectually honest or safe in a busy world. One is compelled to conclude that many of the philosophies based on ignoring differences due to age, sex, etc., are part of the wishful, autistic thinking that perennially finds its outlet in doctrinaire fads and systems immune to reality-testing.

In the more extravert realm of social and political action, the narcistic demand that the self be above moral "restriction" expresses itself, as clinical psychologists have seen in countless case histories, in an apparent concern with "social justice" which is actually more a preoccupation with personal "rights." A more disinterested and fundamental philosopher in this area, John Locke, stated three primary rights of man: to life, to liberty, and to the possession of property. Elaborations beyond these have emerged from politically oriented individuals like Tom Paine, ranging from the ideas of the Encyclopedists of the French Revolution, through those of the German liberals of 1848 and so on. To define individual rights in precise and legalistic form the government of the United States put forward in 1789 its Bill of Rights (ratified in 1791), while the United Nations in 1948 propounded a Universal Declaration of Human Rights.

These are important social documents, excellent in intention, and yet one suspects that to a behavioral scientist attempting to put them on a firm basis many flimsy and inconsistent concepts, too ill-defined to permit measurement, would be found. Even to the more traditional and culturally-local thinking of lawyers one suspects, as argued below that the contractual "rights" would be found to have the disconcerting uselessness of floating in a social vacuum. The basic meaning of the word right is such that it implies a second person who, by contract or otherwise, has the obligation to give the first individual in question that right. A developmental psychologist may divine that this unstated fellow contractor in much of the early clamor was first the father, then God, and then the government. The early Christian Church, and its Judaic ancestor religion, have made much of these rights, in relation to a fatherly (but not necessarily indulgent) God, who fits the Freudian emotional niche of the father Totem. However, whereas no historically known early Christian seems to have dreamt of enforcing his rights by bringing God into court, feelings grew different as the status position was inherited by a king, and spread among an often intolerable aristocracy. Then "rights" usually began to mean something that the king has failed to grant, as Charles I discovered too late, and the English, American and French revolutions passed in succession through a development which asked finally whether the king had those rights in the first place. (King James' *Divine Right of Kings* seemed merely to state the error more clearly.)

Now the position of an evolutionary morality is that definable rights can be set up by contract between the individual and his government; but *there are absolutely no such things as "man's rights" in the abstract.* There has been a tendency for the original father relationship—which expressed an infantile, emotional concept of rights—to split into relation to a government and relation to God, as seen above. But as an anthropomorphic god recedes into an Immanent Cause, and finally into the Cosmos as we now see it, the infantile thinking in the original "rights from father" fades like a pleasant but trivial dream in any man maturely awakening to a day of duties.

A claim—by either individual man or the group—for "rights" against the earthquake, the volcano or the hurricane, is so absurd that we do not contemplate it. Society can actually get only such privileges as it can wrest from nature, and these can then be shared by its members. But it can be questioned whether society can then enter into a contract with its members to provide even such basic services as would guarantee Locke's "life, liberty and property," unless the small print gives it a loophole for its honor in the event that nature beats it. Like the small society of a sinking ship society will find occasions when it cannot honor "rights" to members and the rights have therefore had no immutable meaning from the beginning. (Of what use is it to a first class passenger in the saloon of a sinking liner to threaten to sue the company for allowing salt water in his scotch and soda?)

Yet, the thinking of modern man in these matters is little better than that of his medieval counterpart. It is true that he now sees his rights as a contract with his government rather than as a supernatural gift; but his narcism impels him still to consider these rights as less conditional than we here see that they really are, and to magnify his expectations unrealistically. For example, the standard of living to which a non-working person feels he has a right has risen faster than the community average, for no discernible reason[9]. Roosevelt, as a successful politician (if a poor or insincere philosopher) declared, in addition to acceptable rights to free speech and form of religion, the rights to freedom from want and freedom from fear, and this so-called Second Bill of Rights has been incorporated with additions, in a document "The U.N. Declaration of Rights." Freedom from want is in any case undefinable and one wonders what society's obligations would be toward a hungry man who says he wants a certain film actress more than food. And freedom from fear would be positively dangerous. Most recent research on anxiety and fear (Spielberger, 1966; Sarason *et al.*, 1952; Cattell and Scheier, 1961) shows apprehension and foresight to be inextricably intertwined.

The nearest practical approach to assuring "rights" of this kind would be to return man to the womb. The real model giving a basis for rights to be assigned between an individual and a group is surely that of a simple partnership, in which the guarantees are only as good as the people who offer them. Some interesting recent thinking on rights and duties is that of Johnson, Dokecki and Mowrer (1972) who seek to put the demands of conscience on the footing of a rational contract between the individual and society, regarding agreement to conform in various ways. This removal of any "authority" from the relationship, and the shift to rational agreement, fits the spirit of the times. Nevertheless it is open to the objections already raised: (1) that no social psychologist is yet able to give rational grounds for some of the values which the trial and error experience of society now causes it to demand; (2) that the individual has to conform before his mental age is such as to permit a contract; (3) that his being born into an ongoing society gives him no choice but to agree or get out, and (4) that the forces of temptation are such that a purely rational agreement to resist them is insufficiently potent compared to the guilt response to a categorically, authoritatively imposed value system in a conscience established in early childhood. The Beyondist position is that values should be scientifically rather than merely rationally, established and understood by the adult, intelligent and mentally trained; but that like vaccination, the value system for practical purposes has to be given by authority to the child or those incapable of advanced reasoning.

Society does not know what kind of citizen it is getting, and the unborn does not know what kind of society he is entering. The control of society regarding those to be born into it is at present so weak that no responsible party representing society could enter a contract. Certainly, society cannot now be said to have chosen with whom it will enter a contract. Conversely the foetus has no opportunity to choose the type of society with which it would like to enter into a citizenship contract. But even if this situation were remediable, as by a free migration of citizens accepting and being acceptable to various societies, the fact would remain that no society can honestly, unconditionally enter a contract to provide a certain standard of living, of education, of property, etc. The basic reasons lie not only in the fact that no society can guarantee its own environment, but also in the fact that society is dependent on the quality and behavior of the people who constitute it. Only if the government of a society were given absolute power over its members so that, for example, it could permit only the birth of citizens of good heredity, or forbid its workers to be lazy, or execute everyone who attempted to destroy it, could it venture to enter on a guarantee to its members—even then limited by its own unforeseeable outer environment.

The reality in the rights of the adult individual is that he should be able to make a contract with his society for such privileges as a society of men of his type can hope to provide[10]. Such rights will obviously not stand at some single world wide standard but will differ from society to society, and it is questionable (but *see* Chapter 9) whether a body like the United Nations can logically and morally interfere within truly viable groups in the arrangements between any government, thereof, and its people (assuming dissidents free to migrate). By contrast with the position here reached by a naturalistic examination of the realities of group life it is quite obvious that the pleasure principle has led the majority of citizens, and the more journalistic writers who cater to it, to a view that human beings somehow have "transcendental" rights to this and that from their groups; that they consider these "rights" as justifying one country interfering with another that does not provide them; and that they have what can only be described as a supernatural conception of "the government" as owing them support beyond what the laws of economics (the size of the neighbor's pocket book, in fact) makes in any way feasible.

The belief in absolute and transcendental rights comes to us via revealed religions from the depths of history and pre-history. Had we space it would be interesting to trace the various historical attempts to define "transcendental" or "moral" rights over and above contractual rights. Niebuhr has interesting comments on this (Cranston, 1963) saying that the Stoics (and, earlier, Sophocles in his Antigone), attempted to make this distinction. Anyone who is impressed by the sheer number of those who have taken transcendental rights seriously should note that so wise and progressive a thinker as Bentham wrote "Natural rights is simple nonsense; natural and imprescriptible rights ... nonsense upon stilts." Doubtless men brought up in legalistic atmospheres will put great stock by such phrases as "Whereas recognition of the inherent dignity and of the equal and inalienable rights of all members of the human family is the foundation of freedom, justice and peace in the world," but the social scientist who is a Beyondist will spurn such a farrago of undefined assumptions from the beginning. He will want to calculate what societies with various commitments from their citizens can in turn offer to those citizens, what the probabilities of certain unavoidable social breakdowns of contract are, and what is the least common denominator of rights and duties in a world society. Such calculations need the framework of Chapter 9, Section 9.1.

All moralities consist both of positive injunctions to ideal actions, and imposed restraints against wrong action, and in both senses, but especially the latter, can be perceived by the common man as infringements of his "inherent and inalienable" rights – for which reason religious missionaries are often massacred. Compared to the revealed religions, Beyondism gives in one area and takes away in another. It gives, as civilization gives, real gains in living and in knowledge. It takes away by demanding, again as civilization in general has done, freedoms which the savage possesses, especially, in trimming the cluttered ships of state to make directed, collective voyages of adventure. For example, it requires men to cooperate in giving all information about themselves necessary for research on the social organism (there can be no balking at the census), and in as much as it asks the state to recognize the rights of the unborn and extends morality to genetics, it restricts the rights of citizens to be parents without good cause. The strong narcist, who is often childless, will be the first to protest against taxation and concern for the future generations.

But in deeper matters, Beyondism is a sterner code of belief than the revealed religions have dared to try to maintain. It looks into the panorama of the starry sky and sees, as understanding increases, no increase in probability of finding an all-loving anthropomorphic god. Man has to succor himself. And it sees no individual, personal, immortality with harps in heaven, but only the real immortality of genes, and of cultural actions and contributions that go echoing on for what we can only think of as all eternity. Indeed, it even sees individual death – that ultimate frustration of the narcist but not of the Beyondist – as a desirable necessity. We have to be superseded. As the scientist Pickering reminds us, "In so far as man is an improvement on a monkey, this is due to death." This difference of value may even become a practical issue if medical science succeeds in greatly prolonging individual life. For it may well turn out that there is an optimum length – perhaps a hundred years – beyond which the evolution of societies begins to suffer. Euthanasia may come for both individuals and races, for they need to fulfill their purpose and pass on [11]. Thus there are areas of value in which the clash between a moral system and the narcistic and pleasure principle aspects of human nature is going to be far more acute for Beyondism than for most compromising and palliative moralities of the past.

6.6   THE WELL SPRINGS OF RELIGIOUS DEVOTION IN THE PAST AND IN THE FUTURE

Since the development of the superego is not advanced enough in most men to provide the motive force for the whole moral system, religions of the past have made what we may call "emotional deals" or exchanges. These inventions in human dynamics are like those by which engineers make a stream pump water to a higher level than the stream itself. The neighbor's eye and the policeman's baton are additional aids, but they operate only at lower levels and we may take their action as understood.

A more advanced moral system is, of course, more rewarding *to the group as a whole* — that is how it can be recognized as more advanced — but the gain for each individual in all circumstances is not always equal to this positive average gain and the problem is that of (a) getting him to achieve such identification with the group that he will accept, at times, personal sacrifices, and (b) developing such foresight and capacity to stand deflection strain as will permit the individual to deny himself for the sake of his *own* total future gain. For, as discussed under cultural pressure (page 137) a more rewarding *adaptation* (gain in life success) is often accompanied by a more punishing *adjustment* (emotional demand for control and long-circuiting).

That the successful religions, from the banks of the Ganges to West- minister Abbey, have paid men to adopt a more ethical emotional adjust- ment by imaginary coin is widely recognized. It has been held against them, as counterfeiters, by a host of critics — of whom Voltaire (1946), Marx (1890), Clemenceau (1929), and Russell (1955) perhaps span the range. Heaven, hell, the all-seeing eye of God, the compensated after- life, the bestowal of dignity and status in a church congregation (such as the individual scarcely possesses in a wider society): all these are more positively effective and less expensive than a police force. Well-organized totalitarian Communist and Fascist societies can approach that efficiency, with complete control of education and propaganda, but horde structure, as we have seen, still lacks the capacity to generate supports of con- science for reforms transcending the given society. It is hide-bound in its immediate existing culture.

At the present moment it unfortunately remains true that no social psychologist has systematically and quantitatively investigated the emotional, dynamic exchanges accompanying the successful growth of religio-moral systems; so no brief summary can here be given. One obvious important mechanism is what we shall call *capitalizing by trans- formation under frustration.* This is exemplified by religious develop- ment waiting upon periods of catastrophe or upon society entering culs- de-sac in social evolution which have become rigidly unrewarding. Catastrophes tend to breed depression and guilt, in which new moral restrictions will be accepted, since they are not worse than the current prospects. This the Old Testament prophets knew. Primitives simply placate the gods with larger sacrifices when the crops fail, but alert moral leadership may actually introduce some functional restriction, e.g., an end to homosexual license, at such periods and establish a new dynamics.

And as to social structures that have become unrewarding, numerous writers have pointed out that Christianity grew in Rome partly on the dynamics of slavery. The large population of slaves in the Mediterranean world, hounded by Nero and Decius, crucified by the thousands when they ventured to revolt under Spartacus, murdered in masses by Valerian when their loyalty failed, had little in the way of direct and personal ergic satisfaction to lose even under the most exacting religion. Despairing of any future in this world, they were happy to accept the heavenly millennium of the Church, and in doing so, they accepted as the virtues demanded by a religious morality, the inhibitions on sex and aggression at which the free Roman scoffed. But, since their deprivations had, in any case, accustomed them to what Christ and the saints, in their inspiration, demanded, the new religion was a sheer psychological gift — the gift of making a glorious virtue out of a debasing necessity.

It seems probable, as suggested above, that if historical data on manslaughter and sadistic behavior could be gathered on a larger scale along the lines suggested by the physicist Richardson (1960), we should find that the five hundred years before Christ constituted, for various reasons, an historical acceleration in the trend of aggression. The Peloponnesian War, and the clash of Rome with all surrounding countries, including Israel, illustrate the data. The spiral of frustration-pugnacity-increased frustration until hostility and cruelty reached intolerable levels from which the *volte face* of Christianity offered, as suggested earlier, a compelling chance to escape. Recognition of the inherent viciousness of the spiral of frustration, revenge, war breeding war, is incidentally, older than Christianity, for Aeschylus, in the fifth century B.C. preaches in *Onesteia*, *Agamemnon* and the *Furies* a clear dramatic sermon on the danger.

Whether a fine exegesis would conclude that Christianity in preaching submission, is necessarily opposed to competition, and even to war under certain conditions, is, as we have seen, very doubtful. But there certainly are serious students ready to argue that Christianity aligns with the eight-fold way of Buddhism, and with Gandhiism, in primarily offering an adjustment which delivers one from all strife. At least, many religions have been so interpreted, and the impracticality of turning the other cheek in submission to power has not prevented their being believed. Their historical effect in shutting off the draft from the furnace of competition at that time may well have been good, even though it would be a catastrophe to follow through to the doctrinaire conclusion that the furnace itself must be extinguished. In any case "peace" was a psychological reward for accepting many inhibitions.

The derivation of morality from evolutionary principles cannot offer this escape from competition, or the evasion, in the notion of a loving god, of all real fear, stress or concern about the future of the human race. In fact, although it certainly offers emotional compensations, and a tremendous intellectual satisfaction (as discussed in Chapter 9), yet the spirit and the era to which it belongs deny it any thought of "emotional deals" in the sense of illusions from intellectually inacceptable inventions such as have enabled religions to "get by" historically. Nor can Beyondism offer an adaptation to the racial temperament of each of the world's peoples such as we have argued that all religions do [12]. (It does, however, permit the within-group morality to take its particular adaptive form.)

Perhaps this criticism that proselytizing religions have bent truth to the circumstances and emotional needs of the time (which implies also the use of psychological warfare) may seem uncharitably to overlook the fact that religious congregations are themselves living groups which have to meet the ordinary needs of a group for sufficient human interest to compete for survival. And this fight for survival is such that the folklore recognition that "All is fair in love and war" might well have added to it "and in religion" — especially in primitive times. But it is necessary for a social psychologist frankly to recognize that Beyondism by its nature cannot compete in these terms (maybe so much the worse for its hope of popularity). It necessarily presents an emotionally austere offering, and by its origin in an attempt as honest scientific understanding cannot stoop to emotional deals based on illusion to compensate for this austerity.

Consequently, one can predict no triumphal tide of Beyondist sentiment, but only a slow integral growth through the reasoning of scientific men, and the respect of independent thinkers. Its satisfaction, apart from those which come to the superego itself, are mainly aesthetic, in participating in the magnificence of our unfolding view of the universe. Here it joins with and needs the aesthetic experiences of music and art, as older religions have done in the organ music and the architectural grace of a great cathedral.

6.7   THE OSCILLATIONS OF ENVIRONMENTAL AND CULTURAL PRESSURE AND THE ASSESSMENT OF URGENCY

The pressures—force per unit area—developed on the surface of an inert mass when another body impels it into accelerated motion will depend on the pattern of contact and on the rate of acceleration. Although an organic system has additional features, this gives a simple model of the essentials of what happens when a moral system is applied to a culturo-genetic group previously unaffected by it. The pattern of contact—which is equivalent to the particular points of pressure of the moral system, and which we have so far been discussing—is only half the story of the impact. The reaction—the intensity of pressure and counter pressure—will depend also on the rate of acceleration planned.

The model does not, perhaps, fit moral systems in revealed religions as well as it does that of Beyondism. For mostly they view movement to a static Utopia—the kingdom of god as a community or as a state of mind in individuals. But in Beyondism, once it is granted that evolutionary *movement* is the goal, then an independent decision has to be made as to how rapidly movement should be undertaken. Perhaps there is an optimum rate in regard to human comfort, or perhaps the formula is simply the fastest rate possible?

Since it turns out that the nature of numerous practical moral injunctions is decided by this decision on pace, and since the main thrust of narcism through the social institutions it utilizes, is to *reduce* movement to a comfortable snail's pace, some objective basis for stipulating the correct pace of progress is needed. To ask if such a basis exists in the nature of living matter itself, suffices to bring out more clearly what was relatively implicit in the basic proposition of Chapter 3: that living matter itself cannot be said to have any motivation for progress *per se* at all. Living matter wants only to live and reproduce itself. The chemistry of living matter is a chemistry of persistence in its own form. Only through the interaction with the rest of the universe—with what we have called the labyrinth—does the desire of living matter simply to persist lead it to acquire new forms and to move to greater reality contact, over a wider environment. In that sense, if one wishes to be philosophical, living matter contains the *potential* of wanting to evolve, but only as a bullet contains the energy which permits it to ricochet from the rock in an entirely new direction.

Actually, to the animal and to man untutored to a Beyondist view, the demands made upon him for change appears merely as frustration, and, as we have seen up to this point, the individual's main reaction to his cultural-morality (which, we have seen, learns the lesson of the future sooner than does the genetic nature of the species) is to resist it. Some of the expressions of this resistance have been discovered by our comparisons of countries high and low in the cultural pressure factor (page 136, Table 4.3). Although cultural pressure is not itself an index of progress — indeed to devise a scale of progress applicable equally to measuring progress in all countries would need much debate — it *is* a measure of movement. Consequently, our discussion of movement would be enlightened if we had research data of a longitudinal kind — to supplement the static, cross-sectional data in Table 4.3. An apology is offered for having to lean at present on the skimpy and insufficiently cross-checked longitudinal factor analytic evidence ("*P*-technique") in Fig. 6.2. A plot of the cultural pressures indices (illustrated in Fig. 6.2 for America and Britain) shows that these nations in the Western culture pattern (including Australia) climbed steeply in the Victorian period and continued climbing well into the twentieth century (Adelson, 1950; Cattell, 1966; Gibb, 1956). It is possible that the acceleration is recently in a phase of decline, though we are more likely experiencing a social split in which one section of the population (that alert to Sputnik) is continuing to respond to the educational challenge and another subculture (that symbolized by the guitar and LSD) is trying hard to turn its back on the challenge.

A less precise and quantitative view of cultural change but a comprehensive one, as in Toynbee's definition of Yin and Yang, agrees that there can be substantial historical fluctuations in cultural pressure, largely in response to external challenge. Cultural pressure, though experienced as a stress in demands for psychological adjustment typically accompanies an increase in general community capacity to adapt and survive. Thus the rise in cultural pressure in Fig. 6.2, with its complication and conflict has nevertheless corresponded with a gain in the general population of increased leisure and longevity. However, as Sir Charles Darwin in his farewell volume *(The Next Million Years)* pointed out, scientific advance and certain accidents of history have combined to give humanity in these current few centuries an unprecedented alleviation of and escape from selection for adaptation *as individuals*, which is not likely to continue. Indeed, we may here encounter a special feedback mechanism which could be responsible for the observed historical oscillations. The labors of hard thinking men of science, resting in turn on an orderly society indebted to religious values, have granted the average man surcease from

In America from 1845 to 1935   In Britain from 1837 to 1937

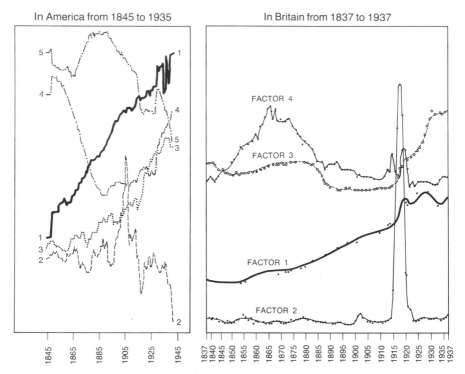

These are the results of *P*-technique (longitudinal factor analysis) on some forty to eighty variables recorded annually. The analysis shows that some five factors are needed to explain the trend and other changes. Only Factor 1 (in the heavy continuous line), which is Cultural Pressure (*see* list of variables on Page 137) is discussed here. The others are included to show the distinctive nature of the time course for the different factors. It is noticeable that growth in this factor faltered in Britain after World War I, but shows some climb in both, though more steeply in America, from that point onward. The vertical scale is that of the standard scores of the factors over the number of years concerned, but is drawn approximately to twice the scale on the American as on the British graph.

Fig. 6.2   Course of the cultural pressure factor in America and Britain.

constant sickness, physical toil, hunger and early death. At one and the same time this has tended to undermine — in the non-scientific fraction of the population — the reality thinking morality which created it and to offer instead, self-indulgence, vagabondage, drug solace, and bizarre expression of individuality as a practicable way of life. It is, incidentally, characteristic of the halcyon periods of low economic stress that they make greater variation possible. This has a valuable function in making available a good variety of cultural and genetic mutations to be subjected to selection in the next period of exacting conditions. However, to achieve the few valuable variants it will ultimately thus profitably retain, mankind has to tolerate a thousand forms of cultural nonsense, e.g., in the various styles of "self-realization" discussed above, which are essentially parasitic upon the scientific effort. The origin of the oscillations of cultural advance which historians believe that they observe may thus lie in a "negative feedback" system whereby the rewards of one era's good moral effort can become the underminers of moral effort in the next.

However, although "Recuillez pour mieux sauter" is sometimes tactically called for, too, a true slowing of progress cannot be morally defended. (Nor perhaps can society risk oscillations which swing too far in the negative direction.) But the pleasure principle from a million protesting voices will question the contrary argument for a principle that all progress, as far as we can control it, should be at a maximum pace. Resistive social philosophies in many forms will spring up to question any attempt to maintain maximum cultural pressure, demanding "What is the authority for urgency?" Since human nature in itself does not want to progress, and is forced to do so only by the impacts of a changing environment, probably no scientist would logically expect any other reaction. The Beyondist adoption of evolutionary progress (as logically justified in Chapter 3) is, after all, an act of intellectual sophisticated enquiry, followed by a basic, superego-directed attempt to adjust to the reality of movement into the future. It goes beyond the obvious "reality" demands of the present. It is not a natural inclination, and the alienated voluptuary, the entrenched conservative, the vagabond, and the criminal present different forms of open, immoral revolt against it.

In asking for an objective evaluation of the urgency for progress in mankind in his present world we shall get into conflict not only with the pervasive passive resistance of narcism but also the rooted anthropocentrisms of Christianity and other revealed religions, which assure humanity that a fond Providence has some special regard for us all. But let us for a moment set aside these lingering illusions and, turning to the vistas of the astronomer and the biologist, frankly recognize the enormous magnitude of the adaptations to catastrophe that may at any moment be demanded collectively of mankind. How impressive and effective, really, is the recent sense of control that we enjoy relative to our environment? When all is said we cling to the surface of a space ship hurtling as yet uncontrollably through the unilluminated "back roads" of space. An encounter with even a trivial meteor of a few hundred miles diameter, falling into any ocean could drown most of the world's population; a sudden deadly mutation in a familiar bacteria could end most of what we value as civilized organization; another world war could leave radiation products that would shorten and cripple life for several generations; or the polluting products of our technology could change world vegetation and world climate to a point where survival would be possible only for a few human beings (provided they be much better equipped with knowledge and intelligence than any we now possess)[13].

The demands of such crises, in relation to our present capacities to meet them, could be so overwhelming that an objective estimate of the dangers leaves no argument whatever for any "luxury" use of "surplus" human resources. Maximally to reduce the chances of complete annihilation we have to recognize that there is no surplus of either resources or time. There is only one pace to increase basic knowledge and genetic progress — the fastest that well-organized societies can achieve. The answers to the questions "Can one argue for any special degree of urgency?" and "At what optimum level should cultural pressure be maintained, in so far as we can control it?" are definitely "Yes" and "The highest."

6.8   THE OFF-BALANCE ENVIRONMENT, THE MASOCHISTIC RESERVE, AND THE DANGER OF THE HEDONIC PACT

Our conclusions on urgency above need a strategy as well as an intention. To jam one's foot down on the accelerator pedal and hold it there is not the surest and quickest way to get through the town. What is the best strategy and tactic for regulating internal cultural pressure and progress in order to be able to adjust to sporadic external demands of unknown magnitude? The answer is partly hidden in social and biological research yet to be done.

First, however, there is the straightforward answer that the standard research designs of social, biological and physical science can be pursued and elaborated. This means, primarily, more efficient pursuit of the "grand experiment" (page 91) with wider genetic and cultural diversifications than we now possess, and more exact recording and analysis. And *within* each group, too, one may suppose that sub-experiments will give more certainty of perspective to the particular group experiment. (Some sub-experiments may be rather far-out "lifeboat" rehearsals for emergency adaptation.)

The idea also deserves enquiry — though it is a speculative one — that societies should seek to maintain what for lack of more suitable existing terms may be called the "masochistic reserve" and the "off-balance environment." By the masochistic reserve is meant a reserve of moral preparedness maintained in periods of success and ease by some kind of self-denying ordinance. At its simplest this may be conceived as the hard training that a team does between games or the dangerous mountain climbing and other exacting sports that seem to flourish between wars. In more subtle ways this perhaps occurs in the Englishman "taking his pleasures sadly!", and in systematic ways in the asceticisms of all great religions.

Such restraints or deferments of ease can often be seemingly "explained" simply as training or preparation, but one finds ancestries which suggest that an intuition of the value of instinctual denial for its own sake is involved. (Hence the use here of masochistic, though this also suggests a satisfaction.) In the flagellation rites of early Christians and others; in circumcision and sub-incision (which masquerade questionably as hygiene but are trauma reducing sexual expression); and in the horrifying sacrificial rites of the Aztecs and Incas — we see an implication that some self-imposed suffering or deprivation is socially preferable to full instinctual expression. In a more subtle and advanced form it exists in the Puritan ethic; in Judaism, as stated independently in say, Kipling's Recessional ("Lest we forget"), and in that brooding quality of Greek tragedy which distinguishes it from the "permissive" or even indulgent blissfulness of Christianity where heaven is admittedly an untrammelled reward once the main renunciations are made.

The question of what this ultimately means in individual dynamics and practical social adjustments could be enlarged upon if space permitted; but let us accept the approximation that it means maintaining *some degree of ergic denial and reserve of sublimated energy* by what can only be described psychologically as a masochistic act, in periods when the cosmic environment alone would not demand such severe pressures. At its simplest it is a rational self-denial of present possibilities of luxury or ease, for future security. Such may be seen, for example, in Russia's denying its population luxury expenditures anywhere near comparable to those in America, while spending as much as America on science and space research. But in its emotional form it goes beyond any ordinary far-sighted self-interest for tangible future rewards. It is a bracing of the spirit, and a vigilance against indulgence, that is carried out essentially for preserving the razor's edge of the mind and the maximization of understanding, regardless of any perceptible task or danger[14]. There is a possibility that this works against the efficiency of some immediate adaptations, as a gyroscope prevents a ship from responding to immediate deflecting forces, but many such adaptations to deflecting forces would be merely to transient fashions. On the debit side there is also the possibility that if maintained too rigidly it will destroy that proliferation of playful new cultural mutations which normally occurs in the relaxation phases and have later use; but properly conceived the masochistic reserve should not do this.

Allied to the masochistic reserve, in the intention of maintaining the momentum of evolutionary movement when actual challenges are not impinging, is the concept of maintaining the "off-balance" environment or of avoiding perfect equilibrium. A society, or a species, is kept in evolutionary movement partly by encountering an environment *which continues to present challenges. The movement also requires that they will be of the right degree of "educational" difficulty.* It has been suggested that one of war's useful functions is in maintaining, while nature's challenges are "off duty," challenges of a relative uniform (human) order of magnitude, unlike the overwhelming magnitude that natural catastrophies, with no measure of man, may intermittently reach[15]. However, there is a sense in which a society can itself maintain such challenges by what we may call an "off-balance" environment, largely by producing changes within itself.

If this is done, it will not be done without danger too. In as much as one readjustment demands another there is even, theoretically, the possibility of a chain reaction—a crescendo of change too fast for any healthy adaptation. (Perhaps France came near to this between 1790 and 1830.) We are already accustomed to the idea that the consequences of the scientific, industrial and cybernetic revolutions—are both "good"—such as more food and longer life—and "bad"—such as excessive population expansion and pollution. Despite the theoretical danger of a complete loss of equilibrium, the possibility has to be considered that a minimum "off-balance" condition from perfect equilibrium should be deliberately maintained, as a man deliberately loses his balance in the act of running.

In this chapter we have been concerned with problems of the interaction of human nature with the demands of moralities, and specifically Beyondist morality, and have considered both the perversities of defenses and the intelligent dynamic and dialectic ways in which human nature may be better joined to a progressive morality. It remains finally to consider one rather far-out possibility, which we shall call the *hedonic social pact.* At the highest level of moral sensitivity, we have considered a society that would, in creating the masochistic reserve, deliberately increase the cultural pressure upon itself. At the other is the possibility of a society equally sophisticated in the perspectives of evolution, but declining evolution as a goal and setting out deliberately to minimize cultural, evolutionary pressures.

Even such a society would find it impossible in the end to avoid the demands of the physical and biological challenges of the universe. Yet if for substantial periods the main pressures were only *those of other groups*, an international "social deal" might long postpone the need to live with evolutionary pressures. These would be welcomed even though the probability would then arise that at the next natural challenge the unprepared society would fail and be eliminated. But such a common agreement among nations to avoid mutual competitive pressures, accepted universally, we will call an *hedonic pact* for its goal is based purely on the pleasure principle (though the *mechanisms* would have to be realistic).

On the basis of history and group dynamics principles it may be questioned whether a culturally and politically homogeneous "one world," whether established by conquest or by the presently discussed mutual agreement to liquidate the uniqueness and competitiveness of racio-cultural experiments, could achieve such stability as to last indefinitely. The failure of the reality principle in the original intention might show up also in an absence of realism in the social alliance mechanisms designed to produce stasis. However, we cannot be sure of this and it is at least possible that a partial or complete hedonic pact could be achieved that would either greatly retard or completely paralyze human evolution for an indefinite period. Indeed, schemes not unlike this are fashionable in those circles of the intelligentsia in which independent thinking is rare and emotional fashions are "culture," and in which "peace" is now the only goal.

What we are now contemplating takes on some of the extravagance of science fiction, and that recent literary extravagance of phantasy has desensitized our capacity to think at once realistically and imaginatively about immediate real problems. The absence of any, or any clearly significant signals from outer space, since listening began fifteen years ago, is, however, a silence that should startle us. The OZMA experiments begun at Greenbank, West Virginia in 1960 on sending and receiving messages from space, as well as other instrumentations, have so far drawn a complete blank. Since there is no evident technical physical reason why the 500,000 planets[16] on which intelligent life is deemed possible should be unable to communicate with us, two possible social reasons, of perhaps equal cogency, should be considered. Both assume a parallelism of development in the physical and biological sciences in any population of intelligent beings, such that when the physical sciences reach a sophistication at which signals could be sent the social sciences would roughly reach a stage at which the engineering problems of a hedonic pact are understood. That same stage would also bring an interest and respect for the significance of social experiments on other planets.

As far as the last is concerned, let us consider the possibility that as living beings increase in intelligence and knowledge they perceive — as is not at all an unreasonable expectation — that even the smallest communication between one world experiment and another could contaminate the experiment and destroy the possibility of truly independent and unique evolution. (We have seen, pages 154, 215, that there is real danger in excessive communication, imitation and dilution of cultural and genetic divergence among racio-cultural national experiments even on the same planet. The need for non-contamination may for some reason be even greater at the next level, and the "cordon sanitaire" of immense astronomical distance may have a real function.) Or, again, there may be some sense in which we are being scrutinized to see if we can yet pass our examinations for entry to an interplanetary society. Thus, apart from the first innocent calls that are sent into space as world cultures enter adolescent stages — where they have as yet incomplete understanding of the social experimental principles requiring isolation — we should expect no messages. (This assumes moral dedication in all evolving groups to a principle close to Beyondism.)

The second possibility is that the particular point of social and technical development we are now discussing is an extremely critical one in another way; namely, in that competing culturo-genetic groups perhaps quite normally tend to seek compromise on competition and slide finally into a mutual hedonic pact in which all will toward differentiation is lost. Other planetary systems may have very different forms of intelligent life from ours; but the probability of a growth in the form of structured societies surely has almost the necessity of mathematics, and it may also be that the probability of ending in the hedonic pact has a far greater probability of occurring than we yet realize. Thus, typically, the technical level for effective interplanetary communication would barely be reached before retrogression would supervene from arrest of natural selection through removal of cultural pressure by an hedonic pact. With so little discussion and research as has yet taken place in a Beyondist framework, we are at the mercy of rather widely indeterminate ranges of probability; but, at least we have been presented with a strange silence from outer space which has one ominous possibility: that of an inherent likelihood of an hedonic pact. The evolution of the human mind, which begins in the lower animals as a servant of body chemistry, poorly developed in its freedom from pleasure principle immediacy, into a magnificent instrument for far-ranging reality thinking, appears to have a built-in risk of arrest which cannot be overlooked.

6.9 SUMMARY

(1) The gap between any existing human genetic endowment and the adaptive behavior potential required in some fairly remote future is unspecifiable, because the latter is still latent in the complexity of the cosmic environment. But, man's culture is an intermediate term, produced by reconnoitering the demands of environmental adjustment some way ahead (as a rule) of the genetic variation. The cultural anticipation of the true future is not entirely correct, and the genetic development will not therefore, entirely follow it — and should not try to do so slavishly. But the culturo-genetic adaptation gap existing between culture and "human nature" can be conceived as an expression of how far human nature itself lags behind the level of moral perception already achieved in the culture. Thus culture exerts a pressure on human nature in what is largely a morally desirable direction, and the concept of "original sin" (as the extent of the innate inadequacy to reach the given moral standards) has an operational meaning.

(2) Revolutionary forces occasioned by the conflict of citizens with their existing culture contain two very different ingredients; one caused, as just stated by (a) the inadequacy of human nature to the *existing* level of cultural demands, and another by (b) the poor structure of the culture as it becomes visible to the most genuine and intelligent reformers. Because of the temporary alliance which ensues between rebels of the first type, against cultural restraint in any form, with the true reforming minority of the second type, seeking insightfully perceived improvements, the tragic cost of violence — of war and revolution — is often substituted for intelligent evolution. Means can be devised by social science for far better separation of these incompatible forces, especially through machinery for a scientifically guided evolution.

(3) The problem of adjustment of human nature to a moral system, considered here first for religio-moral systems generally and then more

specifically for Beyondism, has two aspects. First, man may ask what his genetic make-up can do in bending down the moral system degeneratively to itself. Second, he may ask what wise psychological education and genetic selection can do to bring about the greatest movement of human nature toward meeting the demands of the ethical culture.

The greatest discrepancies, and those most prone to "bend" the moral system, are man's over-endowment in pugnacity and sex (relative to present cultural needs), his narcism, his limited capacity to endure "deflection strain" of instinctual satisfactions, the mental weakness of "autism" (notably his ready falling from the reality principle to the pleasure principle) and the still rudimentary, arrested development of the superego.

(4) Following the standard paradigm of *adjustment process analysis*, we see that the pleasure principle, and the defenses of rationalization, autism, projection, phantasy, etc., which emerge after extended conflict, are capable of powerfully warping intellectual understanding. Indeed, they are likely to develop, (in the rationalizations of social interaction of individuals attempting to avoid frustrating moral demands), numerous philosophical and institutional defenses too elaborate and tangled for logic alone to be able to compel recognition of their fallacies. The possibility is raised of psychologists developing a *dynamic calculus of cognitive warping*, which could be combined with logic machines to evaluate the extent of evasion of the reality principle in social theories. Historically, this becomes a new development of dialectics. It is most needed for theories on paper, for when they are actually worked out in social life, experimentally, unpleasant natural and social consequences, inevitably, correct errors. The "bill" for behavioral maladaptations is exact and inexorable. But in literature, philosophy and political theory — as distinct from the empirical and disciplined area of science — both traditional rationalizations and new, group-sanctioned phantasies, can easily survive.

(5) Specific areas in which these defenses of innate and other maladaptations to ethical culture now operate, e.g., "rationalism" on sexual license; permissive attitudes on crime and against punishment, are studied in the next chapter. But among some broad and basic expressions of such defenses examined here are the non-acceptance of genetic and other individual and group differences, and the narcistic demand for rights without duties. As to the former, differences are not superiorities, except in regard to a particular community's specification of what it most needs. As to the latter, the Beyondist position is that man as an individual in the universe has *no* rights, any claim to such being a delusion from the anthro-

pocentrism of most traditional revealed religions. As a member of a community, man has such *relative* rights to benefits, in a contract with the rest of the group, as a group of that type of person can wrest from the environment. These rights, e.g., to a given standard of living when not working, will therefore, vary from group to group, and cannot be abstractly specified by statements of "inalienable or absolute rights" or narcistic, subjective concepts of suitable levels of "human dignity."

(6) The central well spring of human motivation that may aid moral elevation in society is the superego, which can be cultivated by affectionate and morally disciplined parents, but the growth of which also hinges on genetic contributions. As with other traits with appreciable genetic components and under active natural selection such as intelligence, superego endowment shows a wide scatter in the present population. Moreover, because of what we have called the *counteraction principle*, whereby *within-group* selection for superego endowment may actually be negative, and counteracted only by periodic elevations from *between-group* selection, progress in morality from this basic source is slow.

(7) Because of this immature development of the superego the elevation of moral levels by religion throughout history has been achieved partly by "emotional deals" in which the necessary dynamic forces for introducing inhibitions or higher aspirations are obtained by ergic exchanges, or by the leverage of natural catastrophes, or the use of illusions, requiring belief in the supernatural, such as a compensated heaven.

Beyondism has no such deals to make, partly because it requires the austerity of giving up these same illusions, and partly because its very nature forbids anything but the scientist's uncompromising candor. Thus it offers neither a heaven, nor a Christian escape from competition. What it does offer is the aesthetic-intellectual panorama of science, and the living of a collective adventure less blind and less frustrated by popular incomprehension and uncooperativeness than in humdrum community life in the past.

(8) The narcistic and other forces of defense which seek to evade the challenges of evolution may stop short of disputing the Beyondist goal, and yet deny any urgency in attending to it. Our environment is not inherently designed to present its challenges to human survival in a manner carefully graded didactically to human learning capacity. Thus it may at any time present possibly overwhelming confrontations (as it did with the Ice Ages). Consequently, the question "What level of urgency is appropriate in our plan to advance human evolution?" can only be answered by "The maximum!"

To appreciate the origin and potency of narcistic motive one must remember that living matter and human nature as part of it have no inherent desire to progress but only to continue, the progress being a secondary result appearing through life being forced into the environmental "labyrinth." Two goals which a Beyondist understanding of the situation suggests are (1) the maintenance of a "masochistic reserve" — a reserve of spiritual readiness through self-maintained pressures and restriction on direct instinctual satisfactions maintained through the lulls of environmental pressure; and (2) the engineering of an "off-balance environment," to present a continual training demand. These also perhaps have a danger—that the rate of change may show the character of positive feedback and lead to a complete loss of stability.

(9) When all is said, the greatest danger of distortion of the moral system from inherent action of the pleasure principle, is the possibility of bringing competition itself to a halt, by the "hedonic pact" among all groups.

Although the negative results of attempting to communicate with other intelligent beings in our galaxy are too recent to lead to firm conclusions, they call attention to the possibilities: (a) that the degree of independence required in socio-genetic experiments in different solar systems may be of so high an order as even to demand that there be absolutely no communication among them; and (b) that there is characteristically a high probability of the above hedonic pact occurring at a given level of self-consciousness and technical skill in the evolution of intelligent societies, so that communicating capacity is never reached. This supposes that a well-engineered social pact could bring an arrest of moral and general evolution which only a major (and perhaps, by then, overwhelming) cosmic challenge, could end.

6.10   NOTES FOR CHAPTER 6

[1] Enormous ingenuity has been lavished by good educators on making cultural acquisitions interesting and rewarding. (It is now being continued into creating the B.A. "proletariat of learning" from average teenagers who are not intrinsically "academic" material.) With I.Q.s of, say, 115 or more, sheer intellectual *curiosity* makes possible much learning that, in terms of the child's needs at the early age, would be inapt and irrelevant to him. (Thus here learning does not rest on initial frustration of a need.) But, below this level of intellectual interest there is classroom trouble. A pedagogic "science of classroom learning" makes the acquisition of difficult material possible, but the ultimate principle remains that culture involves a deflection from more innately preferred modes of behavior, and every lapse of a classroom into disorder shows it.

It has been a favorite ploy in the arguments of such "progressive" educators as Neil, Curry, Russell, and most modern writers on education to contrast their virtue in "reasoning" with children with the detentions and canings by which old-fashioned educators (described gratuitiously as "sadistic") used to bring students to acquire rudiments of manners and scholarship. Yet, as one time psychologist to one of the most famous pioneer progressive schools (Dartington Hall), I have heard sensitive young adolescents assert that they would rather be flogged than face again a two-hour conference with the headmaster in which he dissected their erroneous views and behavior in the glaring light of pure rationalism. In short, the idea that reasoning removes coercion — painful coercion — is nonsense. There are varieties of educational suffering.

In this connection the marked drift to anti-intellectualism which several leading educators have recently asserted to be occurring in the young, could well be explained as due to reasoning and intellectual analysis being used as a form of punishment. Like most hypocrisy, cant and "tact" the refusal to recognize that a fundamental problem arises from the existence of a real need for "deflection" in adapting the genetic to the cultural pattern is often the cause of more severe punishment being needed later. Thus, for a missed spanking at six, is substituted a police truncheon at eighteen, and possibly, ultimately, such "education in democracy and realism" as the Germans suffered at Dresden, Hamburg, and Berlin in World War II, the Japanese at Hiroshima, or the British at New Orleans. Parenthetically, let no one suppose that anti-intellectualism is being traced *wholly* to this lie in progressive education. For an equally rooted anti-intellectualism has always existed among those at the *status-quo* maintaining end of the political spectrum, with its massive passive resistance to analysis, logic, or scientific investigation regarding human affairs.

[2] Until the beginning of the welfare-state, the maintenance of a eugenic trend in society was less dependent on a differential birth rate between the competent and the less competent than upon persistence of a differential death rate and final survival rate. Survival of children depended on family provision and care, with differentials in frequency of sexual intercourse having little to do with the final survival rate. When the state takes care of all children, with medical services, the future begins to belong to those with the highest sexual activity and sheer birth rate. Fortunately, even in the welfare state, sexual activity alone may now again cease to be positively relevant, by reason of birth control being simplified enough for it to be provided for all. Sexual activity is, in short, nowadays, both (a) irrelevant to the differential survival rates, and (b) as far as the average is concerned, far in excess of the level needed for the modern two or three child family. The argument becomes, therefore, that "highly sexed" individuals, by natural endowment, will draw no survival advantage from it, and will, indeed, suffer a disadvantage, through their natural interests being deflected from activities of real survival value, e.g., those of the parental-protective drive. Thus, there will be a slow evolution toward types in whom sexual ergic endowment is reduced to a level not so greatly in excess of real reproductive needs.

Some further aspects of the conflict of culturo-moral trends and excessive sexual drive activity will be considered in Chapter 7, directly in relation to sexual mores. Here let us turn to the second "prisoner in the dock," pugnacity, and ask why over-endowment in this erg also conflicts so much with evolutionary morality. The positive value which higher aggressiveness may have for gains by the individual, and even the family or tribe, is soon lost in larger groups because of the greater toll which dissension and inner weakness exact. In the first place it is quite difficult to control and channel surplus purely individual aggression into "useful" socialized forms of gainful aggression between large groups. Thus, it tends to remain as anti-social behavior — murder, rape, and robbery — in the inter-individual field.

Parenthetically, some liberal intellectuals are ready to assert, on the one hand, that it is desirable to throw away the "Victorian" (and Mosaic) strictures on sexual promiscuity ("which is no one else's business"), but strongly object to a corresponding repeal of the social repression for the pugnacious, aggressive drives. By contrast, both the values of revealed religious ideals, and the direct biological and social arguments from a Beyondist ethic would evaluate sex and pugnacity as equally excessive and unadapted.

[3] There are whole sub-cultures, such as Chekov described in "The Cherry Orchard," and as are common in middle class "rentiers" where immense concern is given to minor ailments and where a high degree of "mutual narcism" masquerading as kindliness is built up with no regard to the needs of the outer world as a whole. The same source establishes in many families, (e.g., the American family where adults avoid work by driving children in cars to places to which they should walk!) a pattern of apparent altruism. This "spoiling" has been interpreted by psychiatrists as a mechanism of projection of the individual's own narcism on his children, the individual being determined to satisfy this narcism and now able to do so under the guise of parental "love." But, there are other motives, as recently noted by a teacher, "In the child centered society, parents are often desperately dependent on [the goodwill] of their 'kids,' and intimidated by their criticisms."

Meanwhile the skillful advertising profession also exploits and stimulates narcism: "Treat yourself to the best." Thus it helps to develop an "instant gratification" society. The *Tribune*, in an Editorial (June 10, 1970) on the "Woodstock Music and Art Fair," quotes a hippie participator:

> "'All my life,' said one, 'I've had just about everything I want. And I want to have everything I want for the rest of my life. And I don't want to work because I can't have everything and do everything I want if I have to stay in the same place from 9 to 5.'"

As a statement of what the psychologist is measuring in the narcistic erg, this would be hard to better.

[4] For example, the early steam engine used in pumping coal mines needed a boy to sit beside it and switch its valves on and off. There is a legend that a more lazy boy than usual tied them to the main crank system with pieces of rope, thus discovering the automatic valve opening and shutting mechanisms built into later engines. Be that legend or fact, it illustrates the psychological principle that unless trial and error variations *in the direction of saving effort* were constantly made, we should be burdened for life with more elaborate behavior patterns when simpler ones would actually work. As mentioned in the text, in crossing the North Sea, our ancestors forgot that nouns have genders and that adjectives have to decline, and the English language has benefited ever since by being swept clean of quite unnecessary elaborations which German and French still retain.

[5] Among these movements one might cite that for shorter prison terms, earlier paroles, and abolition of the death penalty. While the Beyondist position on the last is not clear until more research is done on social effects, it *is* clear that frequently, with psychopaths naturally having a high recidivism rate, the period of preventive custody and attempted treatment should be increased rather than decreased.

As guest speaker, the present writer was present at the annual meeting of a state association of practicing psychologists where it was voted, *without a single dissident*, to recommend, as professional psychologists, that the state abolish capital punishment. Such an outcome, *in the present state of knowledge*, points to the psychologists being short in scientific training and personal independence of mind. Similarly, on the parole issue, no psychiatrist or psychologist has developed methods of such reliability of diagnosis that he can guarantee – or even give meaningful odds – that the parolee will not repeat his crime. All that is statistically certain is that after psychologically advised corrective treatment, the parolee still has a decidedly *greater probability of committing the crime (again) than has any random, average member of the community.* As a possible alternate set of values on capital punishment, it might be considered that a man (as in a case this month) who has murdered three women and five innocent children, in a series of well-planned obsessional knife attacks would, if he had any conscience at all, prefer the dignity and surcease of death. Alternatively, if he has no contrition whatever, but boasts of his crimes, he has so defective a brain that he would better be drugged to death like any intolerably dangerous wild animal. What is overlooked in the "humaneness" of prolonged prison sentences for murder is that other human beings have to be brutalized by the custodial task, as wardens living in mortal danger, for a lifetime. Besides, the possibility of escape and further murders is very real. These alternative evaluations are not presented as a definitive argument for capital punishment, but simply to bring out that the alleged modern rational proposals for change are by no means at present being decided on psychological or social scientific evidence. In fact, much that goes on merely offers evidence of a current trend to greater sentimentalism and expression of the pleasure principle.

[6] It is deplorable, to research geneticists particularly, that as long as thirty years after the end of Hitler, objective and untrammelled discussion is still, in many circles, bedeviled by his evil genius. And, since all illusions are, in the end, costly, the societies which allow their thinking for the future to be affected by the trauma of the past are likely yet to have to pay dearly. In America, the second distorting influence has been the melting pot (something of which Britain now has in its own hearth) in which, for reasons of easy social and political absorption, differences of origin were played down – to zero in the official line. This was per-haps socially necessary, but it had its inevitable effect on psychological views as was strongly evident (in contrast to Britain before it acquired a race problem) in the readiness with which professional psychology in America swallowed – hook, line, and sinker – Watsonian-Pavlovian reflexology, admitting no role to heredity whatever in fashioning the individual personality. Fortunately, the growth of experimental professionally scientific psychology in America has been so vigorous and on so broad a front that this early environment has been outgrown; and some of the best behavioral genetics research in the world is now being done in American psychology laboratories.

Because of these advances, it has shocked and alarmed the scientists of free countries to discover, through the recent cases of Shockley (a Nobel Prize winner in Physics), Jensen (a distinguished contributor to research in educational psychology) and Herrnstein, Chair-man of Harvard Psychology Department, that ignoracists in America are not above employ-ing in science political and personally deflamatory tactics. The ethics of scientists is apparently still not invulnerable to the danger of fellow scientists turning on their kind in the interests of a doctrinaire political position.

Shockley's request to the National Academy of Sciences (May 17, 1967) began with the suggestion that the public turmoil over heavy educational expenditures on the backward be resolved by a substantial and impartial scientific investigation of "hereditary aspects of our national human quality problems," i.e., of hereditary and environmental deter-miners (Neary, 1970). He asked for expenditures in basic social science and behavior genetics to be brought nearer to equality with those which the Academy had sponsored in the physical sciences with such success. The request of this eminent physicist (whose work has enormously benefited all mankind) for a hearing on the national problem of genetic quality was turned down by the NAS (business) meeting, 1969, on grounds that the findings "would invite misuse" and that "it is not clear that major social decisions depend on such information." The first repeats almost exactly the words by which religio-political authority tried to stop Galileo and Copernicus. The second recalls one of the common arguments by which authority forbade Vesalius's attempt to advance anatomy by dissecting the human body. The conclusion that "major social decisions" might not depend on such knowledge is also completely remote from reality while billions of dollars are being expended on what has turned out to be the relatively useless and psychologically frustrating "headstart" programs instead of upon birth control clinics. Already, fortunately, the report of the special NSF Committee under Kingsley Davis has reversed (January, 1972) this taboo on behavior genetic investigation of racial differences and enable the blot on the scientific conscience of a great scientific society to be expunged.

It is not surprising, however, that the incident caused astonishment among leaders in other fields. In Congress a Representative from California exclaimed, "I am shocked that men who call themselves scientists are afraid to seek the truth" (C. J. Gubser, Congressional Record, Tuesday, July 15, 1969, No. 117). Indeed, the incident will not only cast doubt among scientists on the basis of election to this supposed elite of scientists, but should also promote some house-cleaning to eradicate evidently real dangers of Siberia being around the corner for dissident truth-seekers.

Incidentally, the need for private aid to science in areas of political debate becomes increasingly evident as one sees such dangers to governments aid to science. Professor Shockley protested to J. W. Gardner, the then Head of HEW, that Kennedy, Thelander, and Prior failed to get support for well-planned genetic researches, while millions were being spent on unproven (and, subsequently, proven useless) teaching innovations. Clearly there existed a danger that the findings of such research might cast doubt on the latter "action research" rashly undertaken for political reasons without basic investigation.

Incidentally, there can surely be no doubt today about the relative effectiveness of removing persistent backwardness, as a severe problem in school systems, by birth control as compared to bringing intensive educational pressure on the unfortunate imbecile or moron. The reasons for arguing the greater potency of genetic measures have been summarized in several places by acknowledged authorities in the field. Thus Reed and Reed (1965) initially calculated that of the fifty percent reduction that could be brought about in mental defect in one generation, half of it could be gained by voluntary sterilization of defectives who would otherwise become parents. In a letter to the present writer, Dr. Reed cites more recent work by Anderson and himself which indicates that social measures could produce a one generation reduction of nearer thirty-four percent, but again half of this fortunate result would result from arresting reproduction in existing diagnosed mental defectives.

[7] The contrast in "fulfillment" concepts is so great that it is obvious that this "humanistic" definition of an ethical goal is almost valueless. When Oscar Wilde, on trial, was asked by Sir Edward Carson if he really believed his statement: "Pleasure is the only thing one should live for," he replied: "I think that the realization of oneself is the prime aim of life." One may contrast Mazzini, the liberator of Italy, "There is strength in repeating to oneself that there is no happiness to be found in this world, that life is self-sacrifice for the sake of something higher" (Letter to the Carlyles). These words could be almost exactly matched in those of leaders of such diverse periods and settings as Christ, Lenin, Nelson, and Madame Curie. When Bernard Shaw said, "What I mean by a religious person is one who conceives himself or herself to be the instrument of some purpose in the universe," he

implied this same view that fulfillment is only to be measured against supra-individual values. And that fulfillment may even come with failure is well expressed by Browning:

> All I aspired to be, and was not, comforts me.
> A brute I might have been, but would not sink i'the scale.

[8] The pre-Beyondist, late Victorian liberal flavor of Huxley's position is shown further by his continuing, "A central conflict of our times is that between nationalism and internationalism, between the concept of many national sovereignties and one world sovereignty. Here the evolutionary touchstone gives an unequivocal answer." As the arguments of Chapter 5 on the subtle balance of powers indicate, the unequivocal answer for Beyondism is *against* a world monopoly of power and values; but to the so-called "liberal" who has still not opened his mind to independent biological reasoning, it is, apparently, axiomatic that we must have "one world."

[9] The rights in the U.N. Declaration now extend to life, liberty, property, justice, and the pursuit of happiness. But, more specifically, they include the right to marry regardless of the fitness of the mate (the Israelite law against marrying Gentiles would presumably be abrogated by this ruling), the right to equal pay for equal work, to equality at law (equally good lawyers?), to form trade unions (but business monopolies?), "to a standard of living adequate for the health and well being of himself and his family (apparently, regardless of its legitimacy or size)," the "right to periodic holidays with pay," and the right to "guaranteed" employment.

An instance of an important social thinker who, though championing the underdog, was very conservative about "rights" is Karl Marx. It comes out clearly that he regarded the collective social class as the important entity, in relation to which anyone asserting *individual* rights was a presumptuous and tiresome enemy of society.

[10] It would be misleading to draw central conclusions from the "hard cases which make bad law," but common psychological mechanisms, as usual, can be brought out more readily in the exaggerated form of "clinical" cases. Any mature person with a moderate experience of public office recognizes types constantly plaintive about their "rights." Characteristically, they have an instant sensitivity to issues of personal rights and an insensitivity amounting to complete callousness on issues of responsibilities. They are, for example, litigious in squeezing the last drop of compensation from business firms and institutions for minor accidents which the healthy man forgets. It is this parody of the real meaning of rights — claiming good rights on a bad theory and for psychologically poor motives — that should alert us to the fact that the psychological forces present in the "clinical" case in extreme form are necessarily present in lesser degree in the general philosophical discussions of rights.

It is one of the tasks of Beyondist moral research to separate this hidden, creeping debasement of the accurate meaning of rights from the publicly accepted expressions. Most universalist religions — as discussed in Section 6.2 above — present as part of their "bait" or emotional sop some of these promises of meaningless rights, for which no cash in the bank will, in fact, be available when the check is presented. "Rights" and "Justice" are even more commonly the slogans of every political movement aiming to destroy some slow-bought gain in austerity of values or cultural dignity, or to give the lazy and shiftless an easy path to what the hard-working and self-restrained have slowly built from primitive poverty.

[11] The optimum conditions for human evolution may well force upon the human

conscience a real choice between longer and shorter life spans. Let us remember that "Hunger for life" did not prevent Christ, and countless saints and heroes since history began, seeking death in the prime of life, for the sake of the human future as their consciences revealed it to them.

The most obvious possible arrests of inter-individual natural selection would be by an immense increase of longevity and by cloning. (The latter — the exact genetic duplication of one individual in large enough numbers to constitute large fractions of the selectable population — is at present a biological phantasy, though it could at some not distant date become a practical issue.) What needs fuller recognition and discussion in regard to present journalistic and political attempts to stir up confusion over "genocide" is that *inter-group* natural selection could correspondingly be arrested by demanding the immortality of a race. The solemn condemnation of all genocide by the politically active group of social scientists who appointed themselves to write the statutes of U.N.E.S.C.O. could perhaps be reasonably espoused as the desirable prohibition of the destruction in war of one race by another; but, in fact, the cry of genocide is raised whenever, by birth control, economic shrinkage, or other natural and peaceful forms of decease, the end of a poorly adapted race might occur. If evolution is to proceed, we must be prepared for races (defined as particular gene pool patterns) *as well as* individuals to fulfill their purpose and pass on.

[12] At this point a little more attention needs to be given to (a) the origins and effects of inexplicitness of basic principles in the revealed religions now holding world wide sway, and (b) the effects of this "vagueness" in permitting adaptations to local cultures and population temperament differences. Socially we must anticipate the beginning of an extensive conflict between traditional religions and evolutionary ethics expressed in terms of scientific thought. In this the traditional religions are going to last beyond their time both from economic reasons — their tremendous endowments — and for intellectual reasons — the defensive maneuvering capacity which this very inexplicitness confers. The generation of great theologians — Barth (1957), Baber (1952), Murray (1965b), Fosdick (1932), Schweitzer (1939), and Tillich (1963) — who interpreted Christianity as such without involvement in arguments such as the present — is probably over, and it showed in the end the surprising degree of personal and subjective judgment that can enter even the most scholarly judgment. They have been followed by a new generation of energetic young Christian theologians — such as Pannenberg (1970) and Moltmann (1967) — resuscitating Christianity in modern, philosophical terms. Naturally, the social scientist wishing to put revealed religion in what seems a true historical and naturalistic scientific perspective desires a more psychological understanding of Christ and his disciples, e.g., recognizing the tradition of the Essenes; the political frustration of the Jewish culture; its mounting sadism, and rigidity, and the Roman exasperation and countersadism. A convincing beginning along these lines appears in the writings of Raymond Lloyd (1971).

And since Beyondism also argues that each local population should enter on its own value experiment it should be instructive also to study the historical adaptations of such a universalistic religion as Christianity to culture and racial temperament. As Darlington points out, its proselytizing efforts initially found a far more congenial temperamental acceptance in the Eastern Mediterranean than anywhere else, for two or three hundred years. Baetke (1962) points out, as many have done, that the passive Masochistic Christ did not fit the temperament and culture of the Baltic peoples in the North West and that "Christianity did not answer questions (e.g., those of salvation by one's own struggle) which the early Teutons had already posed." Had the North Atlantic empire of Canute survived, covering England, Scandinavia, Ireland, Greenland and perhaps New England, it is possible that it would have developed "aggressive individualistic Christianity" (concerning which Charles Kingsley comes to mind as the best exponent). Such a super-Protestantism, untrammelled by what the Normans later brought from the Mediterranean culture, might have interpreted the personality of Christ very differently, stressing his courage, independence and ambition. Still greater differences of temperament than these of the Western World apparently affect the reception of Christianity so adversely as to make it uncongenial as in Mongolian Asia and Japan. Social psychology needs to get busy with the hypothesis to which many such facts point; that the boundaries of congenial acceptance of religious values are partly set by genetic temperament.

But the main point here is that there is not a religion today that does not fall short of expressing objectivity and explicitness of principles. This comes by reason of being inspirationally based on and wrapt up in the personality of an individual in a particular racio-cultural group. The secret of the success of Christianity across much of the world may be a *comparative* universality of an *emotional* character. But, if so, it is a universality of human feelings, not an embracing of any broad scientific and philosophical vistas related to understanding of the biological and physical universe. Indeed, it could be—and has been—accused of being the most anthropocentric of religions. It has brought success at the cost of making man more important than the universe and the only measure of its remote creator. In so doing it fails to convey the inexorable demands of the cosmos beyond man which Beyondism recognizes.

A great Greek Classicist, C. M. Bowra (1966) has said, "Christianity has alleviated much of the tragic view of life which was implicit in Greek thought—the assumption that the gods were inscrutable and inexorable in their dealings with men." This exposes a Utopian expectation of "eternal satisfaction," and possibly even a pleasure principle strain as present in Christianity. Here there is an unmistakable contrast with the Greek recognition of the "inscrutable and inexorable" environmental demand, which comes far closer to the spirit of Beyondism.

[13] Even so "trivial" a matter as reversal of the earth's magnetic field—which has occurred in prehistory—could put severe demands on us in terms of any stresses to which civilized societies have yet been exposed. Dr. K. McDonald and R. Gunst of the U.S. Coast and Geodetic Survey show that earth's magnetic field may disappear about 4000 A.D. High velocity electrons and protons streaming from the sun will then no longer be trapped in the Van Allen belt, but will strike the earth causing an increase in mutation rates, eradicating at least whole species of animals and plants, and causing major climatic changes, probably flooding most cities on coastal plains.

[14] There is a sense in which the patient, cautious and receptive pursuit of science itself has a masochistic—or at least a fearful and reverent psychological adjustment. The scientist is not a conqueror, enforcing his views on nature; he accepts, listening receptively to what nature is prepared to dictate to him. Here one is reminded that the Haldanes, with their two or three generations record of fine scientific contribution, had under the eagle crest of their Scottish family the single motto, "SUFFER!"

[15] Severe traumas, such as natural catastrophes can sometimes present, may actually initially destroy morale—and one suspects from the deterioration of the stone instruments and other cultural performances as the ice age approached that something like this occurred to Neanderthal man in the last ice age. On the other hand, there can be little doubt from group dynamics experiments, e.g., as in the work of Julian, Bishop and Fiedler (1966) that group competition commonly improves the quality of morale, and, seemingly, the quality of interpersonal relations within groups.

[16] The so-called "Greenbank formula" from the conference (of that name, in 1961) of experts used the probability formula, $N = R_r f_p n_e f_i f_i f_c L$, in which R_r is the proportion of stars like our sun, f_i the proportion with possible life, f_i the proportion of life-inhabited planets that might have evolved intelligence, and so on. To these experts of OZMA, it seemed that N might be anything from 50 to 50,000,000, from which 500,000 has here been taken as a rough central tendency.

CHAPTER 7

The Departures of Beyondism from
Traditional and Current Ethical Systems

7.1 TENTATIVE BUT CRUCIAL ILLUSTRATIONS OF VALUE INNOVATIONS IN BEYONDISM

An ethics based on scientific, evolutionary foundations leads to rules of *inter-group* conduct provocatively different from those somewhat unquestioningly assumed today to follow from revealed religions, but the canons of *within-group* morality so reached are remarkably close to Christian and other revealed ethics. (These two realms are not the totality of moral areas, but that of the response of world citizen to world citizen we may leave to Chapter 9.) Nevertheless, some vital differences of values and reasons for values exist even within the intra-group behavior injunctions. Our aim now is to bring out whatever contrasts may exist with traditional religio-moral systems, first, in this chapter *in values themselves*, and secondly, in Chapter 8, in *the social and political action* that would follow. By this contrast Beyondist ethics can be further worked out.

Two dangers accompany these chapters in terms of accuracy of communication—dangers that have not troubled us before. First, concentration on contrasts, though enlightening, is apt to divert attention from equally important features which the systems have in common—features the understanding of which is not necessarily already established. Secondly, Beyondism calls for founding the more specific ethical values upon *research* on the above principles; but most of the necessary research has yet to be accomplished. Consequently, one must recognize that the truth of Beyondism does not stand or fall by the soundness of any specific socio-political injunction reached tentatively in these present pages. The

295

ethical inferences must be considered as examples, planned to be illustrative of method. In some cases they may yet prove as misguided as Leonardo's recipe for an airplane.

In taking a sample of modern problems it has seemed best to take those which have also tended to recur through history. These include wars "to make the world safe for democracy" (Athenian democracy); the presence of distressing poverty, despite systematic charity (any century); difficulties in assimilation of markedly different races (the Pharaoh's finally forbade importation of Nubians into Egypt); pollution (which kept medieval Europe on the move) and so on. It is hard to name an important modern problem that, with suitable change of detail, is not old. But journalists, who by their name have to talk about things of the day; the young, who like to think they are newer than they are; and intellectuals, who like to think their crucifixions by the mob are unique, do not make the best uses of historical education. Perhaps the fact that so many problems are perennial underscores the need for a change from the moral foundations of the last two or three thousand years to a derivation founded on science, which, in its major growth really is new, and therefore may at last offer solutions.

It is unfortunately inevitable that the differences in recommendations of Beyondist and traditional religions will excite marked antipathy in the latter, even though we have argued for retention of their values until more reliably demonstrable values can be established. Only fools knock down the old before they have plans to build the new. It is not by noisy "protest" and destruction that the new Jerusalem will be built, and, indeed, some walls will be the same. And, whereas "the warfare of science and religion" (Draper, 1898; Julian Huxley, 1953; Wells, 1930) disturbed the intellectual a century and a half ago; and the liberal could almost think of churches as the work of the devil, yet today any man of moral feeling is more disturbed by the imminent danger that the fall of this blind Samson will bring down with it the pillars of civilization. For, as we have already reasoned, while the rising tide of alienation, of crime, and of sheer low cultural morale has many specific causes, such as were present in most former historical fluctuations of morale, there is nowadays more basically and more ominously, suggestion of a final decline in the supporting religions' authority for morality. By education and the mass media, the scientific criticisms of revealed religious truth, once entertained by the intellectual few, are now the property, and also often the excuses of the vast insensitive majority. Let us note, however, that all substantial religions today are substantial also in the sense of being educational institutions, and businesses like any other vested interest. Thus they have a natural momentum and resistance to dissolution which precludes any graceful, rational re-organization, if and when new truths make their doctrinal basis no longer tenable. Beyondism, whatever its inherent truth and promise, is socially the frailest of seedlings in a forest of moribund, but enormous, trees.

In my book, *Psychology and Social Progress* (Chapter 5), nearly half a century ago, I was far more concerned, in recognizing these "political" realities, to consider the resistances, rather than the assistances, which traditional revealed religion could offer to a science-derived morality (and which I then called the *ethics of cooperative competition*). Today, I am more optimistic that—except perhaps for such self-sufficient and highly organized churches as that of Roman Catholicism—we can count on active interest from church congregations peopled by *individuals* who are more educated to feel that they should adjust to scientific change. It is therefore possible that individuals of high moral concern will step over their institutional dogmas more freely than ever before, and some whole churches, such as the Unitarian, may prove capable of making a rational evaluation of Beyondism, and of incorporating those parts which they feel careful inquiry can sustain.

Although we must not, therefore, overlook considerable agreement in an approximate sense, it behooves us to be explicit on the main differences from both traditional revealed religions, and such modern re-hashes as Humanism and Existentialism. Five major differences may be summarized as follows:

(1) In Beyondism, ethical values are intended to be subjected to continuous revision on the basis of research. It may be objected that this is not new since it is already systematically done in the Catholic Church by papal encyclicals. Indeed, the evolution of dogma (which Cardinal Newman more happily called the development of doctrine) is a close functional parallel to research, but (a) for true scientific research is substituted exegetical, philological re-analysis, and (b) for the basic evolutionary principle is substituted the preserved sayings of a half-legendary figure. These differences are momentous.

(2) The design in Beyondism for cooperation of diversely oriented cultures, cultivating and stressing their diversities, constitutes a complex universalism of markedly different though inter-locking congregations, very different from the simple (and, may one add, sentimental?) "homogenous" universalism sought in the imperialism of traditional religions.

(3) There are few traditional religions that do not make a virtue of

poverty and dependency or decline to see possible evil in unbounded expressions of compassion. Beyondism has a more conditional and complex attitude on these. First it desires to see poverty eradicated instead of persisting through palliatives. Secondly, though it agrees that love (agape) and compassion are precious, whereas erotic love, pugnacity and narcism are cheap and redundant, it considers that love as pity (agape) can also be subject to many perversions.

(4) Although Beyondism sees all men as bound by a *common purpose*, this is different in the implications to be drawn from it from the universalistic assurance to mankind that they are "all one family." It is different in the economic interpretations[1] regarding mutual support and in the biological implications. Very probably one of the first recommendations of Beyondist research may be that mankind should plan to diverge into several distinct non-interbreeding species — to avoid the danger of having "all eggs in one basket."

(5) Beyondism is incompatible, as an objective search for truth, with attempts to proselytize by "emotional deals," such as all traditional religions have practiced, or by propaganda having the character of psychological warfare, i.e., deliberate deception in the name of persuasion. It has to be completely open to investigation and experiment. This does not prohibit, however, bringing out all the real emotional satisfactions available in Beyondism, in a skilled educational "persuasion."

As to the first difference above, it is one found not only in the contrast of Beyondism and revealed religion, but is part of a vaster contrast between two cultural eras — the static and the movement-oriented. As Bury (1920) best brings out, before the Renaissance men lived essentially in a static world, aware of a rhythm of generations, but expecting and aware of nothing in the nature of a continuous trend. "Progress" did not exist in Egypt, or in any modern sense in Rome or in medieval times, and arose only dimly in the Renaissance. One may argue that even if they gathered other perspectives than those of the ages in which they were born, the great revealed religions *needed* fixed dogmas at the time of their reigns. Only thus could they hope firmly to withstand the mortal storm of human waywardness. Caught in the present widespread *implicit*, popular acceptance of evolutionary change they are at last forced to hesitant, clumsy and *ad hoc* modifications — as is the Catholic reaction to birth control. This charge of maladaption to movement, incidentally, is no criticism of any religion's firmly standing by deeply established principles when attacked by modern popular movements totally ignorant of history and lacking the wise perspective time alone can offer. But the fact remains that it is the essence of Beyondism to maintain positive evolutionary change in the light of deliberate scientific investigation— and this is no part of traditional religions.

However, this vital readiness of Beyondism to grow continuously is also the main source of its vulnerability. Just so is the mammal more vulnerable than the crustacean or the insect, who have only brief periods of growth and danger between long periods of armored rigidity. Here the traditional religions are safe because of the very rigidity of their dogmas. And on the opposite side, the wider social recognition of change as normal to our era brings the danger of expecting too much of a too easy and too unstudied "change for its own sake." There is a type of incontinent mind, especially common in journalism, on T.V. and the stage, which can spawn to the public more excited and impractical ideas in an hour than good scientists can bring home the fallacies of, by a year's tidying up. One can justifiably fear that as soon as it is realized that ethical habits are a domain for hypotheses and invention, a tremendous smoke cloud of ill-considered, fragmentary ideas will arise to blind the public. (*Brave New World* and *1984* are among the least bizarre of the arbitrary propagandas we may expect.)

The misfortune of too glib a public discussion is that ethics is really the most complicated field of science, requiring the greatest deliberation, insight and good sense for its investigation. Above all, it requires experiment and statistical analysis for which the individual journalist or speculative writer is unqualified, so that "popular" discussion has from the beginning a built-in futility. Steady advance, and the development of a real science of morality, can come about only by deep study of large group experiments, requiring research institutions for gathering data, and calling for steady perspectives while generations of scientists make their contributions. Hopefully we shall soon learn enough about the psychology of research itself and have sufficient instrumentation for detecting pleasure principle and narcistic components in thought to be able to separate the charlatans, the sophists, the Cagliostros, and the creators of Piltdown men, from the Darwins, Einsteins and Newtons. Thus, by vocational selection of researchers, and the growth of tests for the quality of thought, the subject can benefit from the deep originality of individual thought, without letting mere cleverness take over the house of wisdom.

With this brief indication of the gains and dangers in moving to an ethics which is built to grow, let us turn to the second above-listed difference from traditional religious values—that of universality. The universal brotherhood of Beyondism is based on a sympathy and love in the understanding that we belong to diverse, competing experiments, sharing the common human fate of transience, but bound by a great goal. Because of its acceptance of different values in different groups this might be designated *federated universalism*. The universalism of revealed religions, on the other hand would abolish competition and establish an egalitarian uniformity of values and aspirations everywhere—hence the expression *"homogeneous" universalism* above. It has been suggested above that the universalistic quality appeared in religion by an historical growth out of a within-group morality first developed only *within* groups by natural selection[2]. Whatever the details of the dynamics of its persuasion may turn out to be, there is no doubt that universalism flatters all kinds of minds with a sense of their liberality and expansiveness. It quickly appeals even to the easy-going man in the street who is apt to deplore—in the field of cultural specialization—everything from customs declarations and language difficulties to armaments' costs and the draft! In particular homogeneous universalism is the dream of those who perceive or misperceive it as an assurance of universal peace—and who value nothing beyond the goal of peace. It finds resounding support among ethical reformist writers, such as Comte, Wells and Russell—and even politicians, as in Wilkie's recent slogan "One World."

Yet if the Beyondist position is sound—and we have encountered nothing yet to contest its basic logic—a simple, sentimental, homogenous form of universalistic ethic would wreck human progress very quickly—in two ways. It would abolish the stimulus and test of inter-group competition, and it would reduce that local group variation in ethical and other values—that cultural adventure and inventive racial mutation—which is the indispensable condition of diversity for evolution by natural selection. Just as the scientist aiming to discover some new and effective product, tries out his various mixtures in a carefully segregated and labelled array of test tubes upon his shelf, so must evolution keep some self-contained, inwardly-developing apartness in its treasures. For evolution has no alternative but to proceed by diversification and selection, culturally and biologically. In the usual goal of homogenistic universalism we are actually being asked to applaud the crowning disaster of all the test tubes crushed in one confused mess in the sink. By contrast, the universalism of Beyondist ethics is very real, but complex. It requires insightful modification by checks and balances, more fully discussed in Chapter 9, Section 9.1.

One must repeat, however, that as far as the rules of *internal* behavior of groups necessary for evolution are concerned, they much resemble those of the great universalistic religions, especially those of Christianity. Indeed, the social psychologist wishing to turn first in his research to the causes and consequences of the most vitally important virtues and vices cannot do better than to start with those which stand at the center of the discussions of the Great Christian divines. Condensed to a distant view, we see the Judaic religion and the Mosaic code, carried further in the beatitudes of the Sermon on the Mount, and, brought by St. Paul to the essentials of faith, hope and charity. These joined with the well-thought-out four virtues from a different stream—that of Greek and specifically Aristotelean—wisdom, justness, temperance and courage. They became for medieval scholars the seven virtues, in the shadows of which stood seven deadly sins. The almost certain alignment with these virtues of the values inferred from evolutionary ethics bids us, despite their salient importance, spend no time in discussion.

Only among the sins is one which perhaps has special relevance—that of malicious envy. For it is in the nature of Beyondism that it will require —certainly more than such secular religions as Communism—a toleration of diversity and apparent inequality of endowment of men, races and cultures. The faster a pennant is carried and the more certain its direction, the more it streams out in the wind. Only in an army in confusion are the vanguard and the rearguard inter-mixed. The most vigorous Beyondist societies are those in which excellence of all kinds is cherished and sympathized with, rather than resented and cut down. Societies with an egalitarianism based on envy and jealousy have paid heavily for it. Men of psychological intuition—and we can go back here to Chaucer's judgment that envy is the worst of the deadly sins, because it is against the good—have recognized that though envy of property or status is of little moment because to the wise man these are trivia, jealous hatred of intellectual or moral excellence is indeed deadly to society.

The third of the main differences listed above between Beyondist and traditional ethics—that on poverty and charity—deserves a section to itself, as follows.

7.2   RELIGIOUS, COMMUNIST AND BEYONDIST CONTRASTS ON THE VIRTUE OF CHARITABLENESS

Within-group morality, as derived from group survival, demands a high degree of altruism and readiness to mutual sacrifice among citizens. Yet the values which it requires in regard to suffering, poverty and failure are very different from those traditional in revealed religions. It is no exaggeration to say that the latter have actually perpetuated poverty, as Communists and socialists explicitly claim, though it may quickly be added that socialistic welfare is doing the same in a deeper sense. The latter systems have abolished begging, and that appeal to the exercise of immediate compassion which the universalistic religions seem to have liked. But there is a sense in which their immediate effect is really only to spread poverty more thinly, and their long term effect to make it a greater threat to the positive advance and adventure of societies. Some of these matters belong to "social machinery" which we reach in Chapter 8, but here our concern is more with the values themselves.

Let us begin by recognizing three types of poverty. First, there is ill-luck, which is distributed as randomly as shot on a battlefield, and the victims of which unquestionably deserve the full compassion and total assistance of comrades. Let us call this Type A or *extrinsic* or *accidental poverty*. In older societies, such as that in which Christ moved, the vicissitudes of fortune and the inequalities of opportunity were so great that possibly a majority of the poor were of "Type A." In many modern societies, with education and medical assistance, universal and free, on the other hand, there is no question that—at least downward from the *average* "good" earning level—a strong correlation exists between poverty and inherent characteristics of the individual designatable as "incompetence" (to be defined more operationally later). The representatives of this *inherent inadequacy*, or Type B poverty tend to be below average intelligence, and/or lacking in normal foresight in looking after health and property, and/or deliberately evasive of work responsibilities and prone to emotional impulses which land them in a cumulative downward spiral of difficulties. For example, it is common for this type to beget a family of a size out of any relation to the financial and characterological capacity of the parents to look after it. Thus, even if given outright by society what would for another type be an adequate annual allowance (though with no relation to their real earning capacity), Type B will remain in poverty where others will not.

Once an average standard of effective living is achieved, the dedicated scientist, writer or artist will usually cease to strive for material gain as such, so that in the "above average" range of income decidedly less correlation of comparative "poverty" with incompetence will be found than in the lower half of the range. Indeed, the "poverty" of, say, the minister or the academic man compared to the wealth of a business man of equal intelligence is purely relative (though it does not prevent the less mature of writers and academics from generalizing that all poverty is undeserved, and that artificial, authoritarian re-distribution of property is "progress").

The normal degree of disregard for material surplus in the culturally interested person can, however, pass over into carelessness and poverty, as in the hippie movement and the "holy men" of India. This we may call Type *C* or cynical or schizoid poverty. When it reaches true poverty, rather than "simple living," entirely new psychological, narcistic and attention-seeking motives, as in the exhibitionistic cynicism of Diogenes, have usually entered in. Aldous Huxley, in his penetrating comments on religious poverty, dirt (which is "material") and disease (presumably also material) in India has said the subtle things needing to be said there (1926), and has brought such observations into a political framework. It is, however, not only far Eastern religions which have these values, for Christianity through medieval times, in such saints as St. Francis, and in its inherent argument "They toil not, neither do they spin," has assumed that men may have such need to be constantly examining their souls that they have no time to work, to create, to advance human knowledge. (One must remember, however, that the Middle East in biblical times gives every evidence of being a large mental hospital.) The person concerned with intellectual culture today need not and must not espouse poverty; for laboratories do not grow out of the ground; libraries and hospitals need expensive factory productions, and ill-nourished people do not think effectively.

The implication of the above analysis is that in Type *B* poverty, which is by far the largest class, we are speaking simultaneously of poverty and inadequacy for the culture, and therefore, our concern is with compassion in a broader sense as it concerns both of these. The evidence that Type *B* poverty in the modern world is associated with relative "incompetence" is very strong. Research on the intelligence levels of occupations shows a decline from the professions (generally overworked by more demand than supply) through skilled workers (generally with a good market) to unskilled (somewhat in excess of demand) and so to the lowest levels in unemployed (Cattell, 1971). But the same holds for the broader definition of competence which includes personality as capacity to work with concentration and to adjust to fellow workers. Repeatedly unemployed persons have a measured personality profile nearer than that of the general population to clinical cases (Cattell, Tatro and Komlos, 1965). It is relevant to remedial measures to ask how much of this is cause and consequence of poverty, and how much is innate or acquired.

The answers to both are undoubtedly "both," but the reality of innate components is shown by the fact that children of persistently unemployed (but well fed) also fall significantly below the intelligence level of children of employed in approximately the same social status.

The "rationalist" approach, in Socialism and Communism and the welfare bureaucracies of most modern countries have essentially continued to handle such poverty as an economic problem only. They have removed the sentimental charity to beggars of traditional religions, but a Beyondist must point out that there is still a sense in which they have "swept the problem under the mat," perhaps in the long run spreading the disease they seek to cure. (This is still more striking in inter-group charity.) Common sense observation, before documentable scientific evidence, has long asserted this—though the assertion has stupidly got entangled in political arguments really concerning quite different economic levels (those between the "haves"—the so-called right—and the "have nots"—disguised as a progressive left). As Hardin (1964, page 28) says, "Ever since their appearance in the sixteenth century, poor laws have been under fire as poor laws: there was more than a little suspicion that they actually increased the poverty that they were supposed to ameliorate."

So much for illustration in a concrete setting. Let us return here to the question of values, where the main point is that whereas traditional religions have lauded compassion and charity as unquestionable virtues, of which there cannot be too much, the psychologist with an evolutionary perspective is compelled to cry "Halt, let us examine this assumption more critically." The inborn, instinctual capacity to feel pity and succor the young, the hurt and the helpless has its obvious biological origins in evolution as something contributing to the survival of the family and the group. In just a few of the situations where this purpose has been perceived to be negated by compassionate behavior men have learnt to behave differently from animals. For example, most primitive peoples when the birth rate is too high for the food supply set out newborn children to die of exposure; men in the last war abandoned their drowning companions rather than risk the certain destruction of the rescuing ship, and Eskimos, when grandparents become too old to eat tough food or move about, abandon them.

Instances of where the spontaneous primitive emotion does not perform its purpose, however, abound more freely in civilized life in what may properly be called perversions of the parental, protective erg—if we properly define perversion by just the same standards of uselessness or harmfulness as we apply in defining, say, sexual perversions. Our society waxes lyrical (especially in dog food advertisements) over-feeding fifty million dear household pet dogs while fifty million children (more remote) go hungry. It finds its "conscience" opposed to sterilization of those who would pass on terrible and irremediable defects. Its tender feelings lead it to cripple medical research by anti-vivisection campaigns. Both life and literature abound in the cult of misguided pity. Perversions, i.e., acts missing the aim of compassion by substituting some immediate indulgence for a more remote and greater alleviation of human suffering, that would be immediately pilloried if they appeared in sex, self-assertion or even curiosity, are assumed to be sanctified because pity is the motive. Literature, from Dickens to Dostoyevsky, abounds in the sanctimonious, tear-jerking tale of "oppression" and neurotics pore over it — and vote against necessary social disciplines. What would society become but a mutual narcism club if it followed the values in Dostoyevsky's *The Thief*? As Lord Snow points out (1959) the Communists, who are determined to eradicate poverty (even though their neglect of genetics may wreck their intentions), in spite of their pride in Russian literature, have felt impelled to disown Dostoyevsky as an enemy of society and the party, as they see it [3].

Every man is a failure, judged by some positive standard in himself or others, and almost every man finds himself in poverty at some time or other through Type A misfortunes. What does a Beyondist ethic say on this? Haldane, a scientist and a completely free questioner of values, says: "In most human societies it is regarded as a duty to help our weak and unfortunate fellows. This may be a fallacy. I do not think it is." The Beyondist answer is clearly that our instinct developed, and rightly developed in terms of Type A misfortune (temporary and unmerited), has real evolutionary function there, but is exploited — mainly by fools and knaves — for perversions in certain forms of Type B distress. Haldane comes to recognize this in a second thought "It is illogical that sound people should be taxed to support the unsound."

In justice to the intelligence of the leaders of traditional religions it should be recognized that they have not always taken the common view of their congregations (and possibly of St. Paul) that love and charity are the solutions to all things. Dean W. R. Mathews (of St. Pauls) writing at a ripe old age of service tells us that Christian ethics is not satisfied that "Your heart will tell you how to act." The real determiner of what is moral in our culture comes from an outer (see Catholicism), not an inner message. It is, of course, a fact of human nature that people love "kindness." A recent "popularity" survey showed that this is the surest single determiner of personal popularity. Every party politician works this for all it is worth, no matter though his promises would bring, by some higher morality, harm to the individuals concerned. And we are fast approaching a point in politics where a woman who does not own a car to take her five illegitimate children to school is considered in "poverty" and fit for the public payroll.

From a broader evolutionary standpoint, *objective* justification for a society being more indulgent to its citizens than is (from other considerations) good for them, must come, from some auxiliary argument and though such are hard to find, possibly it may be that societies of higher mutual indulgence excite more loyalty and therefore, survive better as societies. On the other hand, it may be quite wrong that indulgent societies are stronger, and social psychological research has by no means said the last word on this. Indeed, an eccentric statesman, appealing to a group of more than average superego development, has been known to raise morale by promising nothing but "blood, sweat and tears." Indeed, to make "relief of suffering" a primary goal is to veto many important goals, e.g., much scientific research, when suffering is interpreted as the overprotected sensitivity of some groups would define it. No important advance has been made in human affairs without discomfort, suffering or death, whether one is talking about drilling mountain tunnels, setting up nuclear power stations, practicing medical operations on animals, or travelling to the moon. One does not make an omelette without breaking eggs.

The insights of literary men well before the era of modern psychology (Blake: "Damn braces; bless relaxes") have questioned the community functionality of indiscriminate "love." The motivational dynamics of a progressive community is extremely complex; but it is certain that compassion cannot be given title, as in some traditional and secular religions, to the unqualified designation of being invariably, pre-eminently ethical. The true Beyondist position would seem to be that though *all* forms of human misfortune merit compassion, the unintelligent use of compassion in Type *B* misfortunes is a social problem, since it perpetuates them. The fact that love is rarer than other motivations, such as lust, pride or fear does not mean that it cannot err just as profoundly. Nature is neither compassionate nor cruel, and human societies, since they have to come to terms with the cosmos, can make a bank of human kindness only to the extent that they can afford indulgence compatibly with realistic control and defense against nature.

7.3   THE RELATION OF BEYONDISM TO MODERN ECLECTIC MOVEMENTS, AS IN COMMUNISM, HUMANISM AND EXISTENTIALISM

Clarification of the values of Beyondism by bringing out contrasts with revealed, dogmatic religions has had to be brief and approximate because of space. But clarification through contrast with more recent value systems, such as Humanism, Communism, Existentialism, suffers instead through these entities being far more vaguely defined in the minds of their adherents than are the dogmas of traditional religions, precisioned by one or a thousand years of philosophical and legal interpretation.

Yet in the immediate future the competition of Beyondism for emotional acceptance is likely to be decidedly more intense with the modern "religions" of Humanism, Communism, Existentialism, etc., than with the retreating revealed religions. Despite all their emotional "deals" the revealed religions have had a forthrightness about their values. They have marched under banners emblazoned with quite uncompromising doctrines concerning sin and virtue. But in Humanism, and its somewhat remote relative, Existentialism, the values hide in an excess of literary discussion. What one needs to recognize at the outset is that though these are movements professedly shaped by the intellectual himself, and claiming to be rational and progressive, yet actually their main substantial values are a conglomerate of traditional emotional (largely Christian) values. They make no disturbing, fundamental demands for readjustment, except for a willingness to accept an ill-defined eclecticism. The breadth of their appeal springs partly from their expedient alliances with political and social movements. Thus, especially in Humanism, we see a sprawling encampment on our line of march which is relatively invulnerable to scientific analysis and attack because it is so vague as to its fundamental principles.

Originally, the *humanities* (and the term "Humanistic" as an occasional derivative) referred simply to classical studies, history and literature. That meaning we will leave untouched and set aside from this discussion. But meanwhile, mostly by association with the revival of the classics at the Renaissance, and their battle with medieval dogma, the term "Humanistic" took on liberal and rationalist overtones. The literary companions of Lorenzo de Medici were inclined to trace the spirit of their times back to Plato; but in fact their version lacked Plato's other-worldliness. For them Humanism meant the humane liberal acceptance of individuality, moral or not and the belief that man is good and indeed a lesser god in his own right. (Though Lorenzo, Ficino, Poliziano and Mirandola were duly followed by Savonarola, to whom Lorenzo confessed his sins!) Today it means a ripe urbanity of viewpoint, rooted in the whole sensitivity of literature, and acting as a loose set of ethical values and guiding principles in legislation, education, international conduct and many other fields.

Its values include, centrally, an appeal to rationality, humaneness and restraint, but this does not save it from being a hash of possibly discordant fundamental principles. The rationality, however, is very frequently based on the assumption that we command — without the humility of scientific investigation — all the facts necessary for our personal reasoning. And its restraint has been accused (notably in the political liberal movement) of hiding a conservative tendency not to get involved in any disturbing fundamentals. Granting that we cut loose entirely from the *narrow* academic denotation of "Humanism," merely in connection with "the humanities," and avoid, on another flank, confusion with the equally definite (and substantially different) notion "Humanitarian," the social existence of a Humanistic movement, as we use it here remains real enough. However, not being a recorded doctrine, it would need some sort of "social opinion survey" to fix its boundaries and nature. Perhaps one can indicate these boundaries best by indicating a span of writers — such as Russell, Robert Bridges, Wells, Huxley and so to Sartre and Kafka (on its existential fringes) who are everywhere recognized as "Humanistic." Thence we proceed to Fromm and many modern journalists, and so to others (Leary, Ellis, and the writers of Hippiedom) who represent a somewhat rank outgrowth. Existentialism as such, it is true has the benefit of being more precisely defined by acknowledged prophets, philosophically competent, such as Kierkegaard (1941) and Heidegger (1949). However, it is far weaker and less important as a social value movement, or a generator of ethical values, and we shall center discussion on Humanism as the bigger brother. What they have in common is the placing of intuitively felt human values — an inner truth — as of *more primary importance than any checks on the group outcome.* The last includes the structure of institutions which an ethical system generates, and the capacity of the disciples, by deeds, to meet the demands of an external reality. It is precisely in working *from without inward* that Beyondism differs from them both.

To judge the nature of Humanism by its acts we may note that its "voting record" includes strong support of "permissive" and "Progressive" education (the capitalized *P* denoting a specific historical movement), of reduction of punishments for crimes (most of punishment being alleged to have nothing but a retributive aim); the advocacy of free love and sexual satisfaction in or out of marriage; an opposition to engendering guilt (and an expressed doubt that "sin" is a useful concept); a philosophical position of moral relativism, including belief that many religious creeds are "superstitious"; and a notion that all problems that matter are solved when we have affection and equality among all men. It would be a fair condensation of various other attitudes to say, in one phrase, that it favors a substitution of the worship of man for the worship of God.

Among the threads now woven into what has become an altar cloth for many intellectuals one perceives easily enough the traces of important Christian values (though not of the theological beliefs). Such general universalistic religious tenets have been accepted, for example, as are found in Unitarianism and in classical Platonic and other philosophical values (not extending however, to German, French, Scottish and other philosophers of the nineteenth century). It warmly embraces the liberal literary and artistic, but not the independent and scientific, thinking of our day (as far as that scientific thinking has yet extended to values).

Naturally such a plexus of values has a strong appeal with the liberally educated intellectual of our day. It is more blatantly expressed by the journalistic camp followers of culture, and has been the creed of energetic young reformers for three generations. In such circles the origin in "rationality" is alone stressed; the borrowings of Humanism of its central values from a diluted, dogma-free *intuitive* Christianity is not admitted, and any dependence on economic history (in a Marxian or any other sense) is simply not recognized at all. An historical analysis needs to investigate this last, for there is an obvious connection of the Humanistic trend with the spread of prosperity and the dynamics of a resultantly released narcism, as glanced at in the last chapter. It has particularly been observed (Horn and Knott, 1970) that student activism today really draws its idealism from this source. The new, more radical "cutting edge" of Beyondist thought has not reached the average student, and we witness an expression of the ironic fact noted by Bolitho (1929) that a new reform movement is often embarrassed by students embracing the cast off values belonging to the preceding wave of reform [4].

No space can be given here to the parallel systematic contrast with Communist values, but this can be dispensed with because the contrasts are made at a dozen points throughout the text (e.g., pages 54 and 304). Like Humanism, they owe much to Christianity and the Judaic religions; but as far as Christianity is concerned Communism is definitely pronounced a heresy, as it would be also with respect to the political ideal of democracy if doctrinally defined.

Nor can space be given to completing, by a three cornered comparison, the corresponding contrasts of Humanism and Beyondism with revealed religions. In a Gallup poll in America, showing a 98% belief in God (falling to lower values in Europe) it was explained that in intellectual circles in the latter "old fashioned atheism" is dying and that a "sophisticated Humanism" is replacing it. Curry, an explicit propagandist for Humanism, contrasts traditional and Humanistic religion by saying that the basis of virtue is not trust in God but "Love—an instinct you see in every child or animal." And the contrast of the two positions is quite consistently rounded off by his dictum: "The world must be made to fit man, not man the world," (1937). Here, and in some other features, a scientifically based religion is actually closer in values to inspired religion than to modern eclectic Humanism; for it considers that the universe has a lot to teach man, and that he would be absurd trying to shape it to his pygmy mind, instead of stretching his mental stature to its demands.

7.4   THE CONTRASTS WITH HUMANISM ILLUSTRATED WITH RESPECT TO CRIME AND PUNISHMENT

Beyondism, by its origins in science, is an intellectual movement of a very different character from Humanism. It springs by fundamental, logical steps from the basic theme of evolution, on the one hand, and the newly emerging laws of behavior in groups, on the other. But Humanism at its ripe best, is, like Franklin's *Poor Richard's Almanac*, a digest of the urbane wisdom of the ages. Unfortunately, that wisdom in the process of literary [5] transmission, has come to terms in countless ways with the worldly values of the pleasure principle. Moreover, as to offering leadership, it is unable to tell us anything new, because—except for its borrowings from the universalistic religions—it represents what is comfortable and mediocre in the average view of mankind.

Thus when Humanism prides itself on virtues of kindness, tolerance and liberality two questions occur to the Beyondist: (1) How far has the social growth in these attitudes been a real growth in that spirit of self-sacrifice and frustration-tolerance which is vital to a higher, more active social quest, and how far is it a mere "economic" by-product of material progress and easy times, that—duly tested—might lack true moral tenacity? and (2) how far, in fact, are increased toleration and permissiveness (*see* Royce, 1961) beyond some optimum point to be considered as a social virtue? As to the first, there is little doubt that a man in a hurry is usually less polite than one who is going nowhere; or that a man with a severe toothache will tend to be unsympathetic to anyone who complains of mere discomforts. Conversely, a well-fed man enjoying an after dinner cordial will consider the traffic penalties invoked by a hungry hobo, recently knocked down by his car, to be savage and uncivilized.

The chief gain by which the Humanist measures his progress in virtue, compared to the hair-shirted saint of the Middle Ages, is in the breadth of his tolerance, the decrease in severity of punishment he asks for admitted immorality, and the reduction in the ferocity of strife over moral matters. He congratulates himself on the decline of cruelty, and he sees in every zealot nothing but a bigot. He retains not a glimmer of the understanding for the inner processes of spiritual questioning in the *auto-da-fé* of the Middle Ages, or even for the stern attitudes of the Victorian parent to, say, an increase in illegitimate births. Yet before assuming our tolerance means more readiness to sacrifice to others we should remember that the severity [6] of life in the Middle Ages was such that the average life expectation was thirty years, and we should keep in mind that the standards of dress, feeding, health care and leisured recreation of the unskilled worker or even the unemployed today is that known only to a few aristocrats as recently as two or three hundred years ago.

In this balmy summer day of our technological triumphs it is a natural psychological reaction to mitigate the severity of demands both on others and on oneself. There should be no difficulty in social psychological research demonstrating this in precise quantitative terms as it develops indices of severity-vs-ease of living. And it may then be possible to answer the question as to whether, if returned to a high severity of external demand, our supposed Humanistic progress in love of fellow man would turn out to register any real advance at all. Some ghastly deeds of members of this generation, now being verified in court, but carried out under the stress of war, suggest that most of the supposed Humanistic progress in benignity has not shown itself in fundamental values. The Humanist "gain" in values is nothing new but merely a natural expression of the alleviation of general environmental pressures relative to those faced by our ancestors.

The alleged progress in tolerance may turn out to be worse than a merely standing still. It may be simultaneously a purely environmentally permitted mellowing and also a regression in real concern about moral issues. The permissiveness of Humanism in this generation has shown itself prominently in the attempt to abolish blame, i.e., guilt, or punishment.

Yet it is obviously in part—or maybe in whole—in socially misguided directions. As Tuchman has said, commenting on the recent rise of willful crime and the intellectual's "psychological excuse" of "understanding": "Admittedly, the reluctance to condemn stems partly from a worthy instinct—*tout comprend c'est tout pardonner*—and from a rejection of what was often the hypocrisy of Victorian moral standards. True, there was a large component of hypocrisy in 19th Century morality." But on the tendency to explain and exonerate in terms of the delinquent's environment the speaker proceeded: "I find this very puzzling because I always ask myself in these cases, what about the many neighbors of the wrong-doer, equally poor, equally disadvantaged, equally sufferers from society's neglect who, nevertheless, maintain certain standards of social behavior, who do *not* commit crimes, who do not murder for money or rape for 'kicks.' How does it happen that they know the difference between right and wrong and how long will they abide by the difference if the leaders and opinion-makers and pace-setters continue to shy away from bringing home responsibility to the delinquent?" (1967).

That delinquents—and prospective delinquents and their associates—in general are more demanding—as voting citizens—of reduction of punishments than are non-delinquents, and that they protest regularly, is easily understood. "Live and let live" is their motto, and some degree of this exists in all people. But the roots in the Humanist, though partly residing in his pleasure principle desire to reduce onerous standards, surely stem more from the hypocrisy of making a virtue out of what actually now comes more easily. He is also guilty scientifically (as also are many so-called social scientists) of forgetting other ramifications of the problem than that in which he is interested—the happiness of the delinquent. For thirty or forty years social science has completely neglected what the current recipe for treatment of the criminal minority does to the rest of the community. Any retributive, as distinct from a therapeutic treatment of criminals the Humanist regards as purely irrational and demonstrably due to deep "psychoanalytic" sadisms in the citizen. Yet in the normal integrated person the severity with which demands are made on the self is correlated with the severity of demands on others [7]. It is quite probable from recent research on the superego that its strength in the majority may be *reinforced* by just punishment of offenders and demoralized by an unreflective system of meeting crime by an indulgent "therapeutic" attitude. Parenthetically, there is no proof whatever that the indulgent attitude *is* more therapeutic. Rates for recidivism, parole-breaking, and fresh crime while on parole remain much the same as before the psychiatrist came into the picture. And the "psychoanalytic authorities" who clamor fashionably in journalism for "the new treatment" have done least to assist or carry out the strenuous and objectively-statistical, basic research on personality on which (Horn and Knott, 1970; Scheier, *In press*) a genuine therapy could be established.

The difference between the Beyondist and the Humanist viewpoint here is simply that between a wholistic approach embracing both the delinquent *and* the rest of society, and an existential preoccupation purely with the inner life of the *delinquent*. The rehabilitation of the criminal and a sympathetic understanding of the difficulties which overcame him (including, often, his own earlier arrogance and callousness) is as fully embraced in the Beyondist as in the Humanist ethic. The divergence in emphasis that is important for illustrating the distinction of moral foundation between Beyondism and Humanism is seen in that the latter simply applies a set of "felt" values — actually having the better part of their origin in revealed religion — to any treatment of an individual, with the implicit faith that if the individual is so treated the group will "look after itself." By contrast, Beyondism considers the psychology of *both* the group and the individual from the beginning, and derives the treatment of the individual not from an "intuitive," "revealed" and merely historical brewed potion of values, but from whatever scientific evidence can be obtained as to the relation of such inter-individual values to the survival and progress of the group.

Consequently, in the case of such deviants as the criminal and the neurotic the social scientist with Beyondist values makes as much investigation of their effect on *other* members of society as the Humanist does in tracing the "guilt" of society in permitting the existence of criminals and neurotics. Psychoanalysis, for example, (which is often the only psychology of personality dynamics apparently known to many writers) is full of what society does to the neurotic, but ignorant of what the neurotic does to society. The studies on performance of a hundred groups of ten men each made by Stice and the present writer (Cattell and Stice, 1969), in which the group could not have accounted for the personalities of the individuals *because the individuals were measured before being brought together in their groups*, showed that neurotics had a very adverse effect on the general performance and morale of the groups in which they were "citizens." The corresponding burden upon others, in fear, disorganization and loss, though the criminal surely needs no such proof or illustration, and, by an objective ethics, justifies the community in considering more than the question of what the treatment of the criminal will do for the criminal.

The penetration of sociological and educational writings by Humanism, particularly in this sense that "social action should be aimed to ensure the ease and blamelessness of mankind," has been widespread with the decay of a more rigorous revealed religious ethic in the last two generations. The apparent unawareness among "intellectuals" that they are getting addicted to that "dangerous alloy" (Chapter 1) mixing scientific authority with purely intuitive, unexamined moral dogmas is astonishing. In an otherwise good technical book, Emmet (1966, page 29) tells her readers: "The influence of John Dewey in America in particular has encouraged people to assume that critical scientific intelligence and liberal values and goodwill go together. The great figure of Pareto—an intellectual authority for Fascists and thinkers on the extreme right— should have taught them better." Pareto's goodwill, which no one familiar with his work can doubt, was apparently quite vitiated by his disagreeing ethically, or even politically, with a "liberal!" The possibility that Humanism, though much associated with academic writing, has no claim whatever to being scientifically and philosophically evaluated as "goodwill" any more than is the writing of Pareto, seems not to have occurred to the author of this students' textbook.

Naturally, the topic of crime and punishment taken above is only one of several that could bring out the contrasts between the narcistic values hidden in Humanism and the demands for reality contact and concern for the life of the group central in Beyondism. It is sociologically interesting to see how "Humanist" values have entered into psychiatry, and these forms of clinical psychology remote from testing by scientific demands. Ellis (1970) tells us that for such psychotherapists as himself, Fromm, Perls, Rogers and May the goals are "self-interest, self-direction, tolerance, acceptance of uncertainty . . . and self-acceptance." The resemblance to Mirandola's "Fashion yourself in what form you prefer" is striking. It is one more instance of psychiatry's foisting upon the healthy the values of the sick, since psychiatry's chief task with the latter has been viewed as the reduction of superego demands.

7.5   SOME FURTHER DISPARITIES OF "SECULAR RELIGIOUS VALUES" AND BEYONDISM

Another danger of the muddled secular moralities is that they lend themselves to that inexplicit mixture of social science and values which has become the greatest threat to social progress in our time. Admittedly, a dishonest social scientist can present a misleading conclusion even when, as with Beyondism and the revealed religions, the moral principles are explicit enough to be separately stated. But when the moral views are *themselves* already an unknown mixture the goal of presenting intelligent citizens with clear choices in the integration of scientific facts with moral principles is impossible.

As we shall see in later examples, the arena of public economic policies —quite as much as mores in sex, political life, etc.—becomes powerfully modified by changes in moral values. A respected liberal sociologist, Myrdal, whose reliable factual observations we have quoted with appreciation (page 226), unfortunately elsewhere offers a typical example of the undisciplined and unashamed mixture of secular (Humanist, as it happens) moral values with conclusions proporting to come from social science. The educated citizen knows, he tells us (1965, page 50), "how much more has to be spent by the community in the slum districts, on police, fire protection, health, emergency aid, and the like, and how, in spite of these provisions, there is much more crime and prostitution, illegal gambling and gangsterism, houses on fire, debilities and chronic illnesses, as well as epidemics and short-term sicknesses, alcoholism and mental illnesses, and so on, in the slums than in the type of well-kept and clean suburb where he lives. He knows that infant mortality is higher there, as is the rate at which mothers die at childbirth.

"The educated American also knows since several decades that, as a general proposition, the higher incidence of all these unfortunate things among the poor is usually not due to inborn differences in human quality but is caused by the environment. In a queer contradiction of this theoretical knowledge, he posits a sort of general moral feeling that nobody needs to be unemployed and poor unless he is a bad person. Nevertheless, it remains something of a mystery that the majority of Americans show such lack of concern about these facts."

The real mystery here is that Myrdal can conclude (a) that the majority of Americans are unconcerned, and (b) that the "theoretical knowledge" that inborn differences play no role in the difference of earning capacity, control, etc., is real knowledge. There is a significant correlation of measures of psychological effectiveness with capacity to earn well and to spend and save wisely, and *a statistically significant part* of this association, by any good mathematical model, goes with genetic characteristics. In regard to inborn aspects of emotional instability, mental disorder, etc., the exact figures are still to be investigated[8], but in culture-fair intelligence measures alone the correlation of social status with intelligence is in the 0.2 to 0.6 range. (For the *child* and the social status of his *parent* the correlation is 0.2 to 0.3. For the adult and *his* own social status the correlation, is, as Horn points out (Horn and Knott, 1970), decidedly higher — about 0.6.) This is more positive evidence than anything Myrdal presents and contradicts him in showing that poor status is more a consequence than a cause of the psychological disability.

It is precisely because practical politics is misled by "liberal," Humanistic writings like Myrdal's, conveyed with some authority, that the poverty problem with which he reproaches the American citizen remains unsolved. It so remains, in Myrdal's country as in America, because social action has become confused in aims. There are, of course, both environmental and genetic determiners of individual differences in the ability to maintain good material and moral standards of family life and we simply cannot give in precise figures at the present moment the relative variance contribution of each. The best and safest estimate today would be that they are of equal importance, but Myrdal's "usually" conveys to the innocent reader that the problem is purely one of environmental mistreatment of the slum-maker. To compound his offense he then accuses the intelligent citizen of believing the person of low status is wicked, whereas a far more commonly expressed conclusion of the intelligent citizen is that he is incompetent. And this is a sounder basis for devising remedies for the situation than the sentimental sociologist's conclusion that it has no systematic cause whatever, or the paranoid conclusion that it is all due to "oppression" by this or that group.

Virtually all men believe they are earning less than they deserve, and consequently, are ready even without the persuasion of modern pseudo-liberals to ascribe too much of any deficient earning to external environmental influences. But, as pointed out in Chapter 6, there are, additionally, pleasure principle distortions favoring rejection not only of personal moral responsibility but even of the realities of heredity. As Hardin (1964, page 118) has well said, "The human implications of the idea of selection are so upsetting that even today most people, including many biologists, cannot see [i.e., recognize and accept] the more threatening of them." A similar substantially documented criticism of the incapacity of the current Humanist subdivision of "intellectuals" to absorb and constructively handle the reality of inborn individual differences, was recently made by a distinguished research physiologist, Roger Williams, in the *American Scientist* [9].

Since, at least on a purely *logical* basis, it would make better sense for a person concerned about our society either to enhance the importance of personal responsibility by decrying heredity, or, if his taste runs the other way, to emphasize heredity and appear to take responsibility for the individual, one wonders what motivation moves the Humanist simultaneously to try to reject both heredity *and* responsibility. A moment's psychological analysis suggests that he must be defending the reign of the pleasure principle in human affairs against both the demands of the super-ego and the scientific, cosmic realities, as presented in the facts of biology. This defense of indulgence which, by the nature of Humanism, could so easily dominate the whole movement, is soon perceived to go far beyond the particular illustrations used here of by-passing heredity in social science, of advocating indulgence as the nostrum for crime, or of calling for virtually every form of sexual permissiveness.

Actually the ultimate and central danger in its claim to be a desirable moral system is its emotional over-valuation of human stature, achievement and rights; the removal of any perspective that would keep us aware of the superhuman distances that mere humanity has yet to travel, and the smug anthropocentrism that would tend to overlook all other species and indeed all natural forces, in the universe. A recent "prophet for our generation" has announced, "I preach that it is not sinful to be idle; it's being human." Setting "idleness" as an ideal value quite aside, the important and recurrent *non sequitur* in such writings still remains: that no moral criticism of what is *human* can be permitted. For is it not human to err? The Humanist and mystical poet, Thomas Traherne sings:

> No business serious seemed but one, no work
> But one was found; and that did in me lurk.
> D'ye ask me what? It was with clearer eyes
> To see all creatures full of Deities;
> Especially oneself: And to admire
> The satisfaction of all true desire.

Toleration is one consequence of the virtues raised to a supreme position by Humanism. It is a strangely illogical mentality that can, if it has any conception of truth as a desirable goal, regard all toleration as a virtue. Toleration is logically either a lack of moral concern or a lack of certainty on moral values. Uncertainty is admittedly widespread, but the sincerity of a truth-seeker is judged by the efforts he makes to reduce it. One may suspect other motives than truth-seeking in those who worship, prolong and perpetuate uncertainty. It is consistent with the needs of the pseudo-therapists discussed on page 264 above that they believe in reducing "pathological" (forsooth!) concern about uncertainty and unreliability of evidence. For there is great uncertainty, indeed, about whether their own therapy works! The aim of a scientific ethics—

and indeed of any ethics if we accept the underlying religious values —
is to narrow the uncertainty about the ethical helpfulness of all kinds of
behavior. In as much as Beyondism aims by research to find objective
bases for values, that will reduce uncertainty, it follows that it is actually
aiming to reduce "toleration." That is to say, it is working to reduce
obscurity of moral values within any one society and to abolish hypocritical toleration. At the same time it is deliberately asking for true and
meaningful toleration on a grand scale in terms of sympathy for deliberately diverse experiments, each planned with regard for its own consistent,
sincere set of moral values.

There is, of course, even in such a vague socio-intellectual movement
as Humanism (lacking in clear or rigorous doctrine though it is) a recognition by its adherents of the existence both of a more intelligent and sound
and a more debased and "heretical" expression of doctrine. To be sure
that we are not misrepresenting Humanism it is important strongly to represent the former. Russell, who has been accepted widely as an enlightened
protagonist of Humanism, says, "the importance of moral action diminishes as the social system improves" (1955, page 110). In this statement the
liberal-Utopian philosophy of the last nineteenth century is expressed
and confessed (Marx also believed that the Communist society would
ultimately need no government). Such a statement also has the implication,
however, which few liberals would embrace, that the more society
achieves the less sensitive it becomes in its moral concerns! But the
path of evolution is far longer than this straight line to a Utopia. "Good"
societies — good in their time and in a limited frame of reference — will
come and go; but the importance of moral striving will remain. If society
has reached a high level there are still higher targets possible, and the only
constant goal of a good society is to maintain the tension of moral
purpose to become a better society. The implication of much Humanistic
thought is thus actually that mankind has already "arrived." With a little
goodwill and polishing here and there it is assumed that it will have
reached its Utopia.

Naturally, Beyondism and Humanism, as value systems, have, on the
other hand, not only antitheses, but values in common. They share an
appeal for greater altruism, and a desire for deeper understanding of
individuality, and both contrast with the dogmatic position of revealed
religion, in that they expressly call for a reasoned approach. In other
respects, e.g., in their acceptance of new facts, in their attitude concerning
the fallibility of intuitions in the intelligentsia; in their degree of admiration for man as he now is; and in their sophistication about the psycholog-

ical peril of the pleasure principle — they are poles apart. In the secular, eclectic moralities of this century, which are now probably in their heyday, there is no understanding of what we have called the genetic-cultural adaptation gap; and still less of the notion that culture itself has yet to spiral out of sight beyond the values of the contemporary literature of Humanism. And, of course, there is no entertainment whatever of so revolting an idea as the possible viability of a concept of "original sin." But the greatest difference resides in Humanism deriving morality from the felt inner needs of man, and Beyondism from the realities of man's position in the universe.

7.6   THE DIFFERENTIATION OF BEYONDISM FROM COMMUNISTIC AND CAPITALISTIC VALUES

The issues in this section are not exclusively those of Communist, socialist and capitalist "systems," but are inherent in all government control of economics. For when it was said "We are all socialists nowadays" as if it were a modern trend, the statement overlooked that we have been socialists since Sumer or since the first tribal sharing of well being or adversity. However, the economic aspects of social organization become increasingly important, since, with the control of war, and the extension of law and order, the population and cultural selections which operated on biological grounds operate increasingly through economics. Economics is a persisting expression of the reality principle, to which both individuals and nations still have to adjust.

Economic law is thus a branch of natural law; but, as asserted earlier, economics is still definitely an immature branch of natural science, the predictive and insightful development of which will come only when it is recognized as the "exchange behavior" branch of general social psychology. However, the fact that economic mechanisms have been important in shaping culture, and are closely bound to cultural and even spiritual values well outside what is usually thought of as economics has been recognized by imaginative thinkers down the ages: Aristotle, Fei-Tzu, More, Bacon, Adam Smith, Mill and Marx. For it deals with what are essentially energy exchanges in individual and group psycho-dynamics and is thus at the heart of those problems of group viability with which morality is concerned.

The new issues to be examined concern primarily (a) the motivating effects of economics as they affect the values of the individual and the cultural vigor of the group, and (b) the genetic causes and effects of economic behavior. Our treatment of the latter will be more extended, but only for the reason that it is completely neglected by economic theory and practical economic politics. The behaviors which bear on these issues are those of wage-determination, the use of capital, taxation, inheritance of property, unemployment and medical insurance, and the regulation of business enterprise and worker migration. At the risk of some awkwardness in cutting across organic connections we have to deal with general questions of values in economic organization in this chapter and with the actual machinery of social action in the opening of Chapter 8.

Let us consider first the arguments of capitalism on our "right" and Communism on our "left".

The former claims that free-enterprise (a) produces a better natural adaptation of supply to demand; (b) produces stronger motivation, especially in the managerial group on whose resourcefulness the economy disproportionately depends, and (c) ensures greater individual liberty, politically and culturally, because a man with money of his own is more free to disagree with officialdom. In a fully state-controlled economy the dissident is free only to starve.

The latter claims that (a) the government elite knows better than the average man, brainwashed by advertising, what goods and services should be supported (the scientist or cultural leader in England or America has to admit that Communism does not allow prize fighters, movie stars, or banjo-twangers to take such a fantastic bite from the public purse; to limelight at the expense of more serious enterprises; or to take higher status in public valuation); (b) that the motivation of competitive gain may produce *excessive* concern with material production, at the cost of a well-rounded cultural development of the individual, and that (c) the inheritance of private wealth is an injustice, in that it has no relation to the inheriting individual's contribution to society.

To define some boundary conditions for comments on the above let us recognize that (a) by the practical nature of things only a city or national government, and not private enterprise, can build side-walks, organize sewage systems or guide national defense; (b) certain final integrations among industrial enterprises, e.g., the avoidance of monopolistic powers by either management or workers, have to be carried out by and for the group as a whole, by its appointed government; (c) that the old socialist slogan, of Jaurès and others, "To each according to his needs; from each according to his capacities" is nonsense, since "needs" are boundless. It has, indeed, been abandoned in Communist countries, where the wage differential according to demonstrated "capacities" is almost as great as elsewhere. However, there remains in most socialism a relative leaning toward "leveling," up and down; (d) that there is evidence of a strong *innate* need in man to possess what he possesses. This last conclusion can be supported by the experimental dynamic calculus on the one hand, and by the ethological observations, such as those of Ardrey in *The Territorial Imperative*, where the possession of land, wife or husband, and much else, is shown to be an instinctual condition of security. Since we have argued, in the "masochistic reserve" and "cultural pressure" principles above that some persistent frustrations may issue in greater creativity, the fact that man is happier with rights to possession is not a final argument for it. One can see in Russia, for example, some drive springing from the restless unease of non-possession. If a research outcome may be guessed, however, it would be that both personal creativity and social evolution will prove to be aided by the independence and growth of individuality and enterprise which a right to individual possessions confers.

Several of the values to which a disciplined capitalism now moves intuitively, (in contrast to the greater "rationalistic" arguments in Communism) can actually draw good support from psychological or genetic understandings. It was the habit of Wells, and other "rationalists" of his generation to say that the "capitalist *system*" was a misleading expression because it is actually a muddle! With more experience nowadays, of the still greater dislocation which attempted government planning can produce, we realize that the apparent confusion of private enterprise hides a more efficient natural selection, eliminating faulty businesses, than prevails in civil service organizations (which, as Parkinson, 1957, points out, are apt to judge their success by their growth in expending public money). The same picture of "confusion" as in capitalism is created by a thousand persons at a busy city crossing; but they may all be efficiently going their self-directed, group-determined ways.

The cultural value of private enterprise and wage differentials in (a) quickly adjusting production to felt needs, (b) rewarding stronger motivation in the majority of areas, and (c) separating livelihood from areas of political interference, are too well discussed elsewhere (see Friedman, 1968; Hayek, 1945, and others) to need comment here. But, again, it is the genetic implications that have been missed or avoided in most sociological discussions, so a brief setting out of the following implications is desirable:

(a) That saving, in the hope of passing on to one's children, is something that would be group-advantageous if spread to *all* members, rather than cut down. Leaving this task to be done by authority of the government does not amount to the same thing, since it destroys natural selection in favor of the more foresighted, and does not put the initiative in the hands of a *large* group of more prudent individuals. (Furthermore, governments save only as much as the average man will tolerate.)

(b) That, nevertheless, since correlation of genetic qualities, such as intelligence, in families is already getting low (0.25) by the grandchild generation, the indefinite handing on of wealth should not occur. (Note, however, that due to the assortive mating, the drop off in desirable qualities from the most gifted generation of a family must not be assumed to be as steep as in the classical model of random mating.) Indeed, through gifted families marrying gifted families, decline may be so slow that inheritance of manageable property may continue to place it in highly capable hands for about three or four generations (*vide* the Rockefellers, the Fords, and the Krupps). While criticizing the unlimited inheritance principle regarding individuals it is important to recognize that it applies just as strongly to institutions, which have little or no parallel biological inheritance to justify retaining capital. What is more pointless and wasteful than money tied up for centuries in institutions and foundations, e.g., for breeding dray-horses in an age of automobiles, for saying prayers for the long departed, for training children in home weaving, generations or centuries after the appropriateness has gone? In England, according to Darlington (1969, page 517), "The educational system has remained stunted, not by poverty, but by the wealth of its medieval and Catholic foundations." (*See also* Cattell, 1933a.) The Communist and socialist need to be reminded that in giving the property of the people to the party they are doing just the same for a soon-obsolete doctrinaire position, in a gigantic and irreparable fashion. Every family, institution, business and government needs (in some fashion to be devised by social science) to be periodically dispossessed of its capital—at least of an appreciable fraction—to see if the needs of society, and the service capacities of the institution, still justify the endowments.

(c) That there exists in differentials of earnings, if related to differences of magnitude of service to the community, a ready-made eugenic mechanism—assuming family planning properly keeps birth rates proportional to means. Acting across the full range of the community from the well-organized genius to the unemployable, this offers the most humane and flexible of practicable individual-difference natural selections. To abolish or powerfully reduce income differentials *operating in these conditions* is to throw away a vitally important human service of economics[10].

Here, as elsewhere, if we allow reduction of selection differentials within groups it will throw excessive weight on the more risky mechanisms of between-group selection. The democratic objection to government by a Communist or Fascist elite has been well stated by Thomas Huxley: "It is better to go wrong in freedom, than to go right in chains." However, in the economic field, as in others, the only thing that competitively established differences are *unquestionably* better for is natural selection. The freely and foolishly advertisement-directed spending of democracies which we contrasted above with more government-controlled expenditure of the common "surplus" in Communism has at least the value that (with birth control and the absence of eugenically uncontrolled welfare expenditure) it offers powerful within-group selection. As to between-group selection a stupid people controlled by a wise elite would outlive an averagely intelligent democracy, and, assuming the former actually to be only one-tenth elite, this would be an overall genetic loss.

Reverting from genetic to cultural effects, which can, of course, be different in direction, one may surely note that the "removal of poverty" by forcible redistribution, by taxation—if the gross national product remains constant—is actually a spreading of poverty too thinly to be much noticed. Also one notes that the provision of full employment commonly reduces the output per worker because it involves employing the less efficient (in intelligence, regularity of appearance on the job, mistake-proneness, etc.). These losses are acceptable in terms of maintaining community morale and the individual's self-respect; but they are nevertheless chalked up somewhere in the bill of group survival, and at times of natural or military catastrophe, might reach the level of being decisive for the end of the culture.

7.7 THE RELATION TO ENTRENCHED BUT IMPLICIT VALUES IN SOCIAL ECONOMICS

The consensus of economists—averaging the political biases of the world—seems to be that a free economy, with adjustment occurring continually through the market, is the soundest condition for the main body of society. With this there must be such government or other tactful touches on the helm, and such support of not-directly-earning activities (research, education, maintenance of morality) as can prove remunerative only for the community as a whole. In a freely competitive system supplying what people need, by competing corporations, the managerial class is constantly under selection for enterprise and inventiveness; the worker is motivated to acquire new skills as they become wanted and to migrate where the best wages are to be found, and the educational system is moved to supply—at least in technical fields—the kinds of education that the community needs and for which students will pay fees. Comparative studies of real standards of living among various countries suggest that where such free invention, migration, and adaptive education exist, there is less misery and want than where they do not.

The crucial question of principle which soon arises, however, concerns the correctness of extending the free competition in supply and demand from goods to people. The writings of most economists get quite hazy here and one sees their personal political or temperamental prejudices breaking through. Oddly enough, Karl Marx and the Catholic Church (though not the Protestant Ethic as interpreted by Weber) come together here (as in some recent papal encyclicals) in asserting that the individual must be placed above the market. But one must insist that the premises and the consequences of thus intruding with values from outside the system have not been clearly stated or understood. In both Marx and the Catholic encyclicals the principle is born of an *ad hoc* intuition. Yet as we have pointed out above, the expressions of economic laws offer one of the most realistic contacts with general biological and cosmic energy laws that we have. If the demand for a certain type of man in a certain type of culture is low—say for a low-browed, massively muscular type in an intellectually demanding, electric push-button society—this will show in the wage distribution. If this wage falls very low, and is only restored to a good family sustenance level by what we may call the unmitigated welfare principle, then the adjustive selection trend in society, both educationally and genetically, is destroyed. This complex matter of supply and demand of persons is looked at more closely in Chapter 8; here we are concerned only with the value judgment.

It was part of the lucidity and integrity of Ricardo's basic approach to economics that he recognized that a worker is, as far as economics are concerned, essentially another commodity[11], and that a producer's contract with his help required that "like all other contracts, wages should be left to the fair and free competition of the market." Further, he supposed that just as the supply of goods would decline adjustively when the selling price ceased to be rewarding, so the supply of persons with certain skills (or lack of skills) would decline when the need, and the willingness to provide wages for that kind of work decline. Incidentally, Ricardo is a shock to welfare state social workers because he recognized that "privations" might reduce the supply of, for example, an excess of unskilled workers. That reduction has in the past often occurred by famine, but by foresight and intelligence it could occur by family planning directed by economic incentives, migration, alert shifts in adult and general education, and other adjustments.

If we begin at the basis of economics with the principle that both men and goods should be primarily exposed to the same law, and each only secondarily brought under its specific modifying laws, the picture becomes more consistent and the directions of moral action more clear. For example, we may surmise that if society permits the production and education of the average man to cost more than his value to the given society, its resulting economic bankruptcy is likely to be part of a total group life bankruptcy. If a viable society is the evolutionary aim, it is morally desirable that every man should repay or more than repay the cost of bringing him to adult effectiveness. He might fall short of this in many ways, e.g., by dying too early, by being lazy and parasitic, or by being born with defects which prevent his ever becoming averagely effective.

Now it may well turn out that secondary principles to this primary one — that a man should be paid what he can in some real sense earn — are quite compatible with the goal of survival of a group. For example, trade unions forcing up the basic wage, or welfare for the unemployed, may have survival value, e.g., in tiding a society over what could otherwise become costly disturbances. But these advantages have not been demonstrated by social science, and the *real* reason for these customs is that Christian and other ethical values, which move us for the plight of the unfortunate, favor these defenses against the strict application of the law of supply and demand.

For example, if a certain educational system is producing too many qualified in area A and too few in B, the earnings for A will fall, and the fashionableness of paying for schooling in A will decline while B will increase. (This could be exemplified precisely in India fifty years ago when there were too many lawyers and too few engineers. Young lawyers complained bitterly of absence of jobs and requested government action to support them.) The primary economic principle will automatically adjust the situation, but the special sub-principle might be that those who, "through no fault of their own" (except following the crowd and lacking foresight) are sufferers might, from humanitarian principles, be given financial help. Use of a "dislocating" sub-principle, however, always requires further adjustments, unless efforts are made to keep it temporary and "contained." For example, a leading economist points out "there is a surplus of unskilled youths seeking jobs because the government makes it illegal to pay less than the legal minimum wage."

The view that production of types of men should be subject to the same naturally acting economic laws as any other production runs counter to the narcism and ethnocentrism now hopelessly entangled in many "ethical" systems. In as much as a large factor in economic laws is simply a projection of natural laws of survival present in the general real conditions of the universe the opposition to the economic principle is pointless or evil. The worker should be willing to be educated to the types of job which society needs.

The objection to allowing the economic principles of various economists to play a major role in social planning is, first, that, as is typical of what we called the "treacherous alloy" in social science generally in Chapter 2, they bristle with values—from Adam Smith, through Ricardo, Bentham, Mill, Weber and Marx—that are insufficiently distinguished and labelled as the non-scientific importations they actually are. Secondly, as also stated above, economics is not a science, but only the fragment of a behavioral science yet to come. It is still so little of a science that it is possible today for a distinguished economist such as Colin Clark of Oxford to find the work of another economist, Galbraith of Harvard, so remote from his own reasoned conceptions of economics that he is compelled to call it "garbage"[12]. Only when economic behavior is geared into general psychology as a special branch designated "exchange and earning behavior" are its laws likely to reach those foundations in biology, personality psychology, and learning laws which can give us confidence in what its applications will do.

Meanwhile, economists are invited by politicians to control unemployment, inflation, business cycles, and national currency values, by an array of levers such as bank rates, price controls, printing money, wage adjustments, welfare expenditure levels, with only a partial understanding of how the results follow and no understanding at all of what the long term moral value of the changes may be. For example, an unquestioned "desirable" in economic terms has always been to keep the economic system in brisk action. Here politicians have followed, for example, Keynesian goals of keeping up momentum, regardless of inflation or the possibilities of many unknown effects on population size and quality in relation to resources, pollution, etc. Others have taken it as a matter of course that unemployment should be kept at a minimum, or that interest on capital should be kept at a maximum, by some economic device which has been found empirically to produce this effect, regardless of secondary effects.

If physiologists were to put drugs on the market with equally poor knowledge of their secondary "side effects," or long term consequences to health there would be a public outcry. Moreover, it is hard to find any more insight or wisdom at present guiding economic reform and manipulation than is expressed in a cautious "middle of the road" philosophy which hovers between central government economic control, at its extreme in Marxism, and the maintenance of the stimulus and vigor of a free market, by the principles of Marshall and Adam Smith.

From a Beyondist standpoint such a goal as a maximally functioning economy in Keynesian or other terms cannot be taken as a final "good." But the argument for re-valuing economic desirables in terms of further, more remote goals is quite different from merely objecting to economic laws and their consequences when the shoe happens to pinch. The pinch may be, by broader principles, desirable, but in any case signs of economic dislocation are always important, for with the added depth of an evolutionary perspective, economic laws must be respected as indicators of the soundness of general adjustment and progress. Intelligent acceptance of certain socio-economic trends, and novel adjustments naturally obeying economic laws has historically been conspicuously absent in most communities. The unforeseen rise of the bourgeois, i.e., as part of a more flourishing commerce, upset the ideals and living styles of the aristocracy. The bourgeois ideal—or at least the notion of *laissez-faire* in industrial and commercial competition—was in turn attacked from such diverse directions as Marx's demand for a regulated economy under the dictatorship of the proletariat, and Ruskin's (and Adam Müller's) indignant demands that a commercialized culture recognize the primacy of the goals of the well-balanced and aesthetic life. Naturally there are as many variants of this immediate discontent with the outcome of economic laws as there are value systems. The hippie, the Hindu holy man, the artist, some researchers in pure science, and all such as teachers and preachers who under-earn, are all bound to show some degree of discontent with economics, but there is no common lesson in or single remedy for their diverse protests.

It should be possible to infer from social scientific research a set of public economic regulations and mechanisms that would be fully consistent with evolutionary morality, but no existing system comes close to it, nor would such an economic system actually encompass *all* the values of the Beyondist society. Similarly the full ideals of other systems, democracy, private individual enterprise, Fascism and even Communism are not sufficiently definable in economic terms alone. In Russia or China, for example, the Marxist blueprints, written in purely economic terms have not sufficed, for example, to eliminate the absolutism of Czarist days or the family paternalism of China, respectively, or, indeed, massive value differences in the two cultures. Haldane reacted to what he felt was an excessive emphasis on such economic characterization by saying "I cannot accept the American and the Communist ideals because they are too exclusively economic." Economics cannot encompass the translation of evolutionary morality, for one thing because most of its goals are short term ones. A trivial but rather striking illustration of this disregard of long term social outcomes is found in the way the pressures of agricultural demand were allowed to starve the craft of ocean-going canoe building among Polynesian colonists, with the result that they were marooned, to stay with whatever fate each particular island could offer. They lost the flexibility in survival we have rated high in evolution of cultures. Intensity of attention to *current* economic demands, as practiced by political-economists, could produce similar catastrophes in our own culture.

7.8   SUMMARY

(1) To clarify and enrich the meaning of Beyondism, so far presented only in logical abstractness, the present chapter brings out some illustrations of departures from the familiar values ordained by currently predominant and traditional moral systems. Its bearings are here given with respect to revealed religions such as Christianity, Humanism, Existentialism, Communism, and the implicit secular values in the economic customs regulating society. (This chapter restricts to more general values since the next chapter proceeds to actual social *measures and mechanisms*.) Nevertheless, because dependable inferences for social progress require that the already explicit *basic* principles be brought into conjunction with *particular facts*, and because scientific laws have not yet been established clearly connecting principles and facts in these areas through social science research, the conclusions and illustrations here must be considered quite tentative. Even so it is illuminating to see how strikingly different the new conclusions regarding social values frequently are from the stereotyped answers of either "conservative" revealed religion or "radical" social writings.

(2) Explicit differences from the principal revealed, universalist religions have been constantly brought out in passing, but as we definitely focus on them here we note four main departures: (i) that Beyondism plans for a constant revision of values by scientific research; (ii) that the

universalistic religions are imperialistic in their universalism, seeking an homogeneous and identical set of cultural-moral values, whereas Beyond-ism calls for a distinct ethical value system "federated" in universalism, in which the only common tenet is a basic assertion of brotherhood in supporting whatever persistent competition is necessary to evolution (as an agreement to disagree, while maintaining mutual respect and goodwill); (iii) that there is a sharp difference in attitude to the treatment of poverty and misfortune. For Beyondism proposes a compassion so directed as to offer elimination of the problem, not a perverted compassion which ensures its continuation; (iv) that Beyondism must decline to gain acceptance at the cost of making largely illusory emotional "deals."

(3) In the third respect an important difference is that whereas univer-salistic religions laud love as a motive, and consider that universal love of man by man can do no wrong, Beyondism (while agreeing that love as *agape* [compassion] is normally in short supply in human relations, and a rarer commodity than love as *eros* [passion]) recognizes that *any* emotion can err, since our past genetic emotional development is inadequate to modern needs. One of the defects and dangers of love (as agape) is that it can be more concerned with human freedom from stress than human achievement and adventure in evolution, and that its aim can slide with-out any definable stopping point from concern with man's happiness to concern with his pleasure and indulgence.

(4) The incompatibility of Beyondism with revealed religions (which latter many modern intellectuals and Humanists consider obsolete) turns out, as regards conclusions (rather than ways of reaching them) to be less uncompromising than with a mass of modern writing which some "intel-lectuals" embrace in "Humanism," "Existentialism," and some ill thought out varieties of "liberalism." Humanism now has the quality of a social value movement (no longer restricted to the academic meaning of Humanistic studies) which, like other secular religions, claims to contrast itself as "free thinking" and "rational" with the "dogmatic" basis of revealed religions. Actually, however, it is not one iota less intuitive and *a priori* in the source of its values. Indeed many of the values in Human-ism and Existentialism are a digest of fragments eclectically and un-critically absorbed from various religions over two thousand (and particularly the last four hundred) years of Humanist thought. In some cases their emotional inheritance from dogmatic religions has favored more the comfortable illusions than the austere truths.

(5) Lacking any doctrinal precision and dogma these "secular" re-ligions are open to steady attrition of values. Also they continually tend to

discard restraints that are onerous or exacting, by the intellectual rationalizations based on the pleasure principle and narcism as studied in the last chapter. Especially this shows itself in an unwillingness to entertain concepts of guilt or sin, and to retain the functions of retributive justice and the concept of contrition that psychologically belong therewith.

(6) These are some among many consequences of the basic difference of Beyondism from Humanism, Existentialism and secular moralities generally. The basic difference is that the former begins with the inexorable standards inferred from the nature of the outer world, in relation to survival and progress, whereas the latter start with human felt needs and intuitions. This does not mean that Beyondism ignores the inner depths of feelings. On the contrary it recognizes that the complexity of the external adjustment requires an educated complexity of inner emotional life, particularly in sharing the tragic sense of our adjustive tasks in human destiny as we see it. Thus it appreciates as much as the humanities do the need for moral values to be enriched and explained in art and literature, but questions only the nature of the values. Part of the basic difference indicated by this rejection by Beyondism of naive anthropocentrism appears in the Humanistic ascription to man of godlike capacities, certainties and virtues. Here Beyondism is closer to traditional religions in recognizing his meager status, while yet it exhorts him to the Promethean adventure.

(7) In relation to Communism, and some socialist philosophies the progressive nature of the Beyondism position is apparent in (a) its recognition that any "Utopian" pattern is born static and dead; (b) an emphasis on the indispensableness of individuality of thinking in producing progress, which is unlikely to reach its full independence without being sustained by economic freedom and social independence; (c) perceiving both differences of earning and saving to have, over and above their social motivational value, a function in maintaining genetic progress within the group ("To each according to his needs" is an inadequate philosophy); (d) recognizing that any lack of within-group selection is likely to throw the burden on between-group selection which is less efficient, and (e) regarding the removal of poverty by "spreading it thin" as merely sweeping the real problem under the mat—the real problem being a partly genetic, partly motivational inadequacy in a section of the population which unless dealt with will debilitate the total group and weaken its chances of survival.

In relation to capitalism the problems are (a) the inadequacy of adjustment of rewards to contribution in activities lacking marketable value (basic research, moral enquiry, education); (b) a possibly excessive persistence of inheritance of property in relation to biological inheritance; (c) a poor direction of reward by the free market, e.g., in the encouragement of trivial amusements and material luxuries, relative to wiser production which some direction by a moral, foresighted elite government or church might foster and (d) probably a greater demand on morale in that it has to accommodate to appreciable individual difference in reward without destructive levels of envy being generated.

(8) In considering the relation of the values in current social economic measures to Beyondism one has to evaluate separately (a) the *correctness* of the present science (alas, still infantile) of economic laws in being able to reach assigned goals, and (b) the *implicit* values in the manipulations which both governments and socio-economic doctrinaire reformers now introduce. Any confident dependence on either the science or the entangled values would at present be ill-placed, both because economics as so far developed bears the defects of incomplete relation to social psychology, and because no serious attempt to state its implicit values has ever been made.

(9) Probably the maintenance of a fundamentally free and competitive economy, with the added government controls necessary to integration, and acceptance of the Keynesian aim of a steady, full draft on the furnace of the economy will prove to be basically consistent with Beyondism. But the latter also requires in principle that the genetic and educational supply of people should be made responsive to the market, and this involves clashes with present values. As much adaptation to economic laws as Beyondism advocates does not mean acceptance of economic requirements as final arbiters, but only the recognition that economic laws belong to, and reflect in human society, the realities of man's relations to his environment. For economic manipulations should have more than economic goals, and in fact current economic practices are ignoring the necessary subordination of immediate economic "desirables" to the ultimate evolutionary goals involved in group survival, genetic selection, scientific discovery, and cultural experiment.

7.9 NOTES FOR CHAPTER 7

[1] Most obviously this difference shows in the arguments about "foreign aid" as defined in Chapter 5. This practice is ethically important enough to deserve a book length factually-economic and ethically-critical evaluation. As agreed in Chapter 5, some of it may be justified on an evolutionary ethics basis as persuasion and education toward better values, e.g., in some far Eastern countries a movement from wholesale government corruption toward the values achieved in Western culture, or as an investment in alliances for security.

But much is motivated—at least in the mind of the contributing citizen—by the false value that relieving Type B (not Type A, where it is justified) poverty is meritorious. Thus, for example, when the U.S. terminated its economic aid to booming Taiwan in 1965, it had paid $1,500,000,000 (*National Geographic*, January 1969, page 10) in "welfare imperialism". A lot of healthy, enterprising, and idealistic Americans could have been born here on such an endowment.

[2] Whatever we may think about the soundness of the ethical values themselves we cannot avoid the conclusion that the spread of homogenistic universalism has been, historically, the result of an imperialistic attempt to bring the whole world under the dominion of the given creed. This imperialism is one more symptom supporting the diagnosis we have made earlier that, psychologically, these religions are really within-group moralities (patriotisms) naively blown up to world size, and failing to develop the essential perception which Beyondism has achieved that "one world" must be avoided.

As within-group developments of idealism toward super-personal goals, directed to the survival of the group, national religions (which historically we see in Judaism, the Japanese development of Shintoism, initially in Arabic Mohammedanism and even, perhaps in Christianity (*see below*) begin with the same emotional adjustment quality as universal religions. The emotional identity of a religion with the culture and life of a particular group gets overlooked mainly because such religions have quickly passed from being the truth for that group to acquiring a missionary zeal in claiming to be the truth for everyone. It is an easy intellectual and emotional mistake to suppose that what is super-personal, transcending the individual, ought to be universal. In any case the result is that by a cultural imperialism allied to, and no more and no less justified than, ordinary imperialism, traditional religions have gone out on missionary attempts to convert. In the case of Islam the identification was frank: the sword and the Koran marched side by side. It was scarcely less frank in Christianity, when it marched on the Teutons, or with the Conquistadors, on the natives of South and Central America. Judaism was quick to claim that its God was the only god, and Old Testament history records compulsory circumcisions in the wake of conquest. The popular following of Christ as a possible savior of the Jewish *nation* is judged by some modern historians (Robert Graves' *King Jesus* (1946) is a fine literary treatment) to have been at least as large as that which followed him for what we now call Christianity. (And after his death a separate Jewish congregation sought to adopt his brother as a continuing leader, so widespread was the view that this Messiah was bringing a purely Jewish salvation.)

This historically characteristic spread by successful proselytizing beyond the boundaries of their cultural group or nation was not achieved, by the universalistic religions, without acquiring some new qualities better fitting them for their new role, and the shedding of features (as circumcision was dropped from Christianity following St. Paul) standing as parochialisms in the way of wider conversion. What leads to the sharpest issue between them and Beyondism, however, is their proclamation of universal love and brotherhood as the panacea for the world's ills. The ways in which this can be interpreted vary from conceptions really quite near to Beyondism to others so remote from it, that precise definition here becomes of crucial importance for differentiation. Evolution sees the ideals of kindliness, of treating one's neighbor as oneself, of self-sacrifice for the individual in order that the group may live, develop as a natural and inevitable growth of within-group morality when one group is in competition with another. Such a morality is functional and self-sustaining because it is reinforced, as learning theory would require, by sharing the group success, and, by the dissolution and elimination of groups whose members fail to pursue that morality.

By the natural rewards of success, groups which reach a high level in this morality reach a prominence, which encourages their members to proselytize, while their good morale tends to put them in a position of acquiring the economic and energy resources to do so. Then a universalistic religion is born, which propagates the teaching that love and mutual aid and common belief should hold among all men.

[3] Poverty, as every social scientist will recognize, is a multiply-determined phenomenon, not to be explained or remedied by any one consideration. In other words, no two people dependent on charity or social welfare will be in that position from the same combination of causes. Unfortunately, this fact allows the rhetorical politician, working on a discontented audience suitably unrealistic and non-statistical in its thinking, to refute (by a single dramatic instance, such as is so powerful with the uneducated and less intelligent voter) every true cause which the statistician knows to be operative. If poor education is alleged, he will point out an instance of a highly educated genius who is poor; if emotional inability to work steadily, he will remember a man who has worked steadily all his life and is "comparatively poor," and so on.

Probably the most constant difficulty in getting enlightened, discriminating social action is, however, the confusion through "religious" values of Type B with Type C. The latter belongs with the monastic pledge of poverty, chastity and obedience. (This poverty may actually beget wealth in the institution and the community as Henry VIII realized!) It is in the disciplines of Tolstoyian simplicity, in the more truly poetic and philosophic of the modern hippies, and hidden among the vagabonds of all eras, that Type C individuals confer protective coloration on Type B individuals and thus create social problems of confusion.

[4] One is reminded of the sharp observation of Browning's biographer (Ward, 1967, page 81) regarding Browning's early critics "It is not a habit of the intelligentsia to stand out against the fashions of their own age, bold though they may be against fashions well and truly buried . . . ."

[5] It may seem a partial contradiction to this emphasis on the literary rather than scientific roots of Humanism that one of the best known Humanists is Julian Huxley, a scientist. However, it is very relevant, if personal, to point out that Huxley's brother is Aldous Huxley, and that they represent the urbane and scholarly tradition of uniting scientific and literary values. Aesthetically attractive though the present writer has always found this combination, he would argue that this late Victorian tradition implies, as in the atmosphere of Oxford, a gentlemanly subordination of the austerities (and the starker imperatives) of science to the values already long cultivated or brewed in the humanities. It is hard, especially in some English and European intellectual circles, to break from this urbane tradition (the alternative to this tradition being wrongly perceived to be the brutality of Stalin or the crassness of some American business values). But the human culture derived from science is *not* this Humanism, though it is equally concerned with what is intellectual, scholarly, of value to humanity and, above all, possessed of intellectual integrity.

[6] Again, in considering the stress levels under which societies live, it is important to distinguish between what is here called *severity of deprivation* (or primary drive total frustration) and *cultural pressure* (or frustration of *directness* of satisfaction with demand for long-circuiting). The former has steadily decreased since medieval times, but with over-population may soon cease to do so. The latter has mounted rapidly in the last two or three centuries and particularly steeply in the last generation. It is the *severity-vs-ease* dimension with which we are here concerned.

[7] This is not the same as saying that personal level of social *success* correlates highly

with exactingness of demands on others. The evidence points to superego demands on the self being greater in the lower than the upper middle class, whose advantages and gifts permit them to "rest on their oars" in the social race. This latter status is accompanied, as Horn points out (Horn and Knott, 1970), by a tendency to be more permissive and less exacting with both others and the self. We are speaking here of *demands* on self and on others being correlated, not "achievement" and demands on others.

[8] Both neurotic and schizophrenic characteristics and incidences are negatively correlated with social status. Incidentally, because of the correlation of schizophrenia with lower social status some psychiatrists in Moscow once gravely informed me that it must be a disease of malnutrition! I naturally had to point out that a common feature of borderline schizophrenia is inability to hold a job, which does not help "social status" or nutrition. The appreciable hereditary element in schizophrenia is more fully documented by every year of research (Fuller and Thompson, 1960, page 270; Rosenthal, 1970) and one must conclude that whatever the nature of this weakness is, it is *one* of the behavioral genetic diatheses that become causally associated with *relatively* poor capacity to earn a living.

It is not appropriate here to enter into the technicalities of the assignment of relative contributions of hereditary and environmental conditions to the variability in a given trait, which can be read elsewhere (Vandenberg, 1965). But progress is constantly being made in unravelling the relative contributions in different circumstances, and for most of us the surprising finding resides in heredity being important in many differentials where "common sense" would not have expected it.

[9] Professor Williams concludes: "The excommunication of heredity from social studies — [i.e. from] our attempts to understand and solve human problems — is a 'disaster area in our thinking.'" He contrasts, for example, in his own physiological research field, the scientific attention to individual differences and the value of medicine's recognition, from Hippocrates, that "Different sorts of people have different maladies." To which one might add Osler's: "It is as important to know what kind of man has the disease, as to know what kind of disease the man has." (I paraphrase for brevity.)

[10] Actually this argument for economic differentials is for a two-way effect. Not only could economic regulation aid eugenics but eugenics could aid economic advance. At a purely statistical, descriptive level there is clear evidence of substantial correlation of earning capacity of parent and offspring. This may surprise some economists, but not any psychologist who realizes the extent of inheritance of such qualities as intelligence and emotional stability, which affect earning capacity. The transmission of education and opportunity also contributes to the correlation, but it does not seem to be the whole explanation of the correlation. It would be against the laws of biological inheritance — and still more against what we know about the luck and haphazardness which enter into everyday business success — to expect a *high* prediction on this basis of poverty in the offspring from parental poverty. But for whole *groups* of people the statistical dependability rises. Consequently, the argument is that by increasing the birth rate of the middle class, rather than of the dependent, society would raise the *real* earning level, and the real wealth and standard of living within a generation or two, while the converse is likely to bring the whole of society down to a poorer level. Thus to the Beyondist, the socialist or Communist solution of "buying off" poverty like a blackmailer, within each generation, is by these standards unethical, for it favors the greater reproduction and increase of types sooner or later bound to re-create relative poverty.

The eugenic approach to the poverty problem is at last beginning to come under fairly searching scrutiny by scientists and deserves some thorough research by the government departments concerned. Whatever the "upbuilding" capacity of a differential birth rate, set to be largely determined by earning differentials may turn out to be over a few generations, there seems little doubt from certain periods of history, e.g., the decline of Rome, that a *reversal* of the desirable differential can produce noticeable results in half a dozen generations. Probably never in history has there been a period in which dysgenic trends could take effect so rapidly as in the welfare state — if morale should fail. Two or three generations of disregard for genetic quality might lead to such a breakdown of the economic and cultural level of society as would be well nigh irremediable.

[11] One must remember, however, that a commodity in economics is very broadly and abstractly defined as "whatever is produced by human labor."

[12] *Chicago Tribune*, June 10, 1970.

CHAPTER 8

The Impact of Evolutionary Values on Current Socio-Political Practices

8.1 THE RECONSTRUCTION NEEDED FOR A SCIENTIFICALLY RATIONAL POLITICS

As we turn from values now to actual socio-political action it has to be confessed that the elements of uncertainty in the last chapter due to having to lean on a social science which does not yet exist are still further magnified. When the essence of the present chapter is ultimately developed in some "Beyondist Guide to Social Construction," of the future, containing actual blueprints for action, it will need to rest firmly on a still larger volume, "An Encyclopedia of the Quantitative Social Sciences." Although there may thus be doubts at present as to what degree of this or that political measure is desirable, it should nevertheless be clear by now that the whole conception of what are important measures is very different indeed from current discussion. The entrenched warfare of reactionaries and radicals today is surely concerned with the wrong aspects of the wrong questions. In essence they are disputing points in the construction of a stagecoach when the social scientists would like to talk about the calculations needed regarding jets. The clamor, meanwhile, of the young for "responsibility" and "relevance" in the social sciences, though partly misguided, in failing to recognize that the greatest advances came from disinterested *basic* research, at least does well to serve notice to the politician and the scientist that they have now to get together — and on new issues and new solutions that the scientist proposes.

Unless this is done our age faces a serious danger of general breakdown of social morale. When issues become too confused, and the political machinery for handling them is remote from what is demanded, societies typically suffer a loss of morale. This has been evident before in old and decadent cultures, burdened with inconsistencies that no one can solve. While Western culture is probably not as badly off as many, the disturbances of the last decade show that the system is baffling to the idealism and energies of youth, and increasingly burdened with parasitic growths.

In science we would not ask a vote from a large number of citizens as to whether a particular virus strain is the best for testing the oncogene theory of cancer, nor whether simple structure or confactor rotation better define the concept validity of a measure of anxiety, though both cancer and anxiety are public concerns. It is recognized that for a meaningful decision we need to be sure that (a) the judges are equally adequately informed and educated on the facts and principles needed, and (b) the consequences of the decision in terms of their lives are understood. The conditions in contemporary political decisions are very far indeed from these, and neither a successful business organization nor scientific research could safely be left to the institutional machinery by which decisions are made affecting the life of a nation. The fact that this machinery is probably better than it used to be cannot disguise from an intelligent citizen that it is so obsolete, and remote from what a genius — or even an average academy member — in social science could devise that it is scarcely worth taking seriously. By the present political rules, half understood issues are submitted in unnecessarily confused combinations of "package deals" to judges educationally unequipped to deal with them and subject to strong biases. Meanwhile none of the participants in the decisions are provided with a scientifically-based, clear framework of value dimensions by means of which the real disagreements of the parties concerned could at least be focused in objective, quantitative terms.

With such conditions the miracle is that three thousand million people rise in the morning, keep themselves tolerably washed, dressed and fed, work hard enough to provide a minimal education for the next generation and get themselves to bed again without worse disasters in terms of war, famine, crime, and plague than we currently experience. But the present day politician walks a tightrope, and the chance of a complete loss of balance by some surge of collective human passion and irrationality is appreciable despite even the modern "organizational revolution" (Boulding, 1953). It would be an unrealistic social scientist who would venture to predict at any time that some form of massive catastrophe, or, worse, massive stagnation and regression, is not an immediate possibility. This is no condemnation of the politician, who, considering the irrationality and defective information of those with whom he has to deal, performs marvels of intuitive and artistic skill. As Bismarck said, "Politics is the art of the possible." Chaos and anarchy are the natural state of man, and any form of order and constructive movement in the forms of society to which we have so far stumbled is a matter for gratitude. Thomson's (1964) documented study of the process of political decisions in the twelve days before World War I is a sufficient example to indite the inadequacy of traditional political machinery.

The science that might help us transcend this rule of thumb — political science — is, as I have argued with technical appraisal elsewhere — at the moment, the most backward of the social sciences, pursued with little method than that of historical, non-quantitative observation and pursued more by tendentious "rights" and "lefts" than by scientists. Only in the last decade, in the fine mathematical enquiries of such as Alker (Alker and Russett, 1965), Rummel (1963), Deutsch (1965), Wrigley (1963), Merritt (1970) and similar pioneers, has it begun to seek empirical laws and more subtle and precise concepts than those with which purely literary trained journalists seem content to operate [1].

For the citizen who is convinced that it is time to move ahead definitely on Beyondist principles there are two questions to be answered: (1) On which of the presently-known, intuitively and historically reached systems of social government can we best proceed for the next two or three generations while social science is building a more efficient system? and (2) what is research already beginning to indicate as the direction of political evolution? As to the first, though it is a desirable condition for evolution that men should tend to be loyal to the diverse systems they have sponsored, yet it must surely be considered an open question *scientifically* at the present moment as to which of past and present alternative political forms is best. The earlier Greek communities, and many others since, have found government by the "best" people (aristocracy, from aristos = good) best. Dictatorships have been able to bring about rapid changes of culture (e.g., Japan in 1945–1950), hobbled by fewer checks than those which democracies maintain. Oligarchies, such as that of Venice in its high period of economic and cultural splendor, have proved effective, far-seeing and stable [2]. The word "elite" is a bad one in this generation, yet most of the benefits which distinguish our way of life from that of four hundred years ago are due to a highly selected elite of scientists, engineers and medical researchers who chose to impose upon themselves a disciplined and creative way of life totally different from that accepted in the life pattern of the majority.

However, in order not to be misled by a word, let us note that the "great contributors" we have just recognized as a true elite were neither a social elite nor an elite in political power. Whether in the long run there can be success when an elite governs in freedom from democratic controls is quite a different issue. It is, however, at least clear that an openly recruited elite, as a governing group, *can* give long term stability of standards, as in the government of universities and of the Catholic Church (and the twelve generations of the Grand Council of Venice). The question of how well an elite works is not raised here as a merely academic and historical issue. For with the increasing complexity of a world of advanced technology, and, especially, the growing understanding of society itself in complex scientific terms, the day is approaching when it will be imperative in some way to accommodate the will of the majority to an elite of scientific advisors in government, as discussed in more detail in Chapter 9 below.

Democracy, which, on the experience of Greek city-states, was believed by Plato to degenerate naturally into a tyranny, actually covers a multitude of sub-species, some of which may *not* so degenerate. Although it seems to appear wherever strong, individualistic, independent types get together, as in the Greek cities, and in the first parliament, set up by the Vikings, as Scandinavian settlers in Iceland, a thorough social psychological investigation would doubtless show the style of democracy to vary along several distinct dimensions. People who have the conviction that it is primarily an expression of Christian belief in the worth of the individual may find themselves thinking it is primarily a form of the Christian view of charity, and thus finish up by believing that Democracy is a direct social expression of Christianity. Parenthetically, with the inherent uncertainties of interpretation, it is surprising what Christianity is made to cover politically. For example, the late Dean of Canterbury became a Communist, asserting that Communism is not a Christian heresy, as most theologians might style it, but Christianity itself. It is almost certainly more correct to give political democracy an origin outside Christianity — in the Greek city states and the independent spirit engendered in early North European and North American exploration and expansion. Christianity then attempted to capture it to Christian ideology, but it has retained a spirit of its own, e.g., in the Protestant ethic, private enterprise, faith in the independent individual, and local political freedom. It is obviously not an unvarying concept.

Open-minded though one must be to further developments, democracy has given appreciable pragmatic proof of its basic suitability for a progressive social system (de Tocqueville, 1947), and indeed of the kind required by Beyondism. It has proved a more favorable climate for scientific thought than most political systems, and, above all, it offers a machinery for continuous evolution. *It is one of the few systems in which violent revolution, with all its costs and reverberating distortions, is unnecessary for progress.*

The directions in which one suspects that an evolutionary moral value system would want to develop democracy, for the sake of more rational and efficient political procedures, less confusing and time consuming, would be (a) *in clarifying the dimensions of voting values involved,* and (b) *in separating decisions on felt wants or goals, from decisions on means or machinery.*

If we consider a fairly typical array of everyday disputed issues on which political decisions have to be made — e.g., the fraction of the national income needing to be devoted to scientific research, especially in social science and moral enquiry, the desirable magnitude of wage differentials; the operation of medical, unemployment and educational allowances; the freedom of advertisement, especially in regard to socially undesirable products and socially vulnerable groups (such as the young); the taxation of religious and business corporations and monopolies; the freedom or manipulation of economic incentives to operate in choice of work and migration; the adjustment of income and other taxation to give comparative reward to different kinds of social behavior, and the relative role of government and private capital in deciding directions of production — it is at once evident that the average voter is being asked to make decisions both on changes which affect his *wants* and on technical issues of *means to ends.* The democratic regard for the individual requires that due weight be given to the fact that many individuals would like, say, more medical attention. But as to the measures by which this can be brought about, and as to the rate at which increased medical attention will translate into, say, a requirement for reduced unemployment security, only expert doctors and economists can decide.

In the rest of this chapter some illustrations will be given of how the separation of values and means may work out on a few of these questions. But it should be made clear here at the outset that our purpose is not to get deeply involved in the grosser disputes of the day. There is a temptation, from interest's sake to do this; but space forbids the detailed investigation needed, and our purpose remains as in Chapter 7 to bring out Beyondist values by well-chosen illustrative cases.

The proposal here to separate *wants and values*, concerning which every individual counts equally, from *technical means*, concerning which democracy must be prepared increasingly, in a scientifically complex age, to defer to trusted (but hired and watched) groups of specialists, could lead to very substantial technical advances in the decision processes. Psychologists could — even today — devise analytical voting procedures that would make politics a more exact science, and eliminate the frustration and bewilderment which now destroy morale and render agreement on even simple goals hard to attain. As to *wants and values*, the important advance needed is a clear analysis of the dimensions along which wants and values can vary. In individual psychology what has been called the *dynamic calculus* (Delhees, 1968) is making quantitative treatment possible. And in the domain of social values a series of factor analytic studies by Morris (1956), Rummel (1970), Singer (1965), Wrigley (1963), Eysenck (1954), Merritt (1970), Gouldner (Gouldner and Peterson, 1962), Digman (Digman and Tuttle, 1961), and others on social attitudes, as well as our own and others work (Cattell and Stice, 1969; Fiedler, 1965; Borgatta and Meyer, 1956; Gibb, 1969) on group dynamics, is demonstrating that (1) improved means can be found for making clear, and separating the temperamental, emotional-irrational, from the rational cognitive bases for the present poorly organized groupings of people in partisan activities, and that (2) in the decisions which groups and their leaders have to make, much more could be done to separate the decision on wants or needs, on the one hand, from those on ways and means, on the other.

Parenthetically, one must distinguish between the dimensions of values in some ideal sense, as in the work of, say, Morris or Gorsuch, and the dimensions of actual political party action, such as emerge in the factoring of dimensions of congressional voting, etc. Ultimately we need the former, but even the latter would reveal more realistically and closely what we are really voting about than the battered, journalistic stereotypes in which voters — and particularly the uncritical and the intellectual young — get entrapped. Some of these terms, like Communist, Fascist, racist, capitalist, imperialist, etc., are totally misunderstood by the user and indeed, become little better than terms of abuse[3]. Nowhere in the true sciences does discussion descend to such dismally empty tropes as in political science and political argument. Such terms as "right wing," "left wing," "liberal," "conservative," "Fascist," "democrat," "authoritarian," — constantly utilized as powerful emotional slogans even by "educated" students are now shapeless excreta of dead issues, from which the intentions and ultimate values of the partisan are indeterminable. The "right" and "left," polarity while not completely meaning-less, confuses a dimension "have"-vs-"have not" with one of "progressive-vs-unenterprising" and is empty of most of the important value differences that one would want to express. For, as the multivariate analyses of Gregg and Banks (1965) and many others show, it requires at least seven independent polar dimensions to encompass the political direction qualities. Similarly, the research of Alker and Russett (1965) on voting behavior in the U.S. General Assembly again shows that any simple "progressive"-vs-"conservative" conceptualization is quite unequal to describing a five or six dimensional domain of behavior. In other words, the intentions of any voter (and this would be true in *values* alone, before we ever come to *means*) require that separate scores or votes be recorded on at least seven dimensions if "information" is not to be lost. Naturally (though space forbids our illustrating them here), it would clarify the voter's expression of his wants and values if he were educated, say in high school, to recognize clearly the nature of these distinct dimensions along which he needs to record his wants. Meanwhile, in any case, he is largely forced by the party system, and encouraged to believe by the language of the journalist, that there is only one dimension of political expression—right and left.

A generation ago, as evident in the writings, for example, of Shaw (1944) and Wells (1903) (and, it is to be confessed the present writer, 1933a), it seemed that democracy was to be improved principally by raising the intelligence of the voting population (an echo of the nineteenth century perception that "we must educate our masters"). The above writers, Graham Wallas (1914) and countless practical men advocated that the right to vote be given only to those of proven education, e.g., having historical knowledge adequate to the issue, an I.Q. of 90 or more, and perhaps five years of experience after reaching adult years of the problems of a citizen. The trend is now away from this Jeffersonian concept—voting is to go down to seventeen years of age (educational level disregarded) and may yet go to six. Only if we view voting as a statement of *wants* does this even begin to make sense. (Indeed, if this is what we are seeking at the polls, it is a pity the recording cannot go to those in the womb, and the unconceived citizens of two generations hence, whose wants will be affected by the decisions.) So long as the voting on means is done by experts, this shift from a democratic assessment of means to an assessment of wants is an advance that should be pursued with all possible technical aids.

Granted that a democratic assessment of wants, and an expert evaluation of means is broadly what social science would indicate, then we can surely count on science to develop new designs of political machinery far more capable than those we now habitually use of implementing these intentions. Even if we keep conservatively within the method of committee meetings and decisions, many innovations could be introduced which avail themselves better of individual talents and conflict resolution devices. Roberts' Rules of Order are surely only a trivial advance on the procedures when primitive men agreed to prop their clubs warily by the cave side and get together for a parley. Both, in committees and in law courts, the present rules are little more than devices for orderly talk. There are numerous devices—including immediate use of computers; automatic signals of degree of understanding, approval, etc., registering on the table, for all participants; rapid retrieval of information from library storage; use of logic machines, etc.—that could be suggested by group dynamics' researchers for enhancing the speed, reliability and validity of group judgments. Incidentally, the stagnation and primitiveness of legal procedure—greater than that even of business executive committee traditions—is an instance of the natural consequence of an unfortunately unavoidable monopoly. As the Governor of Illinois, R. B. Ogilvie, himself a lawyer, recently stated, any American business operated with the archaic procedures which we accept as normal in our courts would have gone bankrupt long ago. Every question has more than the two sides of "motion" and "amendment." Political discussion and decision-making are undoubtedly scheduled for enormous improvement if progressive, imaginative social scientists are given a free hand.

The principle of separating wants from means—though there are all degrees of interaction of these (*see* the dynamic lattice, Cattell, 1959)—could bring about improvement not only in within-group but also in international and general inter-group decisions. In the application of "counting of heads" to the voting of *nations*, however, we see one of those false transfers, due to verbal analogy, which constitutes besetting weaknesses of the human mind. As argued elsewhere in this book, if voting *is* to be a substitute for war then it must recognize weighted votes appropriate to each of the nations concerned. The justification for the democratic vote among individuals is that the outcome of any war among factions, apart from its terrible cost, is statistically likely to finish up much the same as a sheer count of the number in each faction would indicate. Except in unusual instances of "correlated error," e.g., where, say, we are counting a vote of the have-nots to expropriate an almost equal number of competents, the approximation made in voting of discounting differences of competence probably yields much the same result as an actual power struggle—without the waste. But *among groups* giving an equal vote to countries, classes, religious denominations, etc., of quite different size and development becomes absurd. Already, in U.N.O., it begets all kinds of trouble. As Lindbergh has well asked (1970, page 169): "How can we measure the strength and influence of a nation and give it a peaceful means of representation equal to that it can demand and enforce by arms? Yet unless we can find some means of doing so, how can any representative body hope to run the world? I think that is the reason the League of Nations failed and the reason why such leagues have failed in the past and will continue to fail in the future — until some means of measuring human character and strength is found. Counting heads is not satisfactory. If we carried that system to its logical conclusion, India and China would rule the world. Representation in proportion to geographical area is worse and not even to be considered seriously. What then can we fall back upon? There is no tangible measure. When agreement fails in matters of vital importance among strong nations, war is likely to result. It is the method of decision we have resorted to in the past, and I see no indication that we will be able to leave it in the future. The most we can do is to reduce the frequency of wars by intelligent agreements among groups of peoples, mutually beneficial, and backed by sufficient armed strength to make war in opposition unprofitable."

"The measuring of human character and strength" is precisely what the social psychologist is aiming to do in the work on the dimensions of cultures and populations discussed on pages 133, 135, and 137 above. And though it is in crude infancy, the work will yet supply the calculus that will avoid war.

A more specific translation of these principles into institutions is deferred to Chapter 9, where we consider the practical steps for amalgamation of research and politics. The development of the nervous system of animals, however, gives us general support for the above conception of separate registering and analysis of wants and means in the single organic group. For by a "wants" count the autonomic system collects and adjusts information from the internal milieu and by a "means" evaluation the central nervous system decides the necessary adjustive steps for meeting the internal demands.

In summary of this section, present political organization though better than government as known to Ghengis Khan or Julius Caesar, is still a hopelessly archaic "game" in relation to the real tasks and purposes of a modern society. Although evolution of small advances is usually more peaceful than when large ones are demanded, and although the Beyondist changes of political formulation here envisaged are as radical as, say, the change from a four legged to an upright-posture in the hominids, a well-supported social science might yet hope to work them out constructively in a generation, by (1) the shift of broad democratic decision to wants and of technical decisions to a democracy of specialists, (2) the training of political leaders in social science, (3) the shift of ethical and other values in social reform from a dogmatic, revealed to an evolutionary, scientifically derived basis, and (4) the setting up of a "ministry of evolution" or the equivalent thereof to enable changes of revolutionary magnitude to be made by evolutionary methods. This last is inherent in scientific method, but has been lacking from human social organization, which has been a history of painful revolutions (Petrie, 1911) always more violent, and with more undesirable "side effects" than need have been. Changes of equal magnitude and suddenness have occurred in scientific "revolutions" (Kuhn, 1962; Price, 1961; Riddle, 1948) but they have been handled by a more "flexible" evolutionary procedure. The whole formulation of present day politics as a battle between a radical, allegedly progressive party and a conservative, supposedly dedicatedly reactionary party is an absurdity. Parties representing different directions of progressive action — different within the limits of scientific approximation — offer a more reasonable design for decisions.

8.2   INSTALLING EUGENIC CONTROL AS A FUNCTION OF GOVERNMENT

Probably one of the more startling innovations in an evolutionary morality, especially for the thinking of last generation's sociologists, is the frank demand for treating genetic considerations as realistically as environmental, educational issues. It is true that in genetic matters we have to think in terms of a much longer time scale, but that is no excuse for not thinking at all. Astronomers give our planet a further lifetime about as long as that it has already had — 5000 million years. Even a million years of attention to genetics could transform the powers of man beyond the span of our present imaginations to conceive. Indeed, two or three generations of eugenics could make most of the sordid and essentially unnecessary "dependency" problems of our civilization (at least those that are of genetic origin) quite obsolete.

Unfortunately the scientist who drives ahead in the field of behavioral genetic research and eugenics today is constantly pressured to abandon his objectivity by the cries of racists on the one hand and ignoracists on the other. The root origins of eugenics in the liberal and progressive enquiries of Darwin, Galton, Huxley and others has not prevented its being smeared in different ways in Germany and Russia, and (through the misunderstandings of a generation ago) in a few communities in America. It is not our purpose to take space here to refute merely willful misunderstandings. (Elsewhere, in a technical genetics work which permits clearer exposure, Nesselroade and the present writer have made such psychological analyses of prejudice; Cattell and Nesselroade, 1973.) Anti-eugenists have come nearest to objectivity when they raise the questions (rhetorically, however), "Who is to choose what type is wanted?" and "Do we know enough in basic genetics to dare to apply it socially?" To these the answers are, first, this is the wrong question: when society makes a genetic decision it has to choose a whole distribution pattern—not a type. And in such decisions the community is stating its needs on the basis of the same rights as its exercises in shaping individuals by education. To the second, which we have already met in the discussion of basic principles in Chapter 4, Section 4.5 above, the answer is a decided "Yes" (for reasons given on page 350 and elsewhere).

Almost as inimical to a sane eugenics as these conservative prejudices, are the radical "science fiction" discussions of random manipulation of heredity without regard to group social effects, as in "chromosomal engineering" or "tissue culture" as Haldane called it. Some equally imaginative but more practicable schemes in eutelegenesis[4] and prophylactic abortion deserve immediate study and pioneer application. However, when all is said, most of these, which we approved in principle in Chapter 4, concern, in the practical applications we now have to consider, only small fractions of the population, while the real aim of a wise eugenics is to have an everyday, workable value system that will act to produce the optimum progressiveness of birth rate patterns operating in ordinary family planning settings across all the families in the nation.

Beyondism calls for as much *positive* social action in eugenics as in euthenics (education and environmental improvement). Broadly the aim of eugenics is to reduce the suffering and death which prevail when natural selection has to operate through differential *death* rates, by working instead through differential *birth* rates. By reason of the degree of resemblance of parent and offspring, it can offer to substitute for a tough natural selection among people already born, a differential birth rate, higher for those parents who show better adjustment. The cost of the approximation (parent to child) is well worth the gain in reduction of human suffering. The question then arises whether the recommendation for size of family should be based on a psychological and medical examination by experts, or whether it can be left to some natural verdict concerning success of the parents which life itself offers. The argument for the former is that in life there are "injustices" – errors in assigning rewards and in public acknowledgements of success. The argument for the latter is that a human committee does not know enough to assign correct importance to the thing it measures, and is biased by the climate of the time in deciding what is good. (One can imagine that at the time of the Olympic Games a popular poll would instruct the expert to give more weight in "goodness as a citizen" to physical prowess.)

A limited use of the former approach – expert advice, but as negative eugenics – already exists in Eugenic Clinics where parents with suspected hereditary defects are advised on parenthood. A more positive use is likely – as psychology and medicine advance – in voluntary societies whose members agree to accept advice on family size from an objective appraisal of their qualities of mental capacity and physical health. Indeed, Beyondism should pioneer in this direction. However, across the bulk of the nation, the "life success verdict" itself is more practical and easy to gear to self-directed family planning. Indeed, if a social system is one in which earnings are fairly related to social contribution, i.e., if society pays most where it most needs capacities, then there is a simple and elegant basis for eugenics – a family size proportional to the earnings of the parents. (It happens also to be correct for euthenics, since it means more children in homes with better environmental opportunities.)

The first premise – that of fairness of earnings – cannot be entirely taken for granted in any existing society. It is said to be a goal of Communist or socialist societies, but so far no clear theory whatever exists for paying one calling more than another (though one could be developed). In free enterprise countries supply and demand supply a non-arbitrary basis, but unfortunately demand runs more to prize fighters, manufacturers of whisky, purveyors of trashy fashionable music, and so on, than to clergymen, teachers, judges, and researchers (especially when they work in theory, in abstract mathematics or *basic*, "irrelevant" science). Admittedly, a critic can make much of this discrepancy of earnings and social worth, and also of the fact that after gaining a reasonably secure middle class income culturally oriented individuals strive no more for increased income, i.e., the correlation of income and capacities though good over the lower range would tend to be poorer over the upper range of income and ability [5]. But through the great central mass of society an adjustment of birth rate to income – even over the small range of one to four children per family – would be a most powerful eugenic tool. It would also permit some degree of genetic control through those economic manipulations of various kinds which governments are increasingly able to make.

If we could bring about the basic and revolutionary condition of the adoption by all citizens of a Beyondist set of values, bringing alertness to the rights of the unborn and the need for evolutionary movement, the necessary details of legislation would take care of themselves. For the rest is only a technical matter for psychologists, geneticists, doctors and economists. As Dr. Francis Crick of Nobel Prize fame has well said: "If we can get across to people that the idea that their children are not entirely their own business and that it is not a private matter, it would be an enormous step forward" (quoted in Rosenfeld, 1969, page 161). The willingness of the general public to get seriously concerned, in this decade, on at least the question of *size* of population, augurs well for an advance, perhaps in this decade, into the long delayed concern with *quality* of population. (The fact that at the moment, as argued in Section 8.4 below, the public's concern with size may actually damage the quality trend, is accidental.)

It will have to be part of this growth in the range of moral concern that citizens will accept "invasion of privacy" (a much greater threat to the criminal than to any honest citizen). For, in company with all other needed headway in social research, eugenic research to determine the birth rate effects of various economic and educative measures, will *require testing and documentation of extensive population samples*. It will also be necessary, as it is wherever new moral values are accepted, to support those measures with whatever legal action is necessary to curb the recklessly anti-social. And here, in face of the usual outcries about "intolerance," and "regimentation" the new moral fiber will have to show its quality. Cases will arise – like the frequently repeating case of the borderline mental defective prostitute, now receiving extensive welfare aid for a dozen illegitimate children – where such definite action as sterilization will be called for. It is grossly unfair and demoralizing of their own ideals, for parents who can with difficulty support two or three children of their own to be burdened with taxes for numerous cases such as this. Yet there will always, apparently, be sentimentalists who become peculiarly hysterical on the sterilization issue (though there are many thousands of, for example, university and other professional men nowadays who are *voluntarily* sterilized after begetting a planned two or three child family). One is reminded of Judge Oliver Wendel Holmes argument when his judgment supported the majority opinion upholding the legality of sterilization of the feeble-minded. "We have seen more than once that the public welfare may call upon the best citizens for their lives. It would be strange if we could not call upon those who already sap the strength of the state for these lesser sacrifices, often not felt by those concerned." One might substitute, for "not felt," "much desired," for there is misery as well as profligacy in the majority of these irresponsibles.

Eugenics in general has two aspects: *restrictive* (or "negative") eugenics as just discussed and *positive* (or creative) eugenics, some problems of which we have yet to discuss. Restrictive eugenics gives least headaches to the social theorist, who finds less difficulty in getting a consensus on what is defective than on defining what is "superior" in positive eugenics. But it may give more headaches to the social legislator. Although the files of welfare services and hospitals and colonies for the mentally inadequate, are crammed with instances of families of defectives expanding from generation to generation at public cost, the public conscience has been slow to awaken and its action relatively easily stopped by arguments about "the practical impossibility" of controlling sexual behavior in the psychopath. "You cannot legislate morality" is the misleading half-truth cynically flung at crusaders in any field. Yet it is patent that societies like Sweden, Britain and New Zealand where birth control has been made available to all, and family planning has become a matter of public conscience, have reached new levels of civilized order and assured living standards partly through these measures. And in America the first real reduction of poverty in many generations (a decline in those below the stipulated poverty line from about 24% to 12%) occurred between 1960 and 1970, precisely in the period in which family planning most rapidly increased.

Positive eugenics finds its goals harder to define than in the above "bringing up the rearguard," but the difficulties will diminish with imaginatively conceived research (*see* Pearson, 1909; Fisher, 1930; Haldane, 1928). Although there can be little reliable insight as yet about the desirability of various temperament traits, psychologists studying actual life adjustment and measured creativity would probably agree on the desirability of higher intelligence, good memory, emotional stability (ego strength), freedom from paranoia, and natural strength of superego. A new science of psychometric prediction in these areas is in fact getting on a firm foundation (*see also* pre-natal diagnostic possibilities; Arehart, 1971a,b).

Those who pooh-pooh attention to genetic improvement are apt to say that even the community itself has no right to choose what type of person shall be born; but they completely overlook that such choice goes on all the time at the wedding altar. Incidentally, this selection by the average man may well be less beneficial than the directions that social scientists would work out. For sexual selection, as Darwin pointed out may actually be a backward eddy in the stream of natural selection. Broad shouldered and heavy jawed men, and women with extensive fore and aft projections may not be at all the types that nature itself is demanding at this stage of culture. Nevertheless, there is much that is good in the implicit and unconscious but none the less real eugenic selection of the next generation by the present one through marriage choice. The explicit and scientifically planned eugenic selection by well-informed scientific opinion is a different matter, though it does not escape the difficult issues of ethics which Ramsey (1970) has raised and which we have discussed here in Chapter 4 (Section 4.5) at the level of basic principle.

Much as one would hope to leave as much as possible in eugenic affairs to the self-directing citizen, his goodwill is not enough unless it is educated, and it is likely that as research establishes firmer knowledge about genetics and social needs, more positive guidance will regularly be sought on both marriage and family size by the citizen. Leaving encouragement of a differential birth rate to economic differentials, as we have seen, is an *approximate* procedure, in as much as correlations of contribution and earnings are at present quite imperfect. The caterer to vices and fads is overpaid (as well as most of those who deal directly with money. *Vide* the Communist jibe "Don't rob a bank; own one"). The teacher, researcher, and minister of religion are commonly underpaid, if the wide ramifications of their beneficial contributions were evaluated.

Practical problems in a eugenic program admittedly thus present a series of difficult challenges which must end in semi-efficient compromises. But society has access to enough sufficiently effective measures to show that it is in earnest in the extention of moral intention to reproductive behavior. A powerful *drive to extend effective birth control habits* to those who are as shiftless in family planning as in all things is a first necessity. The social ethics standards which would *make it a normal expectation that the more well-to-do will have more children* is another. A radical re-weighting of taxation to bear more on the childless is a social measure now much needed. In this connection one must recognize a sinister tendency which initial social research indicates. It seems that when birth control is freely available the differential birth rate tends to change from dysgenic to eugenic; but the *total* birth rate may fall below replacement level. As said above, there are two ways of travelling: "first class, and with children" or (in social competition terms) "he travels fastest who travels alone." And altruism is apparently not yet enough, when choice of births is freely made for the average citizen to maintain the nation's population. He needs to have the will to push himself into "hardship." Here again the future is genetically to the morally responsible, in a way that was not true up to this time and before family limitation was widely available.

8.3   THE ECONOMIC EXPRESSION OF ETHICS; IN INCOME, INSURANCE, TAXATION, MIGRATION AND PRODUCTIVITY

Civilized societies, capitalist or Communist, have "settled down" to accepting as a matter of course, particular differentials and magnitudes in earning rates, in taxation, in interest on money, in profit from management, in expectations of leisure from work, in insurance, and many other economic habits. But most citizens — including bankers — would have no idea at all how to defend these particular interpretations of ethical values in economics by any rational principle. An economist can "explain" them on seemingly tolerable reliable empirical economic laws, but he cannot understand them on broader social psychological principles, nor has he any systematic principles whereby he can justify his manipulations in terms of human rights and moral expectations. Meanwhile we are only half aware that there are quite extensive ethical implications in the currently accepted economic habits of society. The question we now ask is "What are they?" And how would an explicit Beyondist morality set out to justify or modify them?

The principle has been stated above (Chapter 7, Section 7.7) that since economic laws reflect more basic energy and real resource laws extending through our environment, these economic laws need to be respected as if more than merely man-made, by any Beyondist morality that wishes to keep integrated with the realities of evolution. (Few social psychologists except Thorndike (1939) incidentally, have seen the importance of studying these economics-psychology integrations.) It was recognized that in stating this, inherent adjustive realism of sheer economic laws one enters sharp conflict with Humanism and with a revealed religion such as Christianity, which at times has denied the right to collect interest and has wished to fix wages socialistically by standards unrelated to the service given. The response of Beyondism is that such adjustments to economic laws are entirely justified for Type A poverty — accidental or associated with temporary fluctuations in the needs of society—but that the charity of Type *B* poverty—due to systematic failure of individuals to contribute to society—is an unfortunate, immoral deviation, best kept entirely temporary.

Ideas of "controlling the economy" are quite old. They were tried in Egypt and Athens. Both church and state proposed schemes in the Middle Ages. Extensive controls were tried and abandoned in the French Revolution, and so on. Mostly they have been—until the twentieth century—attempts to get something from nothing—to satisfy the clamor of a "pressure group"[6]. The important points are that though the economy can be *effectively* regulated only by understanding its mechanisms, it can be *wisely* regulated only by moral principles that fit the total realities of life on our planet.

As to the beginnings of a rationale for wage differentials it was suggested above that one must begin with the courage to accept Ricardo's recognition that a man's offering of his work is, at least basically, a "commodity" on the market. (Let us remind the reader that economics defines a commodity as "any object embodying human labor for which there is a social demand.") And like any other its value is naturally set by supply and demand. (If society is not wise enough to pay the researcher and the priest, and other providers of services not *immediately* valuable, as in the temporarily unemployed who stands and waits, it is so much the worse for the given society.)

Now it is recognized that these wage differentials are sufficient motivation for the sensible man (a) to make the effort of education for more needed jobs when others needing less education are overcrowded, (b) to migrate from areas of low to higher demand, and to pursue various other adjustments valuable to society. It is only through the Beyondist comprehensive inclusion of genetic and environmental determiners, however, that we come to see that wage differentials are also a signal (under (a) above) for *genetic* supply and demand adjustments. In deliberately illustrating this here in regard to the *innate* component in intelligence we are not ignoring the *environmental* educational component. But for brevity of presentation we shall suppose that the aim of our present educational system, i.e., of giving longer education to the more intelligent, has been achieved, so that a correlation of unity exists between genetic and educational contributions to the individual's level of effectiveness.

Now let us suppose a society in equilibrium for which two histograms (graphs showing *vertically* the *frequencies* of people at each intelligence level, as measured horizontally) are constructed. One (Fig. 8.1) shows the

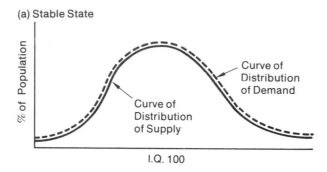

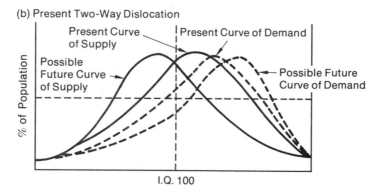

Fig. 8.1 Supply and demand curves of intelligence (Reprinted from Cattell, R. B. *Abilities: Their structure, growth and action.* Boston: Houghton Mifflin, 1971 with the permission of the publisher). Note that the present curve of supply is considered asymmetric about I.Q. 100, though this is not necessary to the argument.

amount of *supply* of intelligence, by birth rate, at each level (left to right), and the other shows the *demand*. Since a population with the initially given distribution of born intelligence has actually generated the jobs which characterize that culture we may suppose that supply and demand are initially in equilibrium. That is to say, there are in the adjusted state, with as many "clever" (demanding) jobs as there are clever people, and simple jobs as simple people. The two histograms (smoothed to "distribution curves") will fit closely, as shown in Fig. 8.1(a).

But from this point of cultural equilibrium onward the two curves can become dislocated by either (a) the birth rates becoming significantly different at different intelligence levels, or (b) the demand in the economy for jobs at different levels of required intelligence-education becoming different through cultural invention, e.g., more computer program writers and fewer ditch diggers may now be required.

The first effect of such a dislocation as is shown in Fig. 8.1(b) will be to change the reward rates (theoretically all being equally paid in Fig. 8.1); and a secondary result may be to create unemployment. Wolfle's examination (1971) of national resources of talent is here useful. If the wages offered are a simple function of the ratio of demand to supply, and if we take this ratio from Fig. 8.1(b) for a series of baseline points projected upward to the pair of curves "present demand-present supply," we shall get the curve derived in Fig. 8.2(a) for wages resulting at each intelligence level. Taking population distribution into account we then obtain a resulting distribution of earning rates as shown in Fig. 8.2(b). Actually this fits in general form the earning distribution that has characterized our society for a generation or more. That is to say, there is no longer either a rectangular or a normal distribution of earnings, but a relative excess of persons either at a low earning level or in unemployed dependence.

A more desirable adjustment to such a situation is not to be reached by attention to education alone. Indeed, the recommendations here coincide with those we have reached in the section on eugenics above, namely, that birth rates should be adjusted to earnings as fixed in the open market. Family planning must respond to the *distribution* of total community family planning needs, not merely to the need to adjust *size* of population. That is to say, the birth rates need to be sensitively adjusted according to the best estimates of the trend of complexity of occupations occurring in our culture.

Instead of such a morally desirable adjustment, however, society is cajoled politically into "charitably" sustaining the economic burden of an "hospitalized enclave" in an otherwise healthy society, occasioned by a largely unemployable group — relative to the demands of new cultural complexities. Indeed, this group now has a state supported high birth rate calculated to *increase* the dislocation. This parasitic burden we shall discuss in other connections below. Meanwhile, any realistic ethical system must regard a man who begets eight children on public welfare as someone as socially dangerous as any criminal[7]. And in general, the attempt by economists to gloss over this *inherent* poverty by paying out in response to the social blackmail ("disgraceful poverty in an affluent society") involved is an anathema to any thinking citizen. The remedy by "affluence" merely hides a more serious disease.

(a) Ratio of Demand to Supply: Price

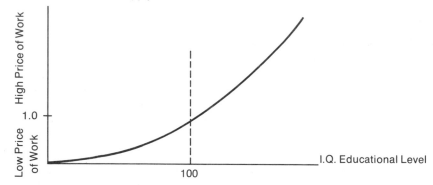

(b) Distribution of Earnings to be Expected from Natural Adjustment to curves in Fig. 8.1(b)

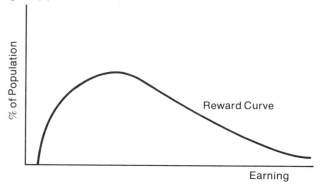

Fig. 8.2   (Reprinted from Cattell, R. B. *Abilities: Their structure, growth and action.* Boston: Houghton Mifflin, 1971 with the permission of the publisher).

A second economic "way of life" that is accepted without enquiry, and which it is vital to discuss because of its powerful effects on evolutionary trends is that of taxation on various differential bases. It seems nowadays to be widely accepted that a larger income and also a larger (total *or* luxury) expenditure should be more heavily taxed than a smaller one. The latter—the tax on *expenditures*—incidentally, has a usefulness lacked by the former in permitting a government to discourage morally less desirable modes of expenditure. One would have thought that the sales tax would therefore be urged, relative to income tax, by economists with broader vision. Indeed, the desirability-undesirability principle of tax planning has the advantage that it can be extended to tax those who do not help produce the next generation, thus in effect giving the proceeds to those who expend on bringing up children.

As regards income tax—especially as a tax merely aimed at reducing the *income* differential (quite regardless of birth rates)—we propose to throw aside traditional habits of acceptance and ask open-mindedly just *what* the rationale and moral values are in setting up any "leveling" taxation. There is, of course, a considerable amount of Utopian writing (and actual practice, e.g., in monasteries, in the kibbutz of Israel, and in some Communist groups) explicitly favoring re-distributions by agreed taxations with the aim essentially of bringing all to the same net earning level. A justification often advanced for such leveling down is that of Russell (1955, page 109): "The main objection to an uneven distribution [of earnings] is that it causes envy and hatred in the less fortunate, leading to fear and correlative hatred in the more fortunate." Good taste and human feeling would certainly forbid the kind of conspicuous expenditure in display which Veblen (1899) criticized, but maintaining moderate differences of resources for living is a different matter, and the Beyondist ethic on envy (page 156) is vitally addressed to assuring this possibility of differences without tension in a good society.

Any argument for bringing individuals of all ages and competences to absolutely the same earning level (or, for that matter, expenditure level) encounters insurmountable objections in any complex society. In an evolutionary framework, there are mainly two objections: (a) genetically, that this denies the economic aid we have just discussed for a differential birth rate adjusted to competence, and (b) environmentally, that it does not provide incentive for facing unusual demands in a job or the uphill pull in education necessary for reaching competence for more complex jobs. (It seems to be demonstrated in Russia, for example, that *free* higher level education still does not attract enough doctors, engineers, and so forth, unless the pay in the job is also higher.)

The aim of producing at least a comparative leveling of incomes through a differential income tax, on the other hand, is widely exemplified in various countries. The fact that it is taken for granted is assumed by many, e.g., Fulbright, quoted below, to mean that it is felt to be just, and that it is good for the group. The historical fact, however, is that the acceptance of the idea of a leveling tax has never been universal, and the current acceptance is quite recent. For example, as Darlington points out (1969) a *single standard* tax for all citizens was characteristic of Rome, consistent with their general ethos of equal national and civic duty by all citizens. Further, he points out that it aided the outward movement of colonists, since those who could not support the tax moved from the center of civilization to regions where it was not collected. This was a pointed remedy for the urban slum! In both Europe and the U.S.A. the differential income tax has no antiquity — an equal "poll tax" being, for many generations, most widely accepted as the model of a tax. Income tax, differential or non-differential, is only two to five generations old.

The rational basis for tax collection of any kind, which in some form is as ancient as civilization, is that certain services — education, postal service, military defense, etc. — being impracticable for private enterprise (requiring essentially a monopoly), and beyond the foresight and willingness of many citizens to organize provision for their own needs, *had* to be centrally organized and supported by taxes. Though instituted as much in democracies as other governments, taxation is authoritarian. For in effect the government says, "Since you cannot spend your money wisely, we will take it and provide what good community opinion thinks you ought to have; and you cannot spend money elsewhere until this tax is collected."

On this basis of payment for common services, which — as concerns indispensable services — is obviously just and necessary, it is difficult to find any reason (assuming that the earning differential is based, in the first place, as analyzed above, on contribution to the real needs of the community) *for charging one person more than another for the same services.* The person of average income does not use more street lighting or free education or free medical service than the person of sub-average income. There is certainly no basis in the general principle of egalitarianism for charging one man more than another for having his children taught in the same school, and so on. Yet a radical United States Senator (Fulbright, 1967, page 223) rhetorically argues that: "In America and other democratic countries, higher income people provide the bulk of the tax money to finance public services of which the poor are the principal beneficiaries; *the redistribution of wealth has become a normal and accepted function of democratic government.* The rich pay, not as a private act of noblesse oblige, but in fulfillment of a social responsibility; the poor receive benefits, not as a lucky gratuity, but as the right of citizens . . . ." (Italics ours).

Here is the whole crass assumption brought out into the open, with all its lack of analytical thought and without a blush for its lack of any fundamental imagination. The oratory in such cases usually conveys that the inference to "the rich" is to some man so rich that he won't miss the tax, and that in any case it must be "unearned" income. The reality that a politician ought to be talking about, however (oratorical tricks like "rich" aside) is not three percent of excessively rich men, but that fifty percent of the population—those above a hard-earned meager average, whom Fulbright and his kind are taxing to pay what they can ill afford to a less industrious or competent fifty percent below the average earning. If one looks at this situation with a fresh mind ("cleared of cant" as Dr. Johnson would say); this is the central fact, though *some* rationale for current tax practice can perhaps be sought in the following: (1) that regardless of responsibility to pay, taxes *cannot* be taken from those who simply do not have anything, i.e., who have neither earned nor saved; (2) that, where the relative have-nots perceive themselves as sixty percent rather than forty percent, i.e., when they have a wrong idea of the median earning (such as Fulbright encourages above), they are numerous enough, by revolution or ballot box, to enforce the heavier tax on the lesser fraction. This implies that the higher tax is a sop to reduce to a safe level that envy which Bertrand Russell and other Humanists consider an acceptable covetousness; (3) that the payment of a heavier tax is a willing act of charity, motivated by Christian or other traditional attitudes to "the poor;" (4) that the tax is a realistic but indirect payment for services in as much as the more able depend on the presence of the less able for the proper working of a total society.

Not one of these can logically face the standards of evolutionary morality and the last is sheer farce (the less able depend even more on the presence of the more able). Let us return to argue this for (2) and (3) but meanwhile return to the brute expediency argued in (1). The reader may find it interesting to try the experiment, as I have done, of asking a dozen or so intelligent and thoughtful men what they consider the justice of taking the larger tax from the man who earns more. After the usual period of speechlessness, characteristic of a custom accepted without thought, my friends usually answer (1), (2) or (3) above, but, pressed, they invariably return to the unassailable position that the man who has spent (or never earned) what he owes cannot be made to pay. Even the remedy which held historically until recently—putting the debtor to work —would not work too well here, since, by definition—in the case of the unemployed—he would cost more to keep than his work is worth in the open market!

As to the third rationale for a graded tax—that it is charity (in thin disguise)—the answer must consistently be what we have already made—that Type *A* but not Type *B* poverty is worthy of it. Under Type *A* one would include the undertaxing of the young, on a scale indicated by an allowance proportional to the typical increase of earning to be expected by a person of the given competence with age, since the professions in particular earn negligibly to say, age twenty-seven. There is no reason why, with suitable corrections such as this, the national income should not be maintainable with a flat income tax rate, for the bulk of it at present comes not from a few high taxes, but from the mass of averagely paying citizens. The proposed reform would in practice only deny cars and families to a minority by requiring them to pay their due debt to society before entering on other expenditures. In this connection sooner or later we have to recognize explicitly the dividing line which any intelligent observer already recognizes—that between free men who earn their way in their society, and the obligated ten percent who are nursed and supervised by social welfare and do not in fact, have citizenship status as it is defined by the contributing majority. The demand to pay a fixed tax would simply be one more criterion of this status.

In the second rationale for a graded tax—that it gives less provocation to the sin of envy—there hides cowardice, but perhaps also two more defensible arguments: (a) That a society which does not treat its members charitably *at all times*, regardless of causes of the misfortunes, will lose certain "defectors" and will not obtain from the average citizen the degree of loyalty found in societies that are infinitely succorant. The briefest consideration of societies that have been particularly tough on their citizens—Sparta, Japan (whose soldiers were always ready to die rather than surrender in World War II), and even some Christian orders such as the Carthusians—strongly suggests that the result of severity is not always a revolt against society. What happens depends on the possession by society of a very positive morale for other reasons. A psychologically rather subtle further possibility seriously deserves research: that although stern loyalty is still compatible with a non-pampering society, certain forms of imaginative creativity occur more freely in a completely compassionate and protective society, since some creative minds are childlike. (b) Not only in this economic realm, but in all problems of succour of individuals by society, one has to consider what is really a special Type *A* misfortune. This is the situation where for a combination of reasons the individual has "sulked" from society and what can only be called redemptive action is called for, e.g., in the modern hippie, certain criminals, the drug taker, and others. It could well be argued that society does well to rid itself of such individuals whose departure is no loss. A professional student of the problem surveying the recent appalling spread of heroin addiction, among individuals who have refused to listen to every plea of reasoning authority pointing out the dangers before they were "hooked," concludes that society really has no alternative but to let these individuals proceed to the last, largest and fatal dose. On the other hand, a Beyondist could argue – and here the conclusion would be the same as the Christian view – that the potential value to society of many a prodigal, redeemed and transformed, is so great as to justify enormous expenditures of affection and patience in a redemptive, rehabilitative undertaking. If we are to approach this objectively it might, on the other hand, turn out, alas, that in a fair proportion of obdurate cases society would actually be better off by spending its resources on new human material. This is the kind of "extreme case that makes bad law," but it brings out, in the economic field, the fact that the individual who cannot make his contribution to society could conceivably be supported on the rational basis of temporary rehabilitative action.

But apart from such exceptions under reason (2) (reasons (3) and (4) are rejected as untenable), it is difficult to find a rational basis for a highly uneven taxation designed to transfer wealth to a less competent half of the population. The operative reason in (2), apart from rehabilitation reason in (2b) is purely tactical: to prevent secession (such as actually occurred several times in history – in Wat Tyler's rebellion, the starving bandits who roamed Russia in the early twenties, and the pirates of the Mediterranean and China Seas).

In questioning endless charities from one fraction of society to another one is not overlooking that such exchanges in special cases may be necessary parts of the functioning of a complex social organism. The charity of parents to children has such functionality; but even that can be overdone. And so we reach the old problem, when does functionality end and the blasphemy which is parasitism begin? In the present case, among citizens all performing more or less required tasks and having equal citizenship rights, where is the rationale for taking from that half which can perform more demanded and paid tasks and giving to the group less unquestionably contributing, or even "slacking"? The body, wiser than political man, gives most nutrition to the muscles most in action.

It would seem, granted that the social organism is in a tolerably effective state of functioning – in education, research, food supply, defense, etc. – that the best criterion we have of the need for reinforcement in any part is the scarcity of people performing in that part, relative to demand. And in an economy with any degree of flexibility this will show by higher earnings there, and by poverty in the types of people least needed. Yet Fulbright, lacking in any clear principles, blandly states above that those earning more must provide "the bulk of the tax" and that "the poor receive benefits," not as charity "but as the right of citizens." He has not demonstrated this as a right, but at least by asserting it he has illustrated the confusion prevailing in popular thought and the need for a basic re-examination of why certain economic customs are tolerated, in the field of earnings and taxation.

If a differential earning is adjustive (and indeed an indication of where we need to produce more people to reduce the differential) then an artificial reduction of this basic adjustive action by a "corrective" tax is a contribution to the inefficiency of the social organism. (And in some cases as dangerous as putting a muffler over an air raid siren, or increasing competing noises.) A "single value" tax, unless it can be demonstrated that some people benefit more than others from the communal ser-vices, would not only be morally more just, but infinitely less costly of everyone's time in the expensive, time consuming business of collec-tion [8].

This last is no small matter, but the main argument against the complex differential, and for the simplicity of a single flat rate for all citizens be-yond school, is that in this way economic custom could be made to assist the spread of (a) more advanced cultural habits and (b) better genetic en-dowment (by the life test), throughout the given society. It would do so through increasing the reward for the first and the survival value for the second. An easy going individual, unwilling entirely to accept the above argument of the falseness of charity, may argue that within the bulk of society — say the middle eighty percent by income — the penalty paid by the upper half to the lower half is so slight as to work no ill. But "trash for over-topping" as Shakespeare reminded us, in times when weights were hung around the necks of the faster fox-hounds in order that the slower pack, might keep up, can be very wearying by the end of a long run — and the slower pack may not catch the fox. The verdict of all care-ful studies on natural selection is that the elimination of one breed and its replacement by another takes place with surprising speed in biological time from quite a small handicap or disadvantage, if steadily continued. The fact is that we do not know how far these artificial handicaps which society now imposes affect, for example, birth rates, health, longevity, or the readiness of citizens to maintain the better earning habits. Does the five hundred dollars per annum taken from the teacher and given to the unskilled factory attendant make just the difference between the for-mer having a two child and a one child family and the latter arguing his children into more effort in school? By contrast, to the dubious practice of taxing by income, the imposition of taxation on expenditures, assisting those who adjust to more desirable ideals and penalizing extravagance and vice is a most valuable economic tool in an evolutionary morality [9].

Another economic practice which needs more research as to its social effects is insurance. When private and voluntary, and still more when governmental and compulsory, it means that those who are careless in living habits, or suffering minor hereditary defects productive of accident and illness are enabled to reproduce at no lesser rate than those who are not. Again the trouble arises from confounding Type A and Type B "misfortunes." Insurance was set up for, and performs a valuable function in, giving peace of mind regarding uncontrollable calamities that may occur to any man. It finishes up, however, by being involved also in Type B "misfortunes": those where the apparent bad luck is systematically related to the qualities of the person. The good driver with no accident in fifty years pays for the drunken driver who regularly smashes up his sports car — or an innocent pedestrian. In this category we must logically, but less happily include the individual with hereditary disease, for which morally he has no responsibility (though his parents have). There is no escaping the fact that health insurance is a tax on the healthy to support the unhealthy, and in so far as the differences are perhaps one-half due to Type B, systematic causes, it is systematically increasing (or preventing the reduction) of disabilities of that kind.

An illustration is provided by the recent report of the National Haematological Society to the effect that keeping one haemophiliac functioning by injections costs $22,000 per year. Now the Beyondist morality is quite clear that any society, for spiritual solidarity, takes care of its defectives. But the cost of a defect (much less than the above clear instance) would normally result in the family concerned having a lower reproduction rate, by a socially beneficial self-adjustment to economic consequences. But insurance removes the connection. The effect may be small, but operating over the millions in a typical national population, it must significantly reduce the elimination of genes bearing various defects. It is not argued, of course, that the main tranquillizing effect of insurance should be foregone because of these undesirable side effects, but only that a problem is created by them and that steps should be taken to reduce the problem. For example, the person with high medical costs through inheritable disabilities should be persuaded to have a smaller family, and the high risk driver should be moved by penalties introduced into insurance to learn to drive or be prohibited from driving. Since insurance must be retained, the argument is for more individually-adjusted, differential premiums. This is not impracticable granted good social recording and psychological measurement—and would continue that humane and graduated implementation of natural selection by economic means which we have already reason to believe would function well in the single value tax and the maintenance of income differentials.

In all three of these fields the cost of not making internal group economic customs maximally functional in a Beyondist sense (and leaving them to the expediency of accidental, traditional social values) will ultimately turn up in having to put the weight of natural selection more upon inter-group than inter-individual survival. Unfortunately, the reckoning between groups is both less efficient per century and more likely to bring the risk of war. Meanwhile, every group which does not eliminate by internal natural selection carries a parasitic burden, eventually largely translated into economic terms, e.g., appearing as reduced power to export, as inflation, or as decline of exchange rate for the national currency—and so on. We have pointed out above that the differentials in this burden may, from an evolutionary standpoint, function usefully in lieu of war, in that the burden when high may so interfere with maintenance of defense that a country perceives it must lose a war and, in essence, surrenders without a war. However, it may also interfere with international scientific advance, as in America in the last decade, where the combined effect of the social burden and war costs began to lose America the lead it had, and the contribution it made, relative to Russia in interplanetary exploration. In that brief period (1960–1970) welfare to families with dependent children increased from $3\frac{1}{2}$ million to nearly $9 million (*Time*, February 8, 1971), a staggering threefold increase of the burden in one decade. President Johnson was prepared to say the true if unpopular thing when he insisted in the middle of this period (June 25, 1965, United Nations Anniversary Ceremony), "Let us face the fact that less than $5 invested in birth control is worth $100 invested in economic support." Incidentally, the parasitic burden is not all accumulated defect or immoral behavior in the ordinary sense, and may include habits of "genteel avoidance" of work. As Hardin (1964) has said, "The recurring inflations which France has suffered are an economic consequence of the excessive burden of pensions and other overhead costs which an aging country has to carry."

If space permitted, a brief analysis should be made of the interaction of economic and moral values in migration. Whenever a country manages its affairs well it is likely to experience a rise in standard of living which then makes it an attractive haven for migrants, often from the less successful of other countries. The question which then arises is whether purely economic laws should hold sway, and permit migration, or whether the idealism of a country in pursuing its own culturo-genetic experiment should override economics. Britain, in terms of immigration from Jamaica and former colonies; the U.S. in relation to Mexican labor; Germany in relation to Italian immigration; Australia and the masses of Indo-China; all face grave decisions. If one agrees with Shakespeare about "this happy breed of men" who made English culture a force for good around the world, he will hesitate to be a party to changing it at its very source, without research to gain far more knowledge than we now have about the social effects.

Migration is *not* immediately a racial matter, but hinges on the fact that cultural habits clearly persist in families over three or four generations, and may persist indefinitely. Consequently, before inviting immigrants on a large scale every culture should weigh the desirability of changing its culture in the way that will inevitably follow. The change may be desirable; it may be undesirable; but it will be real. Also, it should consider the alternative of expanding its own birth rate to supply the needed increase in population, as opposed to experiencing that damping down of its own birth rate that will adaptively follow from an invasion.

With the growth of a fuller understanding of evolutionary ethics, and with far greater knowledge of genetic and social psychological knowledge than we yet have, both cultural and genetic migration and blending will undoubtedly be carried out with far more deliberate and advantageous choice than has been exercised until now. With this will go a firm understanding of the rights of peoples to reject any merely economic determination of migration—as by employers seeking immediate cheap labor regardless of later welfare costs to the community. Instead, with scientific knowledge and a democratic vote, they will choose their preferred immigrants. There is no such "right," as some universalistic religious ethics wish to claim, to complete individual freedom of migration into all societies, uninvited. Each has its own precious experiment to nurture and stand by, in the branching out and general differentiation of human societies.

Our concern in this section has nevertheless been primarily with a revisionary approach to the economic customs determining distribution of support *within* the group. The main issues of *between-group* economic differentials have probably already become clear through discussion of the functions of economic competition in Chapter 5, Section 5.4. The economic standing of a group is very much a function of its psychological qualities. The first correlational analysis of the associates of higher real standard of living in modern nations (Cattell, Breul and Hartmann, 1952) shows surprisingly little causation by the luck of "natural resources." (Bahrein is an extreme exception.) Instead we find two major factors involved, one of which has already been discussed (page 137) as the effect of the intelligence and educational level of the people, while the other seems to be some as yet not easily interpreted temperament association with such culture patterns as the Protestant ethic, such as one would expect from Max Weber's analysis[10]. (Data on percentage of Jewish affiliation was not available for correlation in that research, but in Adelson's longitudinal (1950) factor analysis this percentage also positively related itself to standard of living.) These two factors together — ability level and culture pattern — suggest, however, that the productivity of a community, and its mean standard of living relative to others, rest on moral and intellectual strengths — indeed on such as would be steadily produced from a Beyondist ethics. Data permitting exploring the relation of living standard to the parasitic burden, and to the motivational effect of wage differentials was also not available, so the purely empirical analyses must at present leave our arguments above with the status of hypotheses.

8.4   COMMUNITY GOALS IN POPULATION SIZE, CLASS AND INTERNAL DIVERSITY

Beyondism has definite indications for the size and structure of population, which are in some respects radically different from those toward which presently fashionable views are heading.

As to the optimum population size, though the present popular attention to the real dangers of Malthusian overgrowth is heartening (Pendell, 1951), the concerned reader is being journalistically misled by the astonishing disregard of quality, of evolutionary goals, and of other elements, when discussing the dynamics of population. Moreover, the unfortunate feature about such calls for population reduction as those of, say, Ehrlich (1968) or Paddock (1967) is that they are beamed at middle class Americans, and countries like Japan, which need no preaching to be converted. If population is reduced *relatively*, in countries high in the intelligence-education factor (page 137) compared to those that are deficient, the danger is not merely that we shall have a population "monster," but a "headless monster." Among the world's countries there will, in short, develop a lack of power in those countries where world leadership would otherwise naturally fall.

What one must not forget in the current strong chorus appealing for population reduction (at any rate as it concerns all but a mass of backward countries) is that the very existence of the danger is questioned by some discerning experts, such as the Cambridge economist Colin Clark who writes "With adequate skill and use of fertilizers we find the world capable of supporting . . . ten times its present population." Admittedly, on the other hand, as Wiener (1954) says, the risks in gambling on such forecasts are high. Yet any appeal to history over the last four hundred years tells us that food and other resources, and capacity to handle urban density and pollution, are not fixed, but depend on the quality of the population in producing men of genius. Countries poor in the discipline of scientific education and political good sense may doubtless come to be hard hit by famine and other consequences. But so-called "liberal" writers have no right to try to make our flesh creep in guilt about this, and still less to make us afraid of being overwhelmed by warlike hordes of have-nots. Formidableness in war is much correlated with formidableness in civilization.

The writing has been so long on the wall about the debilitating effect of overpopulation (relative to production) that only the obstinately incorrigible can any longer let themselves fall into this miserable condition. On the other hand, some American and European societies have erred in the opposite direction and are *below* an optimum strength of population, mainly because they have cluttered their lives with material luxuries which they and their advertisers consider necessities. In their one and two child families they should be haunted by the ghosts of children who might have lived. Cutting down by three-quarters the optimistic figure of Clark above, one can, taking the most probable, dynamic view of scientific advance and social attitudes, still conclude that the U.S. can accommodate something nearer to 500 million healthy and effective individuals, *if eugenic attention to quality also guides population policies in this generation.* For the more intelligent and spiritual men become, the less they need in the way of costly amusements and expensive material pleasures to live a full and satisfying life[11].

The central fact remains, as discussed in Chapter 5, that living and education standards being equal a larger population is more desirable than a smaller one, since it favors survival of the given culturo-genetic group in face of natural catastrophes and war, and because it permits more experiment and increase of effectively tried out mutations.

As to the internal structure of a population, which means the form of religious and social class groups with their usual partial inbreeding,

Beyondism similarly questions many common social conclusions which by frequent repetition have taken on the air of rational truths. Thus for some the full realization of the concept of democracy requires a classless society, as in Communism, and Hogben tells us (1951, page 742) that "Science attains its highest dignity [only] in a classless society"[12]. Others insist that a well-developed democracy must be diversified by classes, religious groups, and sub-groups of varied belief. And yet others — mostly dissident intellectuals like D. H. Lawrence, Ezra Pound, etc. — insist that society must naturally consist of a "mass" and a "leading elite."

Social psychology tells us that the notion of definitely discrete social classes is scarcely true — each man belongs to several different groups and (in his lifetime) to different socio-economic classes. The latter, in particular, are more statistical categories than they are organized living groups. As the present writer has suggested elsewhere (1942, 1945a,b) sub-groups have the function for *individuals* of giving belongingness and an asylum from excessive discordance which high variety tends to produce in the main, total group. Besides this function of "asylum" they also have a *group* function of permitting experiments on a small scale, concretely trying out cultural and genetic mutations that may ultimately be valuable in the development of the main group. Social psychology also tells us that under primal threat the smaller congenialities which are the psychological life of classes adjustively break down, producing a broader democracy of shared primary needs, within the life-defending national group — a democracy of basic values which is present underneath all the time in a good society, however sub-culturally stratified.

What is often not taught in school social science courses is that socio-economic classes and religions also favor assortive mating and thus necessarily produce some moderate genetic homogeneity within them. This may not readily be evident to the eye of the amateur, but there have undoubtedly been over the last five hundred years some interesting biological specializations in craft, class and religious groups. It has the effect of widening the variation within society, e.g., raising the representation at upper levels of intelligence, and of reducing genetic regression to the group mean.

What *degree* of such genetic and cultural special development is optimum in a society? Certainly either extreme has grossly evident defects; but by what principles do we aim at the optimum? The Nobel Prize geneticist, G. W. Beadle, argues "Our best course is to assure maximum evolutionary flexibility for future generations by maintaining a high degree of wholesome genetic diversity" (1963). But what is a "high degree?" And do not the sociologist and cultural anthropologist rightly remind us that too great a diversity of culture, e.g., class culture, may reduce social solidarity and the sympathy of man for man?

Here is a vital and unexplored research topic — the optimum diversity of sub-groups for the highest morale and progress. In India and some South American countries one sees a lack of cultural solidarity producing poor consensus of action everywhere, but especially on new, progressive steps. In a few other countries, perhaps one could point to defects from too great uniformity. We know all too little. It is reasonably established that measurable psychological differences exist among classes and religious groups; that assortive mating within them produces greater genetic ranges in society as a whole than would otherwise arise; that the direction of imitation of culture is usually of the upper by the lower; and that in a free society a continuous promotion into higher classes (and some reflux) occurs, most being effected by education (scholarships, etc.). Thus manners and morals have reached highest standards in the "middle class" (actually the uppermost class except for a few aristocrats and plutocrats) and have spread downwards (hence the misnomer "middle class" morality).

The two last instances of dynamics above admit of unfortunate side effects. First, in as much as the prevailing conditions of social promotion admit of some "cheating" (climbing by excessive family reduction, crime, or selfishness) the axis of cultural imitation is awry — from an ideal evolutionary standpoint. Second, the social promotion of qualities with strong genetic components, such as intelligence, tends to leave lower classes impoverished. This depletion contributes to the separation of a "submerged tenth" at the bottom, which is welfare dependent and, in a true sense "hospitalized" relative to the rest of the community. The special aid — guaranteed income regardless of performance — given this class is in turn demoralizing to the low earning but working class immediately above, and the "hospitalized" values may consequently spread upward.

Although segregated classes offer the attractions mentioned above of greater congeniality — as do age classes and school classes — their functionality needs to be re-examined from a Beyondist standpoint. It is perhaps more in religious, political and ethnic groupings, in any case, that examining the problem of optimum degree of diversity becomes most relevant, since classes are only feebly segregated. The issues are well illustrated by a recent letter to *The Times* and the *Telegraph* in England by Lord Elton who criticized the use of the term "to integrate" in a publication "Colour and Citizenship" by the Institute for Race Relations, saying "Everybody knows that Indian immigrants have no intention whatever of being combined into a single congruous whole with Pakistanis . . . . On the contrary (and who can blame them) they are determined to preserve their own culture, traditions, religious and social customs." Integration, he proceeds, is now defined by the Race Relations Board differently from the dictionary as "cultural diversity in an atmosphere of mutual tolerance" which "Philologically . . . is . . . a near perfect definition of non-integration."

It is in respect to religion and race that the diversity issue is most acute, and we cannot set these aside as historical accidents that will disappear. For as men become increasingly sensitive about values they are going to form sub-groups for the fullest expression of values (as has happened often in the further fractionation of Protestantism); and these sub-groups, as Darlington reveals to us, acquire more genetic "racial" peculiarity than common observation suggests. (As also would be expected if, as we have argued above, religious values tend to be partly temperamental, producing some parallelism of race and religion.) The tendency to feel socially and morally lost in very large groups may be a defect in the scope of the average citizen's imagination, which good propaganda, as in Russia, China and pre-war Japan, can overcome. But regardless of the need felt by individuals for the relatively high congeniality of a more homogeneous religious, ethnic or social class sub-group, these sub-groups, we have argued, have a real evolutionary value, and, when, like religions, they cut across countries, a value in checking hostilities between nations.

Nevertheless, as we have argued in Chapter 5, the "tactful" word "tolerance" is a weazle word, used deceitfully by every orator. The fact is that a religious sub-group either believes in its values or it does not, and when community or national decisions have to be made to which all must adhere (as, for example, about Parochial School support, or the maintenance of Jewish holidays) people resist being forced to the lowest common denominator in all beliefs. This is unsatisfying to the individual of keen beliefs and high morale, and, as India or Ireland show, it can reach a point of being unworkable. The answer of the more eunochoid social scientists at the moment, who shape their theories to what civil authority demands, is to make "desegregation" workable by weight of propaganda. They are doing what a heart surgeon does when he tries to knock out the body's deep-seated immune reaction in order to force acceptance of a grafted heart. It *may* be technically achievable, but at what ultimate costs we do not yet know. Intuition tells societies, meanwhile, that it is best not to get into a position of accepting such racially and culturally discordant immigration that such drastic methods are forced upon it. An artist of social design is fully aware that there is an optimum degree of diversity for cultural advance, and that some hybridizations are more promising and acceptable than others, even though science has not done the research work that it should have done to offer firmer guidelines.

8.5   SEXUAL MORALS IN RELATION TO RATIONALIST AND BEYONDIST VALUES

The subject of ethical norms in sex behavior is of such extreme complexity that definition of a precise position would require correspondingly great space, and more quantitative research than psychology has yet accomplished. However, it is necessary—and easy—to expose certain pseudo-psychological reasonings that are attempting to substantiate false positions today. And in any case the whole field presents an opportunity for illustrating "energy economics" principles in morality that have not elsewhere been brought out.

The basic fact that stares one in the face—though few are prepared to see it—is a biological one, namely, that the endowment in sexual activity appropriately evolved for primitive human groups over the last half-million years is entirely excessive for the reproductive conditions and chances of death in civilized society. Here is one of the largest culturo-genetic gaps in human adaptation, with resulting intense moral conflict, neurosis and other products of the sharp disparity. Not only is our endowment too elaborate for the needs of survival, depriving other areas of psychological energy for their expansion, but that same endowment appears too *early* on the scene. The innate triggering by the hormones puts puberty perhaps fifteen years ahead of the time when we are educationally mature enough, in terms of the emotional complexities of our culture, most happily to marry and reproduce. This again is a tragedy—the tragedy essentially of Romeo and Juliet. Possibly some modern societies have a useful compromise solution, in a "trial marriage" norm consisting of early marriage with the expectation that many of these immature marriages will break up in favor of more mature choices, made within a further decade. If this were practiced along with postponement of children until the marriage proves stable, it might—though with the psychological strains of the childless marriage—prove a solution. Incidentally, there is in any case, much to be said—until eugenics can predict genetic fitness more accurately—for children being born of older parents.

In this way genetic defects which typically appear, say, before the age of thirty-five, would be more rapidly eliminated. Indeed, the much-to-be-desired genetic increase of human longevity — as *effective* longevity — will come mainly by the selective reproduction of the long lived [13].

The difference between the well-known position of Christianity, calling for every possible degree of sublimation of the sex erg, and the modern, permissive Humanist position, is very marked. Julian Huxley well expresses the latter (1957) "It is tragic to think of millions of human beings denied the full beauty and exaltation of love precisely while their impulses are strongest and their sensibilities at their highest pitch" (page 207). And proceeds to suggest that "the problem of love ... must be solved *ambulando*, or rather *vivendo*, in living; and the correctness of the solution is only to be measured by the fulfillment achieved." On the sense of tragedy in the situation we are in agreement, but the final Latin expressions do not hide the fact that a solution of ethical expediency is being suggested. For the definition of "fulfillment," with the present poor capacity of psychological moral science to define it, remains at the mercy of one's illusions.

The voice of modern Humanism ("rational ethics") in the realm of marriage and promiscuity is further expressed by the life and writings of, say, Bertrand Russell, or the plea of Rosenfeld (1969), "A man of our time who feels overburdened with his confusions — sexual and otherwise — and his responsibilities — including his marital ones — might see distinct advantages in the more carefree type of world" (i.e., in which the sanctity of the family is not recognized). Knowing many bohemians one must be permitted to smile at the idea that their world becomes "carefree," but one may yet recognize the heightened confusions which Rosenfeld describes. However, it is at least possible that the decay of moral convictions, rooted in the decay of revealed religions, is one of the major causes of these confusions [14].

As a temporary exacerbator of the discrepancy one must recognize that we ride on the crest of a wave of luxury, such as appeared in lesser degrees at ripe stages of older civilizations. There is little doubt — as is more obvious in animal studies — that cold, sickness, high altitude stress, hunger and fear are inhibitors of sex, and in a few instances (Grinker and Spiegel, 1945, observing Marines exposed to fear and stress in World War II) there is evidence of impotence in men and sterility in women induced by stress. Some perceptive literary writers have noted that sexual problems in the young seem almost to vanish in periods where economic stress is more uniform. To this one might perhaps add the comment of a ski resort owner that promiscuity is reduced at times of dangerous competitions, and of Bertrand Russell that (with obverse values!) we might reduce the participation in exciting and dangerous sports by removing obstacles to youthful sex satisfaction.

The evidence being all around us there was actually no need to wait for Freud for the insight, but it was nevertheless Freud who most clearly expressed the role of successful sexual sublimation in the cultural productivity of the artist, musician and mathematician. His analyses at the clinical level are supported by our factor analysis at the community level, where the cultural pressure factor clearly describes the same positive relation of instinctual long-circuiting to cultural productivity. All this is ignored by those who claim that their revolt against Puritanism and Victorianism (not to mention Moses) is "rational" and "enlightened." The crass assertion of an "orgasm culture" by such as Albert Ellis and others posing as representative of psychological science has little relation to the findings of either psychoanalysis or modern experimental psychology. (This is not to say that the psychologist does not meet deviant cases, where clinical disabilities arise, wherein sexual expression needs to be encouraged. And for some defective in altruism, sexual love is sometimes an avenue to love and socialization.) The oratory of Morris (1967) that a physical variation in man made sex more "pervasive" than in the other primates, and somehow produced culture, is an attempt to have one's cake and eat it (which naturally makes a best seller). This mistakes the recent genetic mutations favoring easier sublimation, which make available for culture the desperately needed energy hitherto tied up in the sex drive, a virtue of the sex drive itself, instead of reward from its amendments.

The nature of the genetic constraints which make human reactivity and energy primarily available only for certain directions of behavior is not yet fully understood — but these innate instinctual "channels" are nevertheless real. In the case of a displacement of activity and interest so large as is required in the case of sex it may be that civilization will resort to physiological inventions, if it can find them, that will delay the onset of puberty for a decade, and permit easier sublimation at all ages (an echo of the self-castration of some early Christian leaders). Meanwhile, to the fortunate whom "mightier transports move and thrill," by virtue of greater capacity to sublimate, falls the difficult task of explaining the colors of the sunset to men born blind.

In morality generally it is a higher development to accept a standard and admit inability to reach it (and here the hypocrite is already half reformed) than to commit the supreme sin of denying the value. Yet it is the latter which the rationalists in this field are attempting to do. They rejoice, for example, that chemical and other contraception has divorced sex from procreation, so that sexual license is possible for ever vaster numbers of people. But here they overlook a quite inexorable economic law of biology – that nature wastes nothing (which Aristotle's generalizing powers clearly saw long ago when he said "Nature never makes anything superfluous"). The bill of survival can be met only by *an absolutely functional relation of energy expenditure to what are truly adaptive actions.* Any separation of satisfaction from functionality is sheer suicide. (Biologists are in fact using a biological sex stimulation and false satisfaction as an ingenious way of wiping out insect pests!) However, it is true that (if inter-individual natural selection is not ruined by welfare and other economic machinery as discussed in the last section), the free use of contraception should more rapidly eliminate those lacking in the mutations which transfer sexual energy more freely to general cultural activity. On the other hand a greater emphasis on sexuality in marriage selection itself might have the opposite effect [15].

No matter what our scientific insights may be into sexual sublimation mutations that may, in evolutionary terms, be elegant, we cannot watch unmoved the passing of the sexual romance which constitutes nine-tenths of all literature. When, terrible and fair, the glittering new cultural Phoenix arises, we may yet grow nostalgic for the aroma of the nest of spices, fading on desert air. Yet, turning from verse to statistics, we know that psychology finds that the congenialities which make for life-long romance, are even today very similar to those which make for profound friendship (Cattell and Nesselroade, 1967; Tharp, 1963; Udry, 1966). Such truths, and the diminishing role of the sex drive relative to idealistic planning in deciding when the torch of life shall be handed on, suggest that the spirit of romance will have to spread its wings in wider realms of human endeavor and adventure than those of the sex relation.

8.6 SOME READJUSTMENTS OF VALUES
NEEDED IN EDUCATION

Clearly our purpose in this next look at Beyondist applications – a brief incursion into education – can only be to state the main emphases which Beyondism brings to such a many-facetted and vast domain. Education has responded more than one could have ventured to hope, since 1900, to reformative, rational and scientific approaches. However, one may venture to assert that in perspective the movement that named itself Progressive with a capital P (from Rousseau to A. S. Neil, G. B. Shaw, W. B. Curry, Bertrand Russell and others), will actually prove to have been least positive of the many truly progressive educational-psychological contributions.

In essence it was merely *permissive*, challenging the assumption it claimed to see in traditional education that "every child left to himself would be a wild beast, a violently antisocial being." With wider experience and a more developed source of psychology, it is now evident that this parody need not be far from the truth! More precisely, we have to recognize that *innate human nature is neither good nor bad*—until some cultural standard or direction is introduced to weigh it. It is then evident, so long as there is any evolutionary movement whatever proceeding in culture and ideals that man's nature systematically lags, as in the CAG concept, and is therefore always, both in ethical and other senses, "below standard." Since, in spite of all he says, the untutored man wants only to be himself, it is clear that he will kick against progress, and will himself progress only when kicked by environmental demands.

Shaw came out with the typical "Progressive" objection to character education, which had been a central theme in the English Public Schools, saying "anyone who tries to shape a child's character is the worst kind of abortionist." But this is rhetoric, for every society and religion has a duty to pass on (not irrevocably stamped in, however) the imperfect discoveries that culture has already made. Even if education were to be considered as concerned solely with intellectual matters, education of the emotions would still be important. For the most vicious errors in reasoning are not logical but emotional in origin. Consequently, it is actually more important to extend character education into the intellectual domain, than merely to bring a sceptical intellectualism into the character domain, as the "Progressives" did. Thus, with one of those swings back to sanity which fortunately occur with some regularity, educators come again to the riper wisdom of Matthew Arnold, which inspired the earlier emphasis on education of character and the emotions[16].

Among other unrealisms of the more noisily fashionable elements in education in the twentieth century have been: (1) an excessive estimate of what the school culture can do relative to the home culture. Much of the difference between school classes and schools in various regions has more to do with the kind of children, their homes and social backgrounds, than the goals set by the teachers. Moreover, this home culture has the capacity to persist for generations, and (2) an attempted disregard—egregious among any skilled craftsmen — of how the quality and sources of the raw material need to be watched. Both culturally — in terms of accepting cultural immigrations and in the neglect of need for social work — and genetically — in terms of birth rates — educators have neglected a genuine duty to call the attention of society to the supply problem[17]. Some of the very best, e.g., Thorndike, Terman, Sir Cyril Burt and Sir Godfrey Thomson, did, but most educators preferred the vision of their own omnipotence offered by the nonsense of Helvetius ("L'education peut tout") and Watson (1914). It may be difficult for education to refrain from overselling the role of education; but it has now also to sell the importance of biological factors.

What Beyondism now calls for in the spirit of education has been so well stated by two eminent scientists that an independent attempt here is unnecessary. (The mechanics of improvement of sheer educational efficiency, e.g., by the teaching machines inventions of B. F. Skinner, and classroom reorganizations through sound personality and ability testing, will take care of themselves.) Thus Thomas Huxley, nearly a hundred years ago, tells educators: "The life, the fortune and the happiness of every one of us, depend on our knowing something of the rules of a game infinitely more difficult and complicated than chess. The chessboard is the world, the pieces are the phenomena of the universe, the rules of the game are what we call the laws of nature. The player on the other side is hidden from us. We know that his play is always fair, just and patient. But we also know, to our cost, that he never overlooks a mistake, or makes the smallest allowance for ignorance. To the man who plays well, the highest stakes are paid, with that sort of overflowing generosity with which the strong shows delight in strength. And he who plays ill is checkmated — without haste, but without remorse . . . . what I mean by education is learning the rules of this mighty game." If this spirit of realism can somehow be incorporated in the thinking of young citizens we shall indeed witness a major revolution in social and political direction, and one without violence. It will open up a willing acceptance of new and positive moral values, on an objective foundation, and prepare the way for the world adventure of great socio-genetic experiments which Beyondism envisages. At the same time it will bring about a sufficient sense of the irrelevance of many political organizations and religious systems among which the young are now asked to take partisan stands.

Besides the elusive change in spirit which this indicates for education, it calls also for some changes in content — especially a still larger role for biology, genetics and psychology. This has been well stated by the second scientist we shall quote (the Nobel Prize winner W. B. Shockley who gave us the transistor and various advanced electronic concepts). "The central purpose of our educational system should be to develop a citizen's rational powers and to equip him to understand causal relationships, especially as they apply to man. The greatest obstacle to man's future evolution at the present time is lack of public education on the fact that man is a mammal and subject to known biological laws" (1965, page 104). (To which one might append, as an "Amen" Teilhard de Chardin's "To see or to perish is the condition laid on man.")

Certainly modern man needs to see himself more objectively, as a small but vital part of an evolving universe. He needs sympathetically to accept and adjust to the reality of inborn individual differences. He needs to view racio-cultural groups not as hostile deviants from his own values and kind but as important experiments, to be preserved and advanced. He needs to learn that moral values are continually developing *out of scientific research on group and individual survival*, and that, as such, they have authority. His emotional and character development must be tied to this perception. Indeed, when we come to the final question in the next chapter of the emotional adjustment that is integral to Beyondism it will be realized that the school has a major assignment in emotional education here.

One need which all ethical systems based on scientific or philosophical analysis share, in contrast to dogmatic religions, is the development in their congregations of a capacity to handle psychological warfare. The growing child typically moves from a domain of benevolent instructions, which proceeds faster the more trusting the relation to the teacher, into a deceptive and malevolent jungle of political cant, journalistic slyness, and a language corroded from the precision and elegance of its meaning by the euphemisms of advertising (e.g., the "good life" defined as expending more than $12,000 a year). At this point education has to recognize that the training in scientific method and logic alone, as set out above, is not enough.

By the same strategy as that by which reduction of drug habits is sought through attacking the production sources, education could reduce this problem by special attention to bringing about a high correlation (over individuals) of intelligence and character values. A person of low intelligence who is emotionally disturbed is perhaps unemployable, or even dangerous as a criminal individual. But a society which allows its most highly intelligent individuals to manifest ill-balanced, emotionally perverted and essentially dishonest character traits is in grave danger indeed, for these are the writers, who then become sources of infection rather than of wise leadership. Unfortunately, the evidence at the moment is that the combination of highest intellectual with highest character education is failing. In the early part of this century the classical studies of Burt (1917, 1925), Chassell (1935), and Terman (1926), clearly showed a decided tendency of delinquents to be below average in intelligence, and of highly intelligent children to be on an average of superior character and emotional stability. In some studies the correlation now approaches zero.

Since this mis-marriage of great talent with meanness of character cannot entirely be avoided (as is recognized in the legend of Lucifer) society needs to armor itself far more by training its citizens in defenses against psychological warfare. It might be a good step to begin by teaching young students to distinguish, in their hero worshipping, between the scintillatingly clever and the steady illumination of the wise [18]. Certainly in a democracy it is increasingly vital, as mass communication grows more powerful, deliberately to give school courses that will enable the citizen to think statistically, instead of being affected by the cunningly chosen extreme case; as well as to learn to detect the emotional, sentimental appeal, and to refuse to be stampeded by reports of mass opinion. These qualities are particularly lacking in our socio electorate today, and all the advances of economics, welfare and science will get us nowhere until we cease to be a nation of sheep. The problem is, of course, very old, as one is reminded by Plato's grieving over the young (misled by the sophistries of the sophists), by Aristotle's *Sophistici Elenchi,* and by Francis Bacon dissecting out the "idols of the market place." What is hopeful and new about the present educational possibilities is psychology's better understanding of emotional persuasions and deceptions, beginning with Freud's perception of the defense mechanisms, and the still more precise diagnostic location of emotional twists in thinking that the experimental dynamic calculus now promises to make possible. The better diagnosis of intellectual distortion which the progress of psychology makes possible could lead to an education thoroughly self-conscious in these matters and able to armor the student effectively.

The rather wildly aimed shots of the young revolutionaries of this decade unfortunately indicate a tremendous need for such education. A political writer has recently complained (especially of the young) that "The person with extremist inclinations may succumb with equal readiness to either right or left, depending on which 'ultra' fad catches him first." There is, of course, temperament in this tendency to extremity

(Eysenck, 1954, has shown the common hysterical character in the extreme right and the extreme left), but this perennial peril in political organization arises also from defective education and intelligence. And judging by the prevalence of extreme and ill-balanced views in the distended university student populations today the educational mis-fire we are experiencing is precisely of the kind just indicated—an enormously enlarged intellectual tool kit of information and words (if not always ideas) with extremely poor judgment, lack of training in intellectual discipline, and negligible understanding of methods of scientific and statistical analysis.

Also in this domain—but here it deals with self-deception rather than deception by others—one encounters the cost of what I called some forty years ago (with experiences of Oxford student debate) "tea-table intellectualism." Russell, on reaching ninety (1968, Vol. 2, page 185), has also become suspicious of this and warns against "The habit of affixing easy labels [which] is convenient to those who wish to seem clever without having to think, but has very little relation to reality."

I would be happy if the label I used might also correspond to nothing, but one fears that, actually, "tea-table intellectualism" is very real and has been spread by the educational system from the leisured academic few to become a national habit of conversation and reasoning. Meanwhile, "far back through creeks and inlets making" comes the silent revolution among the few, trained in original scientific thinking, patient in investigation statistically perceptive, breaking with calm realism through the stereotypes and accepted sentimental attitudes which remain the rhetorical stock-in-trade of the self-styled "intellectuals" and the political zealots.

This brings us finally to the danger in education which few educators will ever admit—namely, the danger of too much of it. In some *ideal* sense one may certainly argue that one cannot get too much, but there are two realities which deny the possibility of the ideal so defined, (1) that education for living in an era at time t has to be given in era $(t-1)$, and (2) that when education exceeds the reasoning capacity of the student it can lead to more mistakes than would a "lower" level of education. Partly because of the mind's inherently very limited power to "transfer training," an intensive education often ties up the individual's skills and interests in forms which disqualify him for a changing world. In young countries such as America and Communist Russia, with enormous enthusiasm for education, the remark that a man might have too much school education is received with incredulity, but in older countries, such as France or Britain there are usually some rather esoteric expressions for indicating that a man may have "the defects of his education." Such observations as that James I was "the wisest fool in Christendom" or the old folk saying that a more dangerous fool than one who does not know, is one who does not know that he does not know, illustrate that the perception goes back over centuries. There is nothing to prevent intensive verbal education, for example, producing someone like Mrs. Malaprop. Indeed, investigation might well show that an appreciable fraction of our student population today has been endowed with a larger vocabulary than it can command, just as a young man can learn to play more pieces on the violin than his sense of pitch can sustain. The danger of over-education is well illustrated by Napoleon's comment at Jena, that his opponents had read too many books on tactics. Thus they became fair game for an intelligently original opponent. In the rapidly changing world ahead of us there *is* such a thing as being over-educated in out-worn skills, and made unaware of the limits of one's intelligence. Education to think, using analytical and research methods, and apprehending the limits of one's facts and abilities to reason, becomes the primary requirement.

But, in summary, the chief challenge to education in the coming years is to teach biological and social sciences, and the arts of analyzing arguments, in such a way that a democracy may be prepared to understand the increasing role of expert research in social science in government and the growth of values.

8.7 THE UNSOLVED POLLUTION PROBLEMS OF THE MASS COMMUNICATION MEDIA

The activities of the media of mass communication—newspapers, T.V., radio and magazines—the professional participants in which we shall collectively call "journalists" (because they deal with things of a day) are vastly more prominent in society now, for good or ill, than ever before. An initial major problem here is that in an era in which, as we have seen, a scientific understanding of social life is increasingly important, the education of the journalist remains predominantly literary. Very few scientifically educated men indeed ever take up journalism. (There is economic determination here: graduates in literature, languages, and the arts have fewer places to go than those in science, engineering or medicine.) Literary and scientific sense need have little correlation. Although comparisons may be odious it is commonly conceded that journalism in England tends to have better literary standards than in America, but its feeling and insight for the cultural values of science remains quite poor (Snow, 1959) or inferior.

Since, unfortunately, it is true that the bulk of our population after school ends, reads no serious books (except paper back "best sellers") but turns only a tired eye on newspapers and T.V. and drifts in a jaded state to entertainment by the stage, this is serious. The argument has been firmly made earlier (*see also* Chapter 9, Section 9.6) that the arts cannot be an intellectually legitimate *creative source of new moral values*. Consequently, a second major problem in this area arises from the tendency of the young intellectual and the man in the street to turn to journalism as a convenient source of new values without realizing journalism's fundamental lack of qualification for that service. The highest art of journalism is surely to "hold the mirror up to nature" — the nature of current society and its daily events. The task of deriving and analyzing moral values is increasingly a task for extensively organized social research institutes — or, in an interim period — the established churches of the revealed religions.

Everyone knows that the mass media do not confine themselves to transmission of fact. Even when they do not seek to create new values they actively use persuasive propaganda, either for political parties, or for the unconscious values in their personal background — such as the essentially science-unsympathetic literary educational background cited above. Consequently, although every liberal rightly sided with the "freedom of the press" in the centuries preceding our own, when this was a bulwark of liberty against tyrannical governments or churches, a perceptive person must recognize that functions have changed. This freedom has nowadays the complexion of license, and needs sane controls.

The questions that arise are essentially whether the journalist has a right to the "freedoms" (1) to garble the cultural message into sheer noise, i.e., to pander to the lowest denominator of taste for trivia, gossip, lewdness, violence, etc., just because this "pays." It has been recognized by every sane observer that the media are in fact a prey to every fad and fashion, every sentimentalism and every popular error of thought — all expressed with a loud mouth. (2) To practice as a dispenser of authoritative information on matters beyond the daily news — history, values, economics, psychology, sociology — when the journalist's professional training obviously does not fit him in the least to do so, as countless "leading articles" easily reveal. (3) To "select" the information supplied (sheer "misrepresentation" of facts would of course be unthinkable). A cure for biased selection by requiring an obligatory "stratified sample" is not technically impossible, and without it virtually any direction of persuasion, even of critically intelligent people, is possible, by suitable selection of facts. Men in windowless dungeons can be made to believe that day is night. (4) To defame, without chance of redress, whoever disagrees with them. As to this last there are numerous shocking instances. Lindbergh, who resisted press invasion of privacy because he believed the thrill-hungry press would prevent the recovery of his kidnapped son, has scarcely ever had a fair press since. Vice President Agnew, with extreme temerity, said, "My differences with some of the news media have come not over their right to criticize government . . . but my right to criticize them when I think they have been excessive or irresponsible." And added an appeal "to drive misfits and bizarre extremists from front pages and television screens." Whether Mr. Agnew will be smeared, tarred, and feathered for the rest of his life rests with the not too sensitive consciences of the affronted newspapermen. Chief Justice Burger, in a recent powerful appeal for a civility and decency self-imposed by the press spoke of "editorials shrill with invective" and "savage cartoons." Savage cartoons could yet be honest, but in most countries, and notably in India and South America, one rarely sees a cartoon that imputes a decent motive to any public figure, and one wonders how much this emotionalism and lack of fair play is responsible for creating the type of politician from which such countries seem to suffer. When Froude, looking at various political misdirections suffered in history, declared oratory to be "the harlot of the arts" he pointed to a perennial, still unsolved problem.

More important than these sicknesses of bias, of posing as experts, and of vapidity, perhaps, is the fact that in many countries the press plays a vast confidence trick in claiming to be the bulwark of freedom while it tenaciously exercises a dictatorship of vulgarity. Unless and until a better system of control is devised there is no hope but for an increase in competence, taste and above all, conscience in the individual members of the profession. And no society or segment of society, not even a church, has been able to leave important matters to "individual conscience" without some machinery of lawful control. In as much as the mass media abut on the areas of education, moral direction and politics, they need to develop the controls which these already have. Absolute freedom, being absolute power, corrupts absolutely.

Any constructive reasoning in this area encounters an infantile emotional belief in the mass media, and the less intelligent liberals, that censorship is simply bad *in toto* and in all circumstances. But the fact is that nowhere in nature has a successful organism been built, and nowhere in history has a society survived, without an appropriate, intelligent application of inhibition and censorship[19]. Even on T.V., heroin is not permitted to be explicitly advertised, and rape and murder are supposedly not to be encouraged. In short, society has decided that it *will* exert censorship, and the only question is in what style and in what quantity it is optimum for social health.

Actually, in periods where great issues are not at stake, the main need for control and censorship is of the trivial and trashy rather than the vicious. This brings us back to point one above – the selection of news. Any scientist or educator, concerned to get society to consider vital issues, or any Humanist, concerned to get attention to good literature, has to admit that the effect of the mass media has been to trivialize. It argues that it can survive financially only by putting the sensational and the superficial rather than fundamental and progressive issues before the public. The reading of great books and the time normally available in other generations for individual thought, reflection, and the formation of independent opinion have inevitably suffered grievously with the crescendo of activity in the mass media. In aspiring to a true democracy, one cannot take lightly what is now happening to the mind of the average man. A survey shows that ninety-six percent of American homes have one or more television sets, and that the home set is turned on more than six hours a day (rather more than four hours for each occupant). Bronfenbrenner rightly concludes "the danger is not the behavior it produces, but the behavior it prevents." A survey of *Mensa*, whose members are in the top two percent of the population by intelligence, shows an altogether lower usage. But even the average man used to read some books. Leaving aside what three to six hours of television a day must do to his slim chance to read, when does he have the time to ponder a little in his own way; to digest the confusion of calls upon his values and his vote; and to discover, the independence of thought which his forefathers undoubtedly possessed? In psychology laboratories, one may see rats, dogs and monkeys with brain-embedded electrodes and a radio antenna on top of their heads through which an experimenter can convey prescribed emotional states of mind. Some hundreds of millions of citizens are, today, similarly screwed to T.V. sets, similarly designed to generate a prescribed uniformity of mind and an abdication of individual thinking. What this means for the processes of thought and judgment on which the healthy functioning of societies has hitherto depended, no one yet knows.

8.8   SUMMARY

(1) Although most of the varieties of government that the nations of the world have inherited miraculously "work," through an empirically

derived realism that makes them immune to much individual irrationality, yet they work extremely inefficiently in comparison with conceivable machinery which modern social science could suggest.

Five conditions are necessary to effect changes if need be, of revolutionary magnitudes, by evolutionary methods: (a) A democratic basis for evaluating "wants" broadly, but a machinery to satisfy them by technical means decided by a democracy of specialists. (b) The setting up of a "ministry of evolution," directing change from the top instead of the bottom of society. (c) An abolition of the archaic stereotypes (under terms like left, right, liberal, fascist) which now catch affiliates to particular parties under the conception that an individual must be "radical" or "reactionary" in his allegiances. (d) The psychological selection and social science training of political leaders. (e) The acceptance of evolutionary ethics values in place of dogmatic religious values in social legislation.

The impact of Beyondist ethics upon existing socio-political practices can take place partly through what have been called "the policy sciences" (Lerner and Lasswell, 1951) which may be defined here as the applied social sciences such as economics, city planning, education, in which value judgments are necessarily implied by every decision. Economic laws, "natural" and legal, illustrate this, though they embrace only a part of the principles governing the life of a society. Nevertheless they are geared to biological and psychological realities which govern group survival and can be flouted only at peril of loss of adjustment. This needs to be heeded particularly in international politics where one vote to one country is a travesty of democratic principles.

(2) A major change which Beyondism calls for in the function of government is an extension of moral concern to the genetic quality of the population. It is as democratically appropriate for a nation to define its ideals in genetics as in education—in eugenics as in euthenics. Although psychology and medicine will gradually move toward a capability of making a *direct* evaluation of mental and physical traits, in respect to both *creative and restrictive eugenics*, yet today a reasonably efficient, self-acting machinery, based on total adaptation and contribution to culture, exists in economics. If earnings are allowed to be the index of social demand, a wise government will need to adjust earnings of the small set of occupations (in teaching, religion and research) not directly evaluable by supply and demand on the market. Under these conditions, instituting a positive relation of family size to earnings, throughout society, would suffice to bring about a positive eugenic trend.

A new moral goal always creates new legislation and new possibilities of dereliction and delinquency, and in the genetic field these defaults are constituted by failure of the able and well-to-do to have children and of those on public welfare irresponsibility to multiply. Legislation must support morality here as in any other field, and we should have the courage to face the need for sterilization (when social work attempting to introduce contraception completely fails). In the case of the well-off who decline to raise children, heavy taxation (or large child allowances in income tax) should help the situation. However, the failures of reproduction in the upper income group at least act as natural selection automatically eliminating those of low superego sensitivity.

(3) Natural, inherent economic laws, i.e., those not depending on government manipulation, must be respected in evaluating adaptation, since they reflect realities in a group's adjustment to its cosmic and human environment. Economists have recognized in general that the best functioning system is one with certain group regulations superimposed on a basically free enterprise, self-adjusting market (*see also* Bagehot, 1873). But they have variously hesitated, on account of traditional or Communistic moral values, to follow Ricardo in the implication that a man of a particular ability and education is also in the economic sense a commodity. A theory of human supply and demand distributions suggests that this law is nevertheless operative, and that it should be respected as a guide for genetic and educational policies.

(4) Although the greater taxing of those who earn more is recent, it has become uncritically widely accepted. As far as temporary misfortune or inadequacy of Type A is concerned (childhood, accident, old age) this is a defensible "charity," but the argument from envy and threat is wrong. Those societies will be most successful whose morale and purposefulness is such that optimum differentials can be tolerated. These differentials motivate educational improvement, spread of more effective values, and a eugenic balance in family size differentials. Within the central range of earning the differential income tax, upsetting the ratio of reward to effort, may seem a minor burden, but in biological groups even a small handicap, over a few generations, can bring extinction to the sub-group subjected to it.

The equal tax for all, required by the fact that all essentially get the same services from society, cannot be rejected on any sound principle but only on the expediency that it is assumed impossible to collect from all. The inability to meet this debt should be one criterion of assignment to the perhaps one-tenth of the population considered dependent or "hospitalized." An *enormous* bureaucratic saving would in any case be made by a uniform tax. Other economic customs which need to be considered in terms of ethical value implications are insurance, migration for employment and upset of the natural market for workers, managers, goods and capital by political power. But these are realized and discussed whereas the income tax question has some features of being under a taboo. It is not asserted here that the equal tax in its simplest sense is a proven desideratum, but only that it deserves serious study in the light of new principles.

(5) Current discussions of ideal population size have (a) neglected the arguments earlier (page 366) for the greater survival and evolutionary value of larger populations, (b) extrapolated in too simple a fashion from present cultures and populations, neglecting what scientific invention is likely to do in radically changing problems of food supply and pollution, (c) completely ignored the question of population quality. A monster population *could* be a healthy and vigorous organism. The beaming of hysterical calls for population reduction on the advanced societies, and specifically on the middle classes, is, most unfortunately, calculated to convert world society into a headless monster.

(6) As to the structure of society Beyondism indicates no extreme position such as the abolition of classes or the institution of elites. As far as evidence can at present be read, fractionation of society into cultural and genetic sub-classes has some valuable functions, provided it is on an underlying basis of political and spiritual democracy. Such groups lessen genetic regression to the mean and permit try-outs of diversified sub-cultures, but if carried too far seem to impair morale. Research urgently needs directing to the problem of optimal within-group diversity.

(7) The type of scientific inference from evolutionary goals that may be needed in the aspects of ethical decision that depend on psychological dynamics can be illustrated by the presently much debated field of sexual morals. Here the discrepancy between genetic endowment, in this case in the sex erg, and its realistic adaptive value – the CAG discrepancy – is extremely great, highlighting the problem in all such discrepancies. Rationalism, in the form of modern Humanism, has taken altogether too limited a view of the "effect on others" involved in sex behavior. The need for restriction on sexual freedom is complexly determined, but is indicated especially by the economics of psychological energy, and the dependence of cultural creativity upon ergic sublimation. Though our emotional imagination is inadequate to conceiving that a change could occur in our preoccupation with romantic love it is likely

that, in terms of millennia, comparatively rapid transformations will be brought about here, perhaps with initial aid of biochemical reduction of sex drive, opening up new spectra of cultural-emotional expression to replace this domain of expression.

(8) Apart from the technical advances which learning theory and psychological testing are likely to bring to education, Beyondism needs to emphasize three things: (i) the need for a far more comprehensive and realistic biological education giving perspective on man's place in the universe; (ii) a training essentially in emotional balance and character, ensuring a fair-mindedness and objectivity, without which advances in reasoning are merely dangerous, and at the same time a capacity to defend oneself against the emotional arts of psychological warfare. (Especially, education should avoid a negative correlation of intelligence and character.) And (iii) the extension of the technical development of education toward defining what is ideally needed in the "raw material," i.e., cuthenics and eugenics, need to be geared into more positive mutual connection.

(9) Mass media of communication have been allowed to grow up haphazardly apparently with no other principle than that they are free to say what they like and to cater to whatever public tastes are profitable to them. No other institution in organized civilized societies has been allowed to grow without internal and external controls, and it is only by their mesmerizing popular opinion with the obsolete slogan of complete freedom to comment on policies that they have persisted in this immature structure. Granting the mass media complete political freedom should not be construed as granting them absolute license to undo education and to vulgarize and trivialize the minds of young people. Their unregulated power to put out information selectively, to make more educated viewpoints invisible, e.g., in the interests of advertising, and to destroy reputations, is an anachronism in societies otherwise supplied with the necessary checks and balances of a complex organism. In the recent action of the British, American and other governments on tobacco advertisements, and in some other ways a beginning of the necessary control by social conscience and non-party government has appeared. Until something akin to the accountability in the parallel "information service" of education appears, however, no remedy is visible to the more subtle problem of transmitting mere noise.

(10) Injunctions from Beyondism for practical social action studied in this chapter are to be taken more as illustrative of the kind of steps of inference possible from this approach than as final suggestions. The chief differences of the innovations suggested by Beyondism from the well-

entrenched habits and positions which it opposes reside in the fact that it begins with doubts, with a search for more explicit basic principles and with systematic (statistical) factual investigation. However, since it takes the duty and the ultimate goals of investigation seriously it does not propose to remain lost in doubts. For it follows the scientific philosophy of Bacon that "If a man begins in certainties he shall end in doubts; but if he will be content to begin with doubts, he shall end in certainties."

8.9 NOTES FOR CHAPTER 8

[1] When the newspapers play up the game of "right wing" versus "left wing," and the young form into Fascist and Communist battalions, the investigator of the real dimensions of disagreement can only cry, like the falling Mercutio, "a plague on both their houses." Quite apart from whatever rational and real dimensions can scientifically be discovered for political directions of diverse movement *as such*, the psychologist has found clear evidence that sheer personal temperament dimensions rather than objective data determine an appreciable amount of political decision. The temperament dimension descriptively labelled *tough realism* versus *tender-minded* sentimentality (I factor, premsia, in technical terms; Hall and Lindzey, 1970) seems to decide voting on treatment of criminals much more than any evidence on how effective different treatments are. Similarly, the degree of personal integration of the *self-sentiment* (Q_3 factor; see Cattell and Butcher, 1968) is important in young voters in deciding their degree of acceptance of social values. The *conservativism-vs-radicalism* factor, Q_1, is a personality factor which self-consciously determines the individual to see himself as radical or conservative, and according to his score he tends to "buy the book" of radical or conservative voting, without examination of the real merits of the items. For example, the radical votes for a change to the metric system, and a switch to looser sex behavior, though to a social scientific examination these may have a very different status. Similarly the overprotected $I+$ (premsic) individual is found, as a function of his temperament, to vote against even a defensive war and to vote for abolition of capital punishment, despite the inconsistency in that a take-over by a foreign power might mean wholesale "capital punishment" for democratic opposition.

Another unmistakable indication of the distance that we still have to go in bringing political interests and political categories closer to social scientific realities is the divorce in methods and concepts between students interested in politics and students interested in science. Only very slowly is really scientific thought being brought to bear on political choices and categories. A recent revealing symptom of the state of mind of those who at present follow political science has appeared through observations of the confused disturbances on college campuses over the last decade. The more emotional parades of violence have proved to be recruited more from literary and political science departments than from engineering, science and medicine. The latter have certainly been equally concerned to bring a more progressive and scientific rationality to social affairs, but have tended to investigate in terms of more realistic principles rather than demonstrate. The study of rowdy demonstrations seems to place much present political "science" with the emotional, intuitive and prejudiced (emotionally pre-judged) approaches.

As a more direct personal experience the present writer has to admit thirty years of disappointment in attempting to get political science interested in quantitative and basic social psychological group research, instead of "history" or journalism. After joining the Academy of Political Science and reading its quarterly journal hopefully from about 1940 to 1960, he reluctantly concluded that for most writers and teachers it remains a display of opinion and philosophy rather than a development of the quantitative science which a social psychologist would be interested in trying to build. As far as I can recollect, I never saw any numbers in this journal except at the top corners of the pages!

Perhaps the main thrust of the substantial constructive literature of political science is taxonomic, demonstrating that in the past and in the present virtually every conceivable major type of political government is successfully in operation somewhere around the world. Dictatorships by individuals, "dictatorships" by parties, oligarchies by power groups, oligarchies of selected elites, theocracies, would-be classless democracies in which democracy has a religious quality, as in the U.S., and democracies compatible with class, in which democracy is a system of voting, as in Britain; also various instances of tribal government, and so on through a fine catalogue of types of systems. Indeed, one indication from this data is that the form of political organization might have less effect on the rest of culture, e.g., on personal freedom and economic opportunity, than is often supposed by those preoccupied with politics!

[2] The objections offered to most of these non-democratic systems by thoughtful people are well known. For example, in the U.S. Congress on April 2, 1917, Woodrow Wilson, (at that time expressing no comparable doubts about the working of democracy) had complete doubts about oligarchy. He pointed out that "Cunningly contrived plans of . . . aggression, carried it may be, from generation to generation, can be worked out and kept from the light only within . . . the carefully guarded confidences of a narrow and privileged class. They are happily impossible where public opinion commands and insists upon full information concerning all the nation's affairs." Presumably, the capacity of an oligarchy to plan beyond the whims of popular opinion in a particular generation regarding these undesirable strategic aggressions and defenses would also extend equally, however, to valuable and desirable social and general construction, such as was accomplished, despite meager popular enthusiasm, architecturally in Athens, artistically in Florence, and in the centuries-long building of the great cathedrals by a religious oligarchy.

[3] A very common misuse of slogans and catchwords for example in student and journalistic debate, is presented in, for example, the confusion of "authoritarian" and "dictatorial" in relation to government. The latter, of course, refers today to single person government—what Plato and Aristotle would have called a tyrant, though in intention the "Fuhrer" or Doge may be benevolent rather than tyrannical. An authoritarian government can, by contrast, be democratic, as the Communists would declare the governments of Russia and China to be. The Catholic Church would also declare itself authoritarian but not a dictatorship—since its College of Cardinals qualifies the infallibility of the Pope. "Democratic," as the quite different use in Russia indicates, (East Germany is democratic: West Germany is not!) has also become scarcely a meaningful word for any realistic, precise, international discussion. For example, in the United States and one or two other countries, it means—to the confusion of the rest of the world—merely a political party. In East Germany, it is made synonymous with Communistic. Above, we have defined it basically as a mode of reaching decisions by equal voting power. To many Americans and Scandinavians, again, it seems also to represent a religion, having to do with high valuation of the individual soul and

individuality; and as such, it becomes an historical Nordic culture equivalent of Christianity.

The real social scientist, struggling to apply research to social structure and to discern desirable new directions of political construction finds his struggle toward precision baffled by the present coinage of terms. Though admitting that some meaning can be given, as just indicated, to democracy and that a rough meaning for "practical Communism" and "practical Fascism" may have been given by Stalin and Mussolini, the social psychologist making a fresh approach finds the continued use of these terms inapplicable. He has to cut across massive vortices of debate completely irrelevant to the scientific issues, reminding him, as one said, of trying to reach his laboratory on the afternoon when crowds are streaming to a university football game.

A perennial confusion which waylays the young—and sometimes the mature—is that mentioned above which identifies "progressiveness" with the have-not group and "conservatism" with the haves (say, the Republican Party in the U.S.A.). There is no *logical* reason why, say, the ownership of a factory should make a man less progressive than his workmen. Historically, in Britain for example, some of the most progressive measures in the support of science, of education, of family planning, etc., have been the work of the conservative "managerial" party. (At the moment, for example, it is the Labour Party that is opposing the integration of Britain into a common effective, economic unit with the rest of Europe.) The accidental and illogical basis of have-nots being labelled progressive is that most of the *possible* changes in the distribution of property would naturally take it from those who now have it (and democracy is easily perverted into a system for legalized robbery at the polls). But "progressive" should obviously apply to some truly new designs benefitting the community as a whole, not to a dreary see-saw of alternating mutual exploitations. Yet the confidence trick of calling the less skilled, lower-earning occupational groups (the left wing) the "progressives" is repeated upon each susceptible new generation.

[4] This term is adopted by most writers since 1930 for precise description of what is aimed at by such artificial insemination ("test tube babies" to the journalist) as has a eugenic purpose. Thus a father remote in space or time, chosen for outstanding abilities and social dedication, e.g., a Nobel Prize winner, a great social leader, is enabled (*see* proposals in Muller, 1966, and Graham, 1970) to contribute a relatively large number of children, e.g., mainly, at present, through fatherhood in families where sterility of the husband frustrates the desire for children.

Also among the more "far out" proposals we must consider what Haldane called chromosomal engineering (recently popularized by the Nobel Prize winner Lederberg (1969) under the name "euphenics" with some dissent from other Nobel Prize winners such as Beadle (1963; Sturtevant and Beadle, 1964)). Although attempts directly to change the chromosome structure are likely to achieve technical physiological success before very long, we should in all realism note that even if manipulation of genes were achieved tomorrow, the second and indispensable part of the research necessary for such an undertaking, finding social behavior connections, as described above—would still remain to be done.

A related purely technical advance likely to have substantial utility in negative eugenics is the increasing ability of medicine to recognize chemical and structural defects in the foetus e.g., by amniocentesis and thus produce beneficial abortion (*see* particularly Arehart, 1971a,b).

Here too, in the last resort the genes themselves (in the last case in their earliest foetal expression) have to be related to final social and cultural expressions. In short, chromosomal engineering and foetal detection are real advances, but the greater part of the research task remains, namely, the full *evaluation of the psychological and physiological advantage of the chromosomal change produced*. In view of the immense complexity of having to work out not only the advantage to the individual but also the effect on a complex society of a particular ability or temperament change, one must recognize that "chromosomal engineering" has only accomplished much the smaller, more mechanical part of its aims when the molecular geneticists themselves are through with mapping and manipulating the genes.

[5] While on the subject of critics let us attend to a number of spurious or misunderstood *technical* arguments against eugenics. Prominent among them one finds such as: (1) if a defect is recessive we cannot get rid of it; (2) when a homozygous form is undesirable a heterozygous form involving the same gene is often advantageous; (3) that since any actual capacity is partly inherited and partly a product of education, selection on the basis of the phenotype "produces no genetic selection." The answer to the first is that we can still advance but far more slowly. To the second let us respond by a realistic substitution of "very occasionally" for "often." The third appears mainly in the writings of crassly environmentalist sociologists. A comparatively sober statement of this kind occurs in M. Gordon, *American Peoples Encyclopedia*, where the argument is made that since both environment and heredity entered into achievement, or into actual social or economic status, selection on the basis of this final status outcome would produce no improvement in the genetic level. The answer is that the efficiency of selection when heredity and environment are about equally contributory is *reduced* not abolished. So long as hereditary and environmental gifts are uncorrelated it is still sound and effective to select for genetic effect by operating with final, mixed "performance criterion." If they are *positively* correlated (and negative correlation is rare) on the other hand (and this has been shown to be true in the important case of intelligence and education), adjusting birth rate to actual final *achievement* constitutes very powerful eugenic selection of intelligence as such.

Technical complications reside also in the fact that with certain patterns of genetic origin of traits, selection would be rapid and with others only limited in action, as in a model which Penrose (1948) points out. It seems to be the hobby of anti-eugenists to concoct fancy models of the latter kind to disprove eugenic possibilities, though as Poniatoff shows (in the case just cited), they may lead to scientifically quite absurd results.

[6] As recently as 1918 some governments thought that they could make money by printing it. As Friedman (1968) points out, the heavy printing of new style money by the Bolsheviks at a time when a fixed amount of old style (Czarist) money was in circulation caused the former to lose value by inflation and the latter to climb in its rate of exchange.

[7] It has taken most of a century for a real appreciation of the moral implications of reproductive behavior to be properly appreciated, and most of this advance in sensitivity has been due to the writings of the comparatively small group of scientists who have developed eugenics. For the reader who wishes to get a more thorough grasp of the concepts that have developed in the technical and socially far-sighted literature in this area, a relatively long, yet still too brief note is offered here covering the movement of eugenic ideas.

Although explicit eugenics is less than a century old, the basic propositions appearing in various original thinkers, go back to classical and biblical times. Beginning (as far as records go) with Plato (in *The Republic*) and continuing with a recent crescendo in some far-sighted biologists and social scientists such as Charles Darwin, Galton, Leonard Darwin, J. Huxley (1957), Hardin (1964), Graham (1970), Keith (1949) and Muller (1953), the main argument (at least for "preventive or negative eugenics") has been well summed up by

Dobzhansky (1960): "If we allow the weak and deformed to propagate their kind, we face the prospect of a genetic twilight; but if we let them die or suffer, we face the certainty of a moral twilight." Selection by birth rate instead of death rate is thus the key proposition in eugenics.

Most practical propositions have concentrated more on an economic direction of planned parenthood — rightly as the present writer believes — rather than on sterilization, eutelegenesis, chromosomal engineering, postponement of marriage, and other measures relevant only to small groups, or, too socially radical in nature to be accepted for sometime. As stated in the text, the correlation of income with health, intelligence and responsibility of character is probably positive through 90% of the population, and breaks down only in the very rich and very poor. Consequently fitting birth rate to earnings would ensure a generally positive eugenic trend. In the name of perspective on this, the central fact to keep in mind is that from the abject poverty level to a comfortable living condition the great majority are striving for a common goal of economic adequacy, so that their success tends to correlate with their competence.

The effectiveness of trying to operate simply through an economically-assisted and public-opinion-encouraged differential birth rate, maintained through the bulk of normal families, has, it is true, repeatedly been criticized and doubted (in favor of more spectacular methods) even by some leading thinkers. Thus H. J. Muller (1966) argues "Most genetically less fit individuals would not accept the judgment of their being so themselves, and [would then not] voluntarily engage in less than the average reproduction. Nor, *vice versa*, would the more fit choose to make the career sacrifices . . . by their having larger families." He concedes "this *could* be taken care of . . . by taxes . . . but in a democracy the enforcement of such seeming discrimination would hardly be accepted." Many may think this pessimism to be realistic political psychology. Perhaps it was at the time he wrote but it does not allow for increasing biological education, and, above all, for the newer moral standards that Beyondism can develop.

A supplementary argument for raising the economic level of the submerged tenth, even by outright charity, comes to us in the initial indication (still needing to be checked) that any rise in income produces a *reduction* in family size. The argument is that this works because there is now a standard of living to be threatened by excessive multiplication. From this, one may conclude that the "amour propre" of the sub-normal will work in *favor* of reduction, not as Muller supposed. At the other end of society, as regards the willingness of above average income families to have more children, fitness is here self-defining. Those with insufficient superego development to accept the sacrifices are self-eliminating. Lastly, in doubt of Muller's social judgment above, one must add that he is repeating what seems a fallacious interpretation of the main dynamics of democracy. (Unfortunately, it is *partly* true that envy and jealousy can make democracy a leveling-down operation. But Christianity, Humanism and Beyondism unite in one common moral persuasion here to accept individual differences with appreciation and admiration rather than envy.)

How large an effect eugenics can exert by a positive socio-economic birth rate remains to be evaluated. Meanwhile, at least history tells us that a *negative* eugenic situation can produce notable deterioration in perhaps six to twelve generations, as, for example, in Rome from 150 to 350 A.D. However, probably never in history has there been a period in which dysgenic trends *could* take effect so rapidly as in the modern welfare state — if morale should fail in the larger, unendowed respectable skilled worker and lower middle class section. Three or four generations of disregard for genetic quality through the middle range of society might lead to such a breakdown of society under "welfare" conditions as would be irremediable.

The view that the number of children a family produces is no one else's concern has to go. There is no "God given right to propagate" without an equally "God given right" to starve or to be massacred in world wars. Most economists in political advisory positions are at present doing exactly nothing about the fact that university graduate students subsist on grants which forbid them to have a family until an age which effectively reduces their eugenic contribution to the next generation, whereas welfare state paupers, individuals of few skills, and those unadjusted to steady work can reproduce and have their legitimate and illegitimate offspring taken off their hands.

The *New York Times*, in a recent leader, probably accurately describing the associated misery and vice in the submerged tenth, notes that 33% of the relief "families" surveyed are deserted by the father. It continues, with the journalist's prerogative, to criticize this as an "indictment of an America falling short of its ideals." Actually, *sentimental* "ideals" (notably unreadiness to face biological and economic facts) – such as the above unrestricted "right to propagate" – are the main *cause* of the persistance of this problem for so many generations.

[8] Apart from these serious considerations, it is not entirely trivial to add that the collection of a "single value" tax (over the range of, say, the economically middle 80% of the community) would eliminate such an enormous amount of accounting and inspection time that the benefit to the welfare of the community would be immense. However, if a differential tax *is* to be maintained in this middle range, the social scientist may, at least, point out that a highly necessary income tax reform is a *far* larger child allowance and/or college scholarship allowance for able and stable children *regardless of income of the parent*. (The present practice in many scholarship schemes of denying the earned scholarship of a child whose parents have a middling income and admitting a definitely less bright candidate whose parents have earned less is highly dysgenic.) At present, the cost to the parent who begets an able child suited for and ultimately receiving a university education is far greater (perhaps $15,000 per child) than the production of a less gifted child, whose education may end at eighteen. Economic laws as at present permitted to act thus operate to reduce the production of the bright.

Incidentally, though liberal opinion has hitherto favored scholarships over income tax remission, on the group that heredity is imperfect (though the correlation of mid-parent with mid-child on intelligence is about 0.7), some counter arguments have been overlooked. (1) The "reward" of a scholarship comes twenty years after the decision to conceive a child, which is a little remote to be effective, even with far-sighted parents, whereas tax remission would take effect at birth of the child. (2) In spite of the fact that all psychological investigation shows about education (especially of character) in the home being decidedly more potent than in school, many educators persist in believing that progress in education is entirely a school affair. But economic encouragement which results in more children being brought up in homes with richer cultural backgrou d should certainly be introduced. A larger income tax allowance for children would be such a measure. (3) Finally, a punitive income tax interferes with the process of charitable endowment and support being discriminatingly directed to the values and institutions which individuals themselves perceive as being the source of their success. For example, a man who has succeeded through overcoming alcoholism may direct his modicum of savings to a society to save alcoholics. There is a sense in which this is an empirical and scientific (if still partially blind) direction of saved income to the most appropriate activities for the improvement of society. It is surely more real than the action of a civil service department directing the disposal of taxes collected by a government – swayed either by doctrinaire theory or a "pork barrel" vote.

[9] In as much as the pleasure principle can strongly express itself in individual buying, e.g., by favoring luxuries rather than necessities, and ignoring long term realities (such as the necessity to save), there can surely be little doubt, however, that capitalism needs to move in the direction of more control by sales, by excise taxes and by manipulation of borrowing rates, to tax relatively degenerate forms of expenditure. This is recognized, for example, in respect to tobacco, drugs, alcohol, gambling, etc. But, it needs to be recognized still more when more is spent on, say, breeding race horses and running race tracks or on pornographic literature than on scientific research or up-to-date libraries. Here the real issue becomes one of the degree to which the cultural and political leaders of a country can and should dictate the growth of good, as opposed to decadent and degenerate recreations and expenditures (in terms of course, of the moral values which that society has chosen).

As to the difficult issue concerning the relative extent of the economic control of expenditures and endowments by government and by private individuals, what may be seen as an inconsistency between the above position on the role of private savings (page 326 above) and the present position on taxation of expenditure is not really so. The above position argued that individuals who save should be free to endow what seems to them valuable social service. The present position is that the government should be free to tax relatively heavily expenditures that are frivolous, unhealthy, extravagantly luxurious or degenerate. The anti-establishment fraction of the recent permissive generation will argue that no one knows what is degenerate taste or bad morals. The Beyondist contention is that though this is true of what is derived from intuitive, revealed religious values, social scientific research will soon be capable of more objectively evaluating the relative progressiveness or decadence of expenditures.

[10] As mentioned earlier, the argument was explicitly put by Max Weber that certain religious moral values could also be conductive to the economic success and stability of a community; and he argued specifically that the Protestant ethic of industry, enterprise, and frugality aided this goal. (However, several other ethical cultural values do the same: the industry and frugality of the Chinese, and the values of the Judaic religion, have produced a similar economic differential from their neighbors.) Beyondism would endorse this evolutionary value of a larger population based on wiser economics arguing that inter-group economic competition is one of the most effective and humane methods of engineering expansion and contraction.

An empirical study among sixty-nine countries (Cattell, 1949) does, indeed, show a factor *besides* that of education-affluence (already discussed, *see* Table 4.2) which links average real income to such variables as are shown in Table 8.1. Although the influence which is responsible for the vigor of environmental control expressed in this pattern is not yet clear, there is no doubt about the existence of some independent underlying factor of this pattern in the economic domain.

One may have here a plexus of religious preference, temperament, enterprise, and industry. Incidentally, it has a negative correlation (-0.43) with percentage of population Buddhist. The pattern is offered here as the merest first glimpse of an interesting direction of further research on the contributors and consequences associated with economic level. But, it does offer some support for both Weber's and McDougall's reasonings in this area.

If a fuller array of experimental variables could be introduced (these studies are costly), one suspects that a restrictive attitude to sexual expression, a reluctance to spend on luxuries and to waste time, and other adjustments calculated to direct energy into control of environment and scientific exploration would be found in this pattern. For the present, all that one can note here is that apparently, with equal natural resources and natural gifts,

Table 8.1 "Achievement": An Economic Level Value Dimension in Modern Nations

High real income per head	0.68
High protein consumption	0.47
High percentage of Protestant affiliation	0.46
Low homicide rate	−0.45
Many miles of railroad (and other applied science variables) per person	0.41
High percentage of Nordic population*	0.37
Low frequency of revolutions	−0.35
Low (controlled) birth rate	−0.34

*One may wish to explain this as a geographical accident, or, as McDougall did, as a particular suitability of the Protestant religion to the Nordic temperament. Certainly, the boundaries of religious groups coincide more with racial boundaries than would be expected from chance.

certain value systems inimical to dissipation of emotional energy will alter the whole level of economic production. This observation brings together the considerations in the first part of this section on the inter-individual morality value of economic practices, with their final testing out taking place in the inter-group situation.

[11] The problem of defining and reaching an optimum population covers, of course, more than resources: it includes the problem of living stress from crowding, so long as the present partly neurotic drift to crowded cities prevails, and the problem of pollution. Incidentally, the expression "neurotic" drift is used because an irrational degree of loneliness is a central feature of neurosis (Cattell and Scheier, 1961), and the growth of cities beyond optimum sizes for economic efficiency and health is obviously irrationally determined by a need to be reassuringly surrounded by "life." The stress condition in crowding, if Hoagland's analysis (1947) is correct, will result in evolution toward smaller adrenals and other hormonal changes better fitting us for close living. On the other hand, both this and the pollution problem are more likely to be beaten in the near future by intelligent planning of "rural cities" (M. and L. White, 1962).

[12] As implied above, a true acceptance and practice of regard for the dignity and human worth of the individual has essentially nothing to do with a classless society. Britain as a democracy, has never been classless, yet it stands very high among the countries of the world in its regard for the essential value and dignity of the individual, regardless of his class. On the other hand, when Russell with strong Communist sympathies, visited Russia, he wrote (*Autobiography*, page 142), "There was a hypocritical pretence of equality, and everybody was called 'tovarisch,' but it was amazing how differently this word could be pronounced according to whether the person addressed was Lenin or a lazy servant."

[13] The problem of breeding far greater longevity is puzzling unless we recognize that it can be effected both by (a) direct breeding from older people, as here mentioned and (b) greater breeding in the normal age range from strains known to be long lived. The former is desirable, though on the female side, it requires later cessation (and presumably later beginning) of healthy reproductive capacity too, whence the main selection must come through the male. The latter could be made to operate nowadays, with good family records. But it must have been effective long ago too, and this is puzzling unless we suppose that

tribes of higher culture, in which quite old people could contribute by their wisdom to group survival, did better. This supposes longevity, like superego endowment, to be a product largely of inter-group competition. Lorenz (1966) reports survival guidance from aged members operating even in animal groups.

[14] Naturally, the rationalizations or reasons for evasion of ethical restrictions on sex behavior have been thoroughly worked out over two thousand years. In the twentieth century complete sexual freedom has been considered an adjunct of progressive thought. The recurring expression (in such writings as those of Russell (1957), Lindsey (1925) and, it must be confessed, the present author's writings in his twenties), was "the outworn, rigid, puritannical traditions of Victorianism." This literary "conviction without trial" still goes on, as in a current liberal magazine article on sex by a well-known writer condemning "Christianity, Calvin, Cromwell and the bigots of the present day."

Actually, when criticisms of sexual restriction based on more solid grounds than mere unfashionableness are attempted, the authority quoted by the "intellectual" is essentially Freud. This is a product of superficial reading creating the impression that he proved *restraint* ("suppression") to be pathogenic, whereas in fact he was speaking of *repression*, and even that only as it is bungled in the neurotic minority of the population. The neurotic, incidentally, he described as constitutionally defective (burdened by "the psychosexual constitution"). The view that repressive superego action on sex expression also in the normal person can generate anxiety and neurosis — be it a misunderstanding of Freud or by Freud — has been powerfully criticized in Mowrer's recent writings (1967). The finding by Gorsuch (Gorsuch and Cattell, 1972) and by the present writer (1971) that in the normal range a stronger superego is actually correlated very significantly with *reduced* anxiety also explodes the literary misunderstanding concerning neurosis and control.

At a descriptive level, and in relations to some "symptoms" which the critics of restraint and sublimation describe, the movement for sexual license may be right in saying with Russell that continence breeds an inclination in the young to violent and dangerous sports, and, among religious philosophers, a tendency to fanaticism and bigotry (strong conviction). A different meaning is given to this by St. Paul, St. Augustine and others, e.g., in the latter's observation, "I feel that nothing more turns the masculine mind from the heights than female blandishment and that contact of bodies without which a wife may not be had." Priestly celibacy and that until recently adopted by university fellows in the older universities, rests on the same conviction, which at least proves that neurosis is not a necessary condition of a celibate way of life.

Naturally, the question of the effects on group survival and cultural creativity of various kinds and degrees of restriction on sexual expression is a complex one. It is a problem for multivariate calculation, employing individual and social measures that we are only just beginning to make; wherefore, no authoritative scientific opinion can yet be given. What we *can* be quite sure of is that the arguments of these "progressive" pundits who would liberalize and increase sexual activity on "rational" grounds are no more scientifically defensible than the quite opposite intuitions of St. Paul or St. Augustine. If anything, both the evidence in comparative anthropology from such studies as that of Unwin (1934), showing positive correlation of religio-educational level of culture with amount of sexual restriction, and the psychological evidence (Freud's clinical; our own in the dynamic calculus) suggest that the Christian patterns are more probably right. A wise society would do well to accept the tried moral standards of revealed society until such time as experiment can deliver objective and profitably discussable findings. At the same time, as Somerset Maugham has reasonably argued, common sense might admit some abatement for the late adolescent and the immature of the moral restraints practicable for mature persons.

In relation to the arguments of Morris (1956) — a student of anthropology — quoted above — not arguments for sexual license *per se* but for enormous cultural contributions having stemmed from human sexual *expression* — one must point out that although man seems more preoccupied with sex than most (undomesticated) animal species, the more important correlations to watch are those of sexual expression and cultural creativity among individuals, *within* the species. What we then observe, in Unwin's (1934) and other obtained correlations over a hundred or more tribal cultures, is that *among* human groups, the higher cultural and religious development corresponds with greater restriction on sexual expression. Freud's view of religious, artistic, and other cultural inventions, as based on sexual sublimation, simply gives support to such anthropological evidence on a basis of deeper and more individual clinical observation. Any psychologist is aware, on the other hand, that for a minority of shut-in types sexualization is the beginning of socialization, so that a man may first come to an understanding of his fellow man through his fellow woman. But, this is a minor mechanism, since most obtain healthy social contact with and sympathy for mankind without the prop of either sexuality or homosexuality.

The well-known opposite extreme to the "progressive" is that expressed by certain religions, notably Catholicism, where either life-long celibacy or chastity, even through the present long pre-marital period, are the ideals. Psychologists do not yet understand what the full range of effects from such restraint are, though neurosis, intolerance, sublimation into agape or perversion into homosexuality, and cultural creativity are among the results discussed. An argument for this position is that although the majority of human beings seem incapable of the severe adjustment demanded, yet the cultural gains may be very great through an elite which succeeds. That that "muse of fire" Shakespeare himself referred to sex as "expense of spirit in a waste of shame" and left in his will to Ann Hathaway only his "second best bed," may instance this adjustment of the creative elite.

[15] It has been pointed out above in other connections that sexual selection may act contrary to (the rest of) natural selection. Indeed there are intriguing problems here to which Darwin pointed, and Lorenz more recently, but which human population genetics research has still not got around to investigating. An overemphasis on sexual selection could offset valuable, positive natural selection developments. The burdensomely antlered deer, the steatopygous woman of certain African tribes, extinct birds that outdid the peacock, and the over-breasted female of American Playboy pin-ups alike evidence secondary sex characteristics which, though good advertising in the mating field, in the *total* account of natural selection are disadvantageous, often to the degree of producing extinction. The high role of sexual interest in man, and its increase under certain conditions, suggest that sexual selection might be magnified in its effects in humans relative to other species. It is possible — judging by occupations considered fit for woman — that Western culture has gone further than Russia in generating a child or doll-like type of woman disadvantaged in terms of natural survival selection in an adult occupational world such as sustains the survival of groups. This undermining of natural selection by sexual selection — with which "women's liberation" should be concerned — needs serious thought in any consideration of the total effect of sexual mores.

[16] The philosophy of education — as distinct from its technical structure and practice — has, like philosophy in general, shown a tendency to move vaguely in sterile circles, largely determined by the social mood. For this reason I have found it illuminating — as the reader may do — to go back and compare the summary I made forty years ago with the present position ("Education: the Conquest of Obstruction" in Cattell, 1933b). Although I was at that time inclined to treat psychoanalysis as reliable scientific psychology, and consequently to swallow the "Progressive" philosophy, the positions reached show on the whole enough

consistency with the present to give some confidence in their capacity to stand the test of time.

Probably the major conclusion from the comparison is the optimistic one that real progress can be made in a field when a sufficient number of keen and enquiring minds concentrate upon it. For a remarkably high percentage of the pioneer changes advocated in 1933 have in fact succeeded. They included the abolition of Greek and Latin as the core of secondary education, since experiments on transfer of training gave no support for maintaining them on any grounds of "intellectual discipline"; the increase in higher education for women; and the growth of social sciences in the universities. Freedom has been won, moreover, in the new universities from the modeling of curricula and style (in England) slavishly on the inbred values of Oxford and Cambridge. There has been more complete removal of traditional religious sectarian education from school curricula. Instead of psychology being regarded as a pursuit of cranks, a scientific application of the new ability findings of psychology to scholarship selection (testing) and of learning theory to teaching methods has become accepted as natural and reasonable. In the schools the argument for greater emphasis on science, especially on biology has largely been achieved. Only the doubts I then expressed about using the school for moral character education seems to have rested on a poor foundation, and to have led nowhere.

[17] The right of the educator, as of every other craftsman, to pass on specific recommendations for better raw material for the process he is called upon to carry out will surely not be questioned—especially when he is accused of having grossly failed to reach required standards in his product, generation after generation. In a U.S. Office of Education report for 1970 we are told (with some implied criticism of the teacher) that "half the unemployed between 16 and 21 years of age are functionally illiterate." Other surveys show about three million illiterates in the U.S. population. (This percentage is about par for Western cultures, except for a few outstandingly organized and probably more gifted populations, such as those of Switzerland, Sweden and Japan.) It is time that our education policy makers recognized that this is a greater reflection on our neglect of eugenics, than of education, for three million is just about the percentage below an I.Q. of 70 (which level is accepted in most countries as the borderline of mental defect).

Incidentally, the question of what we are actually doing when we think we are raising the I.Q. directly by education has recently been very thoroughly examined in books by Cancro (1971) and Eysenck (1971). The cost of school education per child—if a fixed standard is aimed at—goes up steeply with lowered I.Q. And time as well as money presents a problem; for thirty years of education might barely suffice to develop certain occupationally required intellectual skills in truly sub-average intelligences. Education of *personality* and *character* in the low I.Q., is, however, a more promising possibility. In the inter-group cooperative competition which Beyondism encourages the relative survival of communities will be determined partly by *the cost of their educating their populations* (*to an agreed common standard*). The main determiner of the size of this burden of costs will be the magnitude of the genetic lag.

[18] It will be remembered that the idol of the young intellectuals of the 'nineties was Oscar Wilde, and of the 'thirties, C. E. M. Joad, both of whom well illustrated in their lives the combination of brilliance with irresponsibility (seen in the statistics of personality measurement in the tendency of surgency (F factor) to go with low superego (G factor), as both Eysenck (1952) and the present writer (1957) have shown). Joad will be remembered as responsible for the Oxford oath ("under no circumstances to defend our country"),

which Churchill showed encouraged Hitler. (He later pleaded guilty in court to common monetary cheating, systematically carried out — of the railroad company.) The mess made of their lives by *some* (but an appreciable fraction compared to the average man) among scintillating intellects over the last three hundred years is well documented. Vanity, shoddy thinking, argument by cunning emotional displacement, and sheer mis-statement are unfortunately highly prevalent in such "intellectual leaders" over the last century as Oscar Wilde, Byron, Maupassant, Shaw, Laski, Hugo, Nietzsche, Wagner — and perhaps one should include even Dickens and Aldous Huxley (particularly over drug usage). Even Einstein, after presenting the theory that wars can be avoided, and around 1930, urging all young men to refuse military service, seemed to see no inconsistency in writing to Roosevelt in 1940, as instanced above, urging development of the atom bomb. These unrealisms, and worse, of many "highly educated" persons contrast sadly with the basic integrity of many others, such as Dr. Johnson, Emerson, Newton, Faraday, Madame Curie, Wells, Zola, Thomas Huxley, Browning, Tennyson, Franklin, and many great scientists. But a general positive correlation of intellect and character in all generations cannot be asserted.

[19] In some countries, where "morale" in a general sense is good, and crime rates low, e.g., Switzerland, New Zealand, Germany, Britain, Sweden, Japan, an impalpable self-censorship has long established positive standards in mass media, though how this subtle mechanism operates or why it is associated with higher score on general morality is not yet clear. It is, however, a common observation among travelled educated Americans that British T.V. and radio (to cite one they can follow without language difficulties) are pitched at such a level that the term "idiot box" scarcely continues to apply to T.V. A similar mild and intangible but real "censorship" is perceptible regionally in America. In the realm of newspaper publishing, the cultural interest of New York and San Francisco has given better quality of comment, while the theocracy of the Mormons in the mountain states has resulted in good reading largely free of the crapulence in the rest of the nation's newspapers. Anyone who calls this "cultural persuasion" by an educated minority — also seen in certain T.V. programs — an objectionable censorship had better ask whether the tyranny of commerce and the dictatorship of the untutored intelligences which censor his own media by means of gross "popularity counts" (for advertisers), are not a far more revolting form of censorship. The journalist may object that, unlike the educator, he does not have a captive audience, and unlike the churchman, he is not paid to elevate. If this means that he will choose to ignore both intellectual and moral quality, then societies that would survive and evolve must find some different and better way of doing what the journalist now conceives he is there to do.

CHAPTER 9

The Integration of the Emotional Life with Progressive Institutions

9.1 THE VARIETIES OF CONSCIENCE AND THEIR INSTITUTIONAL PARALLELS

To clarify the nature of Beyondism we have allowed it in the last chapter to clash with, and strike some sparks from, currently accepted ethics. Yet it must be kept in mind that the social scientific research that would go from principles to specific social actions is still largely to be done. Consequently, when we infer some deviant conclusion, such as the desirability of substantially increasing an intelligent population, of instituting a uniform income tax, or of biochemically postponing sexual activity for a decade, we deal only in tentative probabilities, even though no less uncertain than present customs.

In the end the true nature of evolutionary ethics perhaps comes out most clearly in what it positively demands in new social construction, the description of which is the object of this chapter. However, part of that "construction" is also the internal state of mind of a citizen in such a new world. And although we have stated it as a unique feature of the development of Beyondism—so different from that of revealed religions—that the inner state must be discovered from the external requirements, not vice versa—virtually nothing has been said about that inner adjustment.

That the inner state must possess—though restrained and harnessed to a sense of cosmic realities—the same warmth and enthusiasm for life and human good—as any more emotional religion, is by now self-evident. And this in turn, by the nature of human enterprises, must produce a fierce intolerance for ways that are patently wrong. When a light appears in the murk of our times, and the king of sciences — moral science — is crowned, we do not have to apologize for some glory coming back into our lives. The call of the Beyondist conscience is going to be louder and clearer than anything we have heard for a long time and will call us to great enterprises beyond the understanding of older moral systems.

But let us, again, first be somewhat coldly investigative about the nature of the growth of sentiment structures and conscience within the individual in most societies. The most obvious fact is that these sentiment structures and loyalties mirror within the microcosm of the individual the institutions, and their dynamic relations, within the macrocosm of society. From the plastic potential ergic structure of the inborn drives, roughly adapted to the primitive existence of the last million years, there emerges, by social rewards and punishments, a very different structure which we can map factor-analytically as a system of subsidiated attitudes, in sentiments to home, to school and peers, to church, to hobbies, to occupation, to country, and finally to the vital image which each individual forms of himself. Among those structures is what we have been calling — in acceptance of psychoanalysis checked by factor analysis — the unitary structure of the superego or conscience.

The manner of origin and development of the superego is somewhat different from that of other acquired sentiment structures. It may have a larger genetic component; it begins earlier, in an affectionate and dependent relation of the child to the parent, and it responds to more abstract features of the culture than do some other sentiments. Our concern here is not with its early origins, which are much the same for all ethical systems, but with the specific complexities that arise when the later growth of conscience has to accommodate itself to a more complex world picture, and to new institutions that are likely to appear only in this generation. It will be argued that institutions need to be set up, in accordance with the Beyondist insight, for the research development of progressive ethical value systems: (1) In inter-individual behavior within each group, (2) in inter-group behavior among nations united in cooperative competition toward an evolutionary goal; and in other specific role relations that derive from these. Indeed, fully to describe the main *dramatis personae* on this stage of human relations one must list (1) the individual human beings, (2) groups and their governments, (3) a world government, and (4) the universe itself (standing for some Supreme Being or Principle). However, if we consider also the distinct roles in which men can have distinctive obligations we can get nearly twice as many roles engendering some twenty or more [1] relations of the type:

Citizen to fellow citizen within the group
Group government to group government
Citizen of one group to citizen of another
Individual to his conception of the universe
World government to world citizen, etc.

The obligations attaching to one *role* of an individual as distinct from those belonging to another of his roles imply to the psychologist that a distinct structural outgrowth from the main conscience must grow up to deal with each. In a sense each role corresponds to a distinct self-concept and set of habits. Correspondingly, there are expectations by society which ultimately get concretely embodied in the statutes of the lawyer. Indeed, in the end, in any complex society, these role obligations are by no means vague or tenuous. However, though obligations are equally real between groups (since national and international law recognize such duties and rights, e.g., among corporations and nations) yet the attitudes concerned necessarily reside in the minds and consciences of *individuals*. Take away people and there is nothing in social institutions but forlorn books, of ink and mouldering paper. Consequently, as we now consider the structure of conscience in the individual it must include also the readiness to adopt roles which make *group institutions* work. And, as Shwayder (1965) brings out from a sociological standpoint, human beings are indeed capable of highly organized "stratification" of their mental sets in relation to their different roles in various social institutions and groups. The individual conscience is, indeed, a city of "many mansions." Psychological analysis suggests that in its early growth, from which all these developments branch off, it is a unitary power, founded on correct early emotional nurturing, but, in the intelligent citizen it is indeed an elaborate tree of specific branches of attitudes. While correlation analysis remains to be done, one may yet begin by recognizing three major systems:

(1) The *citizens cultural conscience* concerned with one's own probably national cultural group, operating in good citizenship and patriotism.

(2) The *world conscience* which tells what a man owes to all men, and,

(3) The *transcendental conscience* between a man and his God, or as a scientist might more cautiously wish to say, between a man and his conception of what the purpose of the universe is (Gorsuch, 1965; Aronfreed, 1968).

Parenthetically, the first of these is not merely a man's concern to keep his respectability and status in his immediate community. This is a demonstrably different psychological structure, the self-sentiment (hypothesized to be, technically, the factor symbolized as Q_3 (Cattell, Eber and Tatsuoka, 1970), among experimentally recognized personality factors).

Corresponding to these domains of conscience there are also — as there must ever be with erratic humans — domains of law, and institutions to maintain the law. About the extent and development of isomorphisms of this type — of attitudes with institutions and legal obligations — sociologists such as Cooley (1922), have well written. Corresponding to the within-group conscience there is civil law and government. For the between group or world conscience there now exists international law and a United Nations Organization. These define rights, for rights are always in potential conflict, between individuals, between groups, and between the rights of individuals and the rights of the group to which individuals belong. Concern about the conflict of the individual's rights and those of his own group might seem superfluous were it not that in every generation there are political claims for either the group or the individual having absolute rights relative to the other. As to the third or transcendental conscience, to God or the Cosmic Purpose, there has been in the past the universalistic religions with such uncertain rights and laws as grew up in their congregations, but always including the right of a man to explain his acts to his own conscience. Beyondism gives more content, logic and distinctness to this right of a man to go over the heads of men to this cosmic temple of the transcendental conscience. It is in connection with the individual's groping for this transcendental conscience that those with what Wendell Holmes called "three storey intellects with a skylight" will feel the need for new institutes for moral research. Such national and international research foundations weighing the meaning of new knowledge for ethical construction are a matter for our study in the next two sections.

Now the three-fold structure just defined above, and the further combinations from it to be discussed below, clearly present a hierarchy of interacting demands, not a simple set of rigid injunctions. Wordsworth may be right that "in all ages every human heart is human," but the savage one saves from drowning is not necessarily the man one would choose to take to dinner next week with the family. The society he and his are aiming to construct may be very different from that to which you and yours want to dedicate your lives, and what you have in common is the brotherhood of man in the second conscience, that enjoining co-operative competition. It is just possible that one may not even have that. Universalistic religions have never agreed that literally and physically all men are in the brotherhood, but only those who have adopted the values of their universality. The Beyondist conscience does not include, except as a misguided person who may yet be persuaded, some despiser of the human future of cooperative competition, and universalistic religions have considered those who blaspheme against their gods to be excommunicated to outer darkness. Without this meaning the brother-hood of man might as well include the primates, or any other animal one helps in distress. And just as the major world conscience has these restrictions and intensities of meaning so does the cultural conscience have its real obligations and restrictions in conduct within the broader world conscience. There is or should be no real conflict within these hier-archies of values — they can in principle be clearly worked out as, in a simpler sense, can the notion to "render to Caesar what is Caesar's, and to God what is God's." If there is conflict or confusion — and there seems to be much in the modern mind — it is because the individual has not learnt to clarify the dictates of his conscience.

The place where obscurity has hung longest is on issues between the second, or world conscience, on the one hand, and the transcendental conscience on the other — and this is shown also in the structure of institu-tions. In simple (perhaps over-simple) terms this means the contrast between the love of man and the love of God. The distinction has not been helped by scriptural sayings in Christianity and elsewhere that the love of mankind is the same as the love of God. Indeed, a rather well-known and admired poem of about a hundred and fifty years ago beautifully expressed this, but was pointedly and appropriately answered fifty years ago by a verse by G. K. Chesterton[2]. The primary objection to the love of the cosmic purpose being made identical with love of man, is that man has to be superseded. And though this probably has to come about through the cooperation of men, the argument remains clear if we recognize that it could conceivably fall in the end to other species than man to lead the evolution of mind.

However, whereas the world conscience toward man, assembled in his various groups, is or can be clarified in its injunctions, and, indeed, tied to the expressed aims of an enlightened United Nations Organization, the transcendental conscience cannot be so anchored, which is both its glory and its vulnerability to perversion. One does not have to be a member of some authoritarian religion such as Roman Catholicism, but only a psychologist to recognize that seeming arrogance, and certainly error, can be expressed in the individual's direct perception of what is right. If society is nevertheless, to keep its heart open to the calls of an Ahknaton, Alexander, Christ, Huss, Luther, Galileo or Savonarola, it must risk also the Thugs of India, the Black Mass, the Assassins of Persia, Hitler and the recent murderous Hippie sects in California. Doubtless better education about the evolutionary theme in our cosmos, and psychological sophistication will reduce such perversions, but both the fanatic and those who risk their lives to execute him share the inevitable tragedy of the Promethean will in the transcendental conscience. The transcendental conscience is truly conscience, but except for individual education and reason, it is on its own.

9.2   THE LEADERSHIP OF THE WITHIN-GROUP MORAL RESEARCH INSTITUTES

The transcendental conscience need not, however, remain a completely untutored conscience, as it happens to have been in some less happy cases above, Intelligent men may form their convictions of where mankind needs to go from all the resources of scientific information now available about man and his universe. But in as much as there will be scientists in institutions studying such matters, he who would go beyond the common scientific opinion will need to demonstrate his genius.

Our age is quite familiar with the notion of organized research institutions, but has never explored the organizational requirements when such institutions are to be concerned with *research on ethics and morality*. Indeed, society is still faltering in action and unorganized in its thinking even about ordinary research institutes. It does not know whether to organize research through universities, government, or independent (private) institutions devoted purely to research; it bestows, on all of these together, decidedly less than the amount that would continue to give good return on its money; and it continues to endow mainly the physical sciences when the major need today is really to endow the social sciences.

Throughout this book the novel arguments from several directions have centered on the conclusion that the inauguration of social scientific research institutions directed to finding moral values, i.e., to deciding what action is progressive, is the primary need of our times. At this point the question at last becomes a practical one, and we have to ask "Under what auspices?" and "With what duties?" As to the latter, admittedly the *general* principle of Beyondism itself needs to be intensively studied (notably on the vexed question of recognizing the signs of healthy survival potential). But the main concern of these research institutes will be to reveal what everyday inter-individual life values and actions are the best, i.e., contribute maximally to evolution at each point in the history of a society and under each condition of the social and physical environment.

The scientist is naturally apt to concentrate wholly on the efficient internal working of a research laboratory; but a research institute cannot and does not live on its own. It must be kept going by a community willing to support it. There must develop as a matrix around it the complex technical civilization necessary to its supplies and trained personnel, e.g., libraries, computers, universities. Finally, in the last resort, it must have political support and even military defense. By the necessities of organic growth, these new shoots of research organization develop from the existing branches of a structured society. But here we are going to argue from that isomorphism of institutions within society with sentiments within the individual *that three types of institutions are needed corresponding to the three consciences.* In the first place there must be research institutions in each nation, catering especially to the needs of the community cultural search for ethical values partly unique to its own culture. Secondly, mankind must logically proceed to a world research center, supported by the federation of nations, and enquiring particularly about the best standards in the area of the world conscience as it affects international action. And finally, there must be moral research institutions outside of the first two, dealing with the transcendental conscience, and for good reasons, privately supported, as churches and religious communities have long been privately supported[3].

The institutes for the investigation of within-group moral values, on the nature of which we will first concentrate in this section, will need to cover biological, psychological, economic and sociological questions. They will have neither service nor (except incidentally) teaching concerns, but be maximally organized as permanent groups of highly trained scientists in creative research. Universities appropriately combine research and teaching, for the sake of good teaching, and need so to continue. But there are many arguments (Cattell, 1972) for setting up independent research organizations where research teams bring research to the highest level as an art, as in the Max Planck Institutes, the Pasteur Institute, NASA, and various other such centers now well established.

One reason which is sufficient in itself at this point to justify full, unimpeded concentration on research is that the social scientific study of morality is both new born and faced with extremely complex problems and has yet to find not only its results but its methods. In quite other aspects of its activity, as studied in Section 9.6 below, Beyondism has also to *teach* values. But unlike most other ethical systems it has first to find them. The main principles at least may be said already to be clear, but before it can offer individual and social guidance with any authority a great deal of the science of social psychology has to be brought into being and put firmly on its feet. The present writer has made no secret in this book of the fact that he has been drawing the cash of scientific sub-principles from a bank of social science which at present has negligible capital. The methodological and conceptual levels of anthropology, sociology, economics and even of psychology itself, in relation to the height of our goal, are almost a matter for despair. But often it is darkest before the dawn, and perhaps, with more rigorous training and selection, justifying that adequate endowment which social science has always lacked, some dependable technical guidance may be available by, say, 2000 A.D.

In such a National Institute for Research on Morality one envisages the most imaginative, fertile and maturely disciplined pure researchers in the fields of psychology, economics, sociology and so forth gathered together to follow out the implications of their sciences specifically in the field of values and internal legislation. "How much" they might ask "would an increase of income tax allowance for a fourth child increase the supply of gifted leaders in science and politics a generation hence?" "What is the effect of a one hour increase or decrease of T.V. programs for children on the amount and quality of what children read?" "What is the effect on national resources for scientific work and for defense of going to a four day working week?" "How has an increase in biological education in the schools affected political attitudes?" "What are the correlations between degree of permissiveness in pornographic publication and the growth of anti-intellectualism in students?" "Does the population personality score on introversion and independence of mind rise when a larger fraction of the population moves from crowded city centers to suburbs?" "What are the effects of different wage incentive systems on the level of the gross national product?" "What is the effect of more efficient detection and appropriate treatment and punishment of crime upon the general morale of the non-delinquent?"

As distinct from workers in the physical sciences these researchers will obviously have to bring about great developments in what is called *multivariate experimental method* (Cattell, 1966). For everything in the organic life processes of a society depends on everything else. In multi-variate analysis the sheltering walls of academic specialties must be transcended: economics integrates with psychology and political science with sociology. The finest minds available will barely suffice to bring about the span of technical thinking that will be required. The chains of cause and effect between, say, a change in the role of sex in marriage, and a turn in the curve of economic inflation, may be long, but they are real, and call for understanding in a social science worthy of the name.

And all the time these men who serve the mental and moral health of their society will live with the haunting and never entirely answerable question "What is the specific value which our own culture brings to the world?" Within this question, as we saw in Chapter 4, is contained the problem of deriving the individual moral behavior values which sustain the viability of groups in general and *of that group in particular.* Above all, what are the indices of greater and lesser group survival potential from which all other values are to be derived? Any lawful connection established anywhere in the social sciences will be a precious contribution to the work of these teams. And they must contain the geniuses who can cover all sciences and re-interpret social science data in the light of the abstract conception of morality itself.

The national institute – or institutes, since restriction to one "school" must be avoided – of moral value research will necessarily be responsible to government, since the quite vast resources they will need (by present concepts of research endowment) can only come from government. (If one is nervous about this let him not overlook that a check on government bias exists in the independently endowed and trusteed parallel research organizations concerned with the international and transcendental consciences.) Even with these checks and balances in the outer community it would seem that in this field more selection is needed for emotional stability and fair-mindedness than is commonly exercised with researchers. Essentially they need to be men with no prior social axe to grind, but motivated simply by a vast curiosity concerning the truth, and tested for mental health and emotional stability. For, as we have seen, intellectual brilliance is no guarantee whatever of sanity of conclusion, and once a fabric of theory is polluted by emotional bias it is well-nigh impossible to cleanse it. The fact is that we are speaking of a new type of scientist who, setting aside the stereotypes which prevailed in the warfare of science and religion, will combine the daring imagination and ingenuity of the man of science with the unshakable dedication to ethics of the man of religion.

9.3   THE SETTING OF THE RESEARCH INSTITUTES FOR THE WORLD FEDERATION AND THE FREE ENQUIRERS

Just as the national laboratories must concern themselves with the search for the best inter-individual behavioral values in the given society, and the fullest expression of the experimental values on which each group has staked its destiny, so an international moral research center has to be designed to work out the ethical rules in interaction among groups. There will be constant exchange between the national moral research institutes and the research institute of the federated, united nations, each needing data and findings from the other, in ways to be discussed, for its fullest functioning.

Work aplenty—urgent work—awaits both of these kinds of organization, but it will be clear from the basic analysis in Chapters 4 and 5 that the search for values starts from greater depths of chaos in the case of inter-group relations. It does so in the first place because of the wrong clues initially offered—indeed dogmatically enforced—by universalistic religions. But its difficulties arise also because the indispensability of cultural and genetic experiment for establishing values has not been realized even by many intelligent and progressive men. Thirdly, there are obscurities because the possibilities and natural consequences of behavior among nations have not been so thoroughly observed empirically and developed into common lore as have those of behavior between individuals.

To exemplify the first of these points, we have to note that many reformers and activists in international affairs are so completely emotionally obsessed by the issue of war that they cannot get to the various other issues in international relations needing attention in a complete solution of inter-group morality. Others are still "homogenizers," assuming that the solution of all difficulties is to make all people culturally and racially alike. They are more than a hundred years behind Darwin who pointed out the value of segregated development, recognizing that races find the highest expression of beauty in their own type, and cultures the most congenial ideals in their own creations. (See, for some basic discussion here, Huxley and Kettlewell, 1965.) Others again refuse to recognize that a rational solution of social problems calls for a *dynamic* equilibrium in which expansive and retractive adjustments must continually go on. They see the inter-group future as a morgue maintaining the body of the status quo as it existed among nations in the midtwentieth century. For these would-be directors of our international political life there are no fundamental scientific guiding principles, only a political, Humanistic rhetoric in which such whore phrases as "social justice and equality," "basic freedom" and "human dignity" continue to prostitute their beauty to every impostor.

A world federation of nations is a mere roadblock in human progress unless it sets up an evolutionary research organization, the positive purposes of which must be:

(1) Extracting general laws of group life, by comparative scientific methods, from a central collation of the data gathered in the research institutes of individual countries. This might be styled the work of a "clearing house." But its functions will be more than those of a mere central exchange since comparative data offers greater resources for expert analysis and significant conclusions than from any individual country.

(2) Advising on interactions. Mostly these interactions would be of an economic and cultural kind. But we have had to mention earlier the more baleful instance of war, suggesting that many wars would not occur if the margin of guesswork as to the cost and outcome could be reduced. In its broader and more positive and ambitious sense as symbolized by "the Great Experiment" this "advising on interaction" is only an application of the basic research in working out the optimum rules for general evolution by cooperative competition—the definition of international morality. The meaning of law and justice among groups rests on the outcome of this aspect of research.

(3) Advising on directions of genetic and cultural development in *individual groups*, both for their own good and in terms of maximizing the informational gain for all with respect to finding diverse rewarding directions of evolution. This amounts to exercising some planning of a world experiment to the extent that nations are willing to compromise on sovereign rights.

In connection with the first, the point has been made earlier in speaking of "the Grand Experiment" and its design, that many considerations point to smaller groups than present national groups being better units for experiment. The security from power, and virtually this alone, seems responsible for the reduction of cultural experiments, yet it is admittedly an inadequate test of a culture that does not include its power of self defense. More than one discerning historian has connected the stagnation of Egypt in 1000 B.C. to 100 B.C., and the rapid evolution in Greece in the same period, with the geographical effect of Egypt being one sprawling nation on a great plain and Greek culture a set of forever competing small communities in partly isolated mountain valleys. Probably the solution to greater partitioning of experimental and competing cultures is intelligent alliances, in fact the grand alliance of a world federation capable of giving military security.

In connection with the third and last aim above, it is obvious that there will be recommendations for expansion and contraction of groups; for initiating migrations and stopping migrations; for deciding when economic aid is appropriate, and when it is malfeasance; for advocating and discouraging cultural borrowing; and especially for controlling violence, in the form of war or revolution. But on no account must there be any attempt at imposition of a single value system beyond that necessary to give some order to group interaction and experiment [4]. Primarily, the search will be for rules of international morality in the general sense, but participant nations will also necessarily want to determine the outcome of applications of the rules to the particular developing situations that concern them.

The point made above that though knowledge can and should be pursued for its own sake, yet a research institute — group or intergroup — cannot live without economic support and political defense, holds most strongly for the international research institutes. Full discussion of the framework of power, trusteeship and responsibility appropriate for these institutes would be a considerable undertaking; but the main principles are clear. Nowhere would the spirit of monopoly be more incongruous than in an institution devoted to diversification. Though under a federated trusteeship it must be accorded that special degree of freedom to investigate and discuss which scientific work always requires. Nevertheless, until the time when Beyondist principles are universally accepted one must anticipate difficulties from self-interests and cynicisms, principally, in that citizens of some countries or their governments will either (a) decline to pay contributions to a central, world research agency, or (b) decline, on grounds of privacy or national security, to contribute information, or to open themselves to investigation.

As to the first of these difficulties it is necessary to recognize that we are entering a new age calling on both individuals and nations to adopt new attitudes to disclosing information about themselves. The simple right to forbid "invasion of privacy" — incidentally, long a special aid to the delinquent and the anti-social — no longer expresses the full rights of the situation. In psychological and medical research, especially, progress is absolutely dependent on obtaining cooperation and a candid supply of information from all persons. Valuable attempts by researchers to solve problems relevant to everyone can be hampered by the callous refusals of a minority to cooperate [5]. Attempts to justify non-cooperation are often based on such literary misunderstandings as appear in the confused values of Orwell's *1984*. In individual life the problem has not yet been faced sufficiently to bring out the essential justice of denying, to those who thus sabotage research, the benefit of findings from past medical and psychological research. But among nations it may well be that those who do not contribute to the international research institute will be more explicitly and firmly denied the benefit of its findings. At the class level this non-cooperation might be instanced in the middle classes of France evading, by "right to privacy," income tax investigative machinery accepted for a generation in Germany, Britain, the U.S.A. and elsewhere – and suffering some losses of services in consequence.

However, in the case of nations there are also entirely legitimate secrecies, as on armaments, and on developmental research connected therewith. Similarly, in fields remote from armaments, nations will continue, by virtue of their values, to spend at different levels of generosity in scientific research, and from this work they will differentially benefit, as they should. For example, a country that sacrifices to get more rapid advance in what we have called chromosomal engineering, may hope to beget say, more highly intelligent individuals perhaps a generation earlier than others.

But in the international research institute itself we may expect an objectivity and wisdom in international dealings through the high intellectual and moral quality of the personnel. The latter could be achieved through recruitment and training being specifically directed to this objective. Science has never lacked men with a high degree of internationalism in their outlook. Loyalty, however, normally begins at home, and commonly it grows around people and an institution rather than an abstract principle [6]. In the creation of a supra-national moral research institute new and delicate balances of loyalties have to be worked out. Here there seems occasion for loyalty primarily to a principle – the principle of aiding evolutionary advance – and the legal trusteeship must be honestly consistent with this. The present U.N.O. and U.N.E.S.C.O. constitutions are not consistent with a higher principle of evolutionary advance [7].

An alternative design is to consider the international research institute as strictly the "property" of a concrete united nations organization – a club – and to leave the development of truly basic values to an institute for the "third conscience" – the transcendental conscience. This requires resort to privately endowed research institutes, responsible to no political organization, national or international. If the whole argument of this book, that morality has to derive from research, is correct, the day is gone when the inspired individual—free but lone individual enquirer—can move to new values with any precision by his own unaided efforts.

Here rises a dilemma which deserves more consideration than can be given in this space. Progress in science arises in the last resort as much as in the arts from *individuals*. Committees cannot make research, they can only evaluate it. And no large professional body has yet found a way to prevent the cunning stuffed shirt from ultimately slowly elbowing his way to the front. The genius is generally out in the desert, except where some sharp social demand forces the stuffed shirts to bring him in as the person who can do things. And this state of affairs will hold more than ever in the inspired branch of science, requiring all knowledge for its province, which is ethical research. These John the Baptist's of science will be found "crying in the wilderness." How can this be reconciled with the need in this area for a research team, a laboratory, a computer and financial resources? Fortunately, in spite of the tendency of the unobservant to apply automatically the general community "democracy" to science, scientists have by experience learnt that breakthroughs are made by individuals—actually individuals leading small teams, in the close, intensive action in which the leadership of the leader is granted by immediate appreciation.

This is where the creative force of the individual is preserved in modern science, and where in the last resort it must be preserved as the source of directives from the transcendental conscience in the organization of ethical research institutes. But even with such teams the tune is called in the last resort by the organization that pays the piper, and therefore it cannot be by either an individual or a world government. The research institutes for the transcendental conscience must be privately endowed, and here they fit consistently into the general matrix of a free enterprise economy. What we may call the Free Enquirers will be laboratories financed by voluntary contribution, from individuals and congregations, reporting to trusteeships divorced from major power structures.

9.4   ON ORGANIZING A REVOLUTION OF VALUES BY EVOLUTIONARY METHODS

There can be no doubt that by any standards of major change that has occurred in economics, education and national organization, Beyondism constitutes a revolution. Yet it can be achieved by steady evolutionary measures. For it is the peculiarity of a moral system based on continual research that it changes its values from above, and has no need for rebellion — provided it has also the emotionally educative machinery to keep pace with the cognitive discoveries. Our purpose in this and the next section is therefore to ask how social change in general, and the central change toward Beyondist values in particular, can be brought about without the injuries which society incurs in disruptive revolution. "Authority" must change, but if possible without heat and violence, and this requires new types of organization within societies as we have known them (*see* Kuhn, 1962).

Revolutions of the past have typically been accompanied — like wars of which they are a sub-class — by violence, destruction, vast losses of human energy, and what a doctor prescribing a drug would call "dangerous side effects." The latter include the breakdown of desirable cultural elements in the old regime, the extermination (as in the French aristocracy or the Kulaks of Russia) of population strains which though culturally misguided were genetically valuable, and the substitution of one brand of intransigence, arrogance, corruption, and rigidity for another. The content changes, but the evil style perpetuates itself, as in certain South American revolutions, with a merely superficial change of personnel and political label.

Although it has been our determination to keep an open mind, pending social science research by men of genius, even as to the desirability of democracy, and though one can already see ways (Chapter 8, Section 8.1) in which democracy probably needs to be improved, yet the chances are extremely high that democracy is the only system that can ensure an evolutionary mode of progress. Although one may question the lyrical assertions of such as Hogben that science and democracy are one in spirit (for the Royal Society was founded at the Restoration of kingship, and the purely intellectual pursuit of science has enjoyed good morale under government by elites, as in modern Russia, Renaissance Italy, and the later days of Greek cities), yet it is unquestionable that democracy is ideal for the steady *application* of the changing views of science.

The essence of Beyondism is the substitution of revolution from above for revolution from below. Its social science research institutes should be constantly creating new goals and new ways of living, and these research institutes consequently need an "adult educational" or evangelical force to bring these developments into being in the body of society. Parenthetically, this does not mean that emotional pressures from "non-intellectual" man ought to be or can be ignored. On the contrary, since re-searchers are dedicated scientists, not sybaritic intellectuals, every emotional need, every felt pressure in the body of society, will be sensitively recorded and heeded by social seismographs, and ways found to give valid expression to these sometimes irrational forces.

The neo-democratic design of voting for *needs* for everyone, and choosing *means* through democracy-responsible boards of experts, is an ideal one for a society directed by social scientists. It permits a rapid and steady flow of progressive evolutionary movements, instead of the lagging and lurching pace of a succession of revolutions. As seen above, the two main conditions for the evolutionary type of progress are skillful, adjustive, creative navigation by fully organized research institutes, and the avoidance of that overheating at some points in the machinery which generates intolerable instinctual frustrations, pugnacity, and essentially psychotic outbreaks of war or revolution despite every good reason for avoiding them.

Now, in the past every new movement with as radical a message as Beyondism has had to build itself up as an evangelical movement, as did the old religious movements, or the French Encyclopedists, or in the idealistic political conspiracy of Marxism. But so long as we are speaking of democratic societies, with freedom of the press, and so long as the teaching of science can proceed in our schools with no Bryantesque or Lysenkian barriers, perhaps the values here propounded will flow through ordinary channels and require no "church?" Perhaps the inevitable loss of moral authority in our generation, through that authority having rested on nothing better than inspired religious dogmatism, will—after the usual "diseased" break down into countless splinter cults, or diffuse emotional fads—duly be followed by a steady, inevitable and ultimately strong recrystallization around a scientifically-derived, Beyondist foundation of values. There is, as Winborn and Jansen (1967) and Horn and Knott (1970) have shown with solid psychological evidence, some fine intelligence, idealism and spirit mixed with the sheer rebellion in the student reform movements of this decade. The time is ripe for an evangelical movement at the student level illuminated by a deeper scientific understanding of moral values and the concept of steady research as the basis of change.

It is an attractive and not altogether unrealistic hope that this could transpire. But even assuming that the pleasant halcyon days of the twentieth century prosperity continue, without arrest by some major regression or calamity, this *unaided* spread of the new values could still be slow and uncertain. (The catastrophic regressions could be World War III, or a society lost to all collective ambition through an indefinitely indulgent welfare hospitalization; drug addiction; a complete surrender to the satisfaction of Madison Avenue-created luxury "essentials"; or some rigid dictatorship or return to dogmatic and mystical religions, panicked by these degenerations.)

Catastrophe or sudden degeneration aside, we may look for a period in which Beyondist values steadily impregnate social thinking. But from what source, in the form of an institution, is this to come? The argument for not leaving all to the normal progress of secular education is that this may be slow, and that there has always seemed some desirability in a church-like movement being separate—the separation of state and church. Doubtless the Beyondist position will become more widely understood, and carry stronger conviction, as—in the normal growth of education—the biological and social sciences are more extensively and thoroughly taught in schools. But there are special developments in Beyondism that are too rarefied, too debatable, and—to appreciable minorities—too shocking, to be properly taught in schools generally. It is part of our argument about the third conscience that the quintessence of Beyondist development must come from privately endowed moral research centers, feeling their way with greater freedom, both politically and in terms of assigned tasks, than the more matter-of-fact national research institutes, busy with national morale problems specific to the country, and with more time for basic questioning than the world research institute, occupied with the ethics and machinery of international adjustments.

Finally, there is a third argument—beyond the inherent slowness of school change, and the desirability of church-state separation—for the need of special propagation by a ministry over and above research activity. It is that even the seeming moribund sectarian institutions which now occupy the field of moral leadership have teeth, and are possessed of the ordinary angry reluctance of all living imperialisms to "make way." Mostly, like the Catholic Church, or even the Mormons or the Christian Scientists, they are, in size, monsters, with enormous economic endowments. Like the first small mammal, running by the elephantine feet of the titanic Silurian monsters, the new morality could be crushed with scarcely an effort if a confrontation arose. (The vagabond status and research sterility of the psychological departments in the universities of, say, Italy or Spain, compared to those of, say, Holland or Britain, show, to say the least, what a large old tree can do to saplings in its shade.)

Moreover, it would be a mistake to suppose that vested interests and the deadly resentments of conservatism exist only in old religious organizations. It remains to be seen, but one may guess that a scientist in Russia would run grave risks of professional demotion if he argued, as Beyondism does, for maintaining, for example, a positive genetic differential within society, oriented to new cultural demands. Unwillingness to re-structure thinking, in relation to what have later been confirmed as, originally, sufficiently cogent arguments, blots the record of every form of human activity. The only advantage which science has, in this matter, is not any inherent absence of irrational conservatism, but the conscious ideal and the methodological mechanisms for getting rid of inertia more rapidly than in other fields. Even for scientific men – if we are to talk of the majority – to accept the fact that the earth from which they had observed "the heavenly spheres" was not itself the fixed center of the heavens, took most of two hundred years after Copernicus, and two thousand after Aristarchus. The ideas of Newton on optics, of Harvey on the circulation of the blood, of Semmelweiss on infection, of Pasteur on spontaneous generation versus germ action, of Darwin on the origin of species, and of Willard Gibb (in his neglect by his contemporary American physicists) on thermodynamics, present well-known instances of almost incredible delay in accepting clearly substantiable new concepts.

Getting nearer to the biological foundations of our thinking here one can instance the more than twenty year delay in grasping the importance of what Mendel said in 1880. Also one must record the fact that although Darwinism was accepted by leading biologists in about thirty years, the honest exploration of the social implications of evolution is still not widely achieved. For example, some English scientists (mainly left wing but also in the name of the Church of England) tried a generation ago to make "Social Darwinism" – as they called it – a "dirty" word. While deriding the narrow-mindedness of the bishops – in this matter their allies – these socialist "liberals" had already set up invincible barriers in their own minds to the social implications of Darwin's theory. However, even in general scientific terms the real grasp of Darwinian and neo-Darwinian findings has been much slower than faith in the prevalence of the intellectually open mind in science would lead one to expect. As Haskins (1970, page 366) points out, "Not until the great revival of Darwinism in its modern garb *during the middle decades of this (the 20th) century* (italics by the present writer) could it be considered to have achieved a firm basis in the thinking of men." This is easily illustrated by the total misunderstanding of the earlier – perhaps premature, yet clear –

statements by the present writer in 1933 and 1944, of the basic derivation of Beyondist ethics from science. Admittedly, there are, at intervals,
reorganizations of ideas even in the field of science (Copernicus, Dalton,
Harvey, Pasteur, Rutherford, Einstein) that demand revolution rather
than evolution. Fortunately, they do not demand social revolution,
except in the ranks and hierarchies of learned societies. And here one
is almost forced to accept as much as half a century's delay, in lieu of
bloody executions, according to Kuhn's mordant but partly true statement that obsolete theories do not disappear from the field except with
the retirement of those who hold them.

The implication of this observation — and the obvious conclusion from
much of history — is that revolution requires the young, and the same may
be said of the substitute — more speedy evolution — which is our concern
here. Nevertheless, as pointed out elsewhere in this book (pages 33, 346)
the forces demanding change are by no means simply and genuinely
progressive forces, and it must be said, similarly, that the stereotype of
the young as progressive and the old as conservative is altogether too
simple. On quite a variety of psychological measures mature people (at
least up to senility) are revealed to be more aware of the relativity of
"truths" and *more* open to re-structurings of thought than the young.
Their apparent greater conservatism is largely situational, e.g., they
possess more property and status and stand to lose by change, whereas
the young, taking food and clothes and university buildings and plane
tickets as much for granted as the air they breathe, believe they have
nothing to lose. Even so, the as yet uncommitted energy, and the altruism
unabraded by experience of a cynical world, are and must remain one
of the most priceless ingredients in social progressive action[8]. The
problem, as we have said, is to separate out the neurotic anarchist who,
under the deluding banner of "reform," is revolting against the necessary
restraints and obligations minimal to *all* culture, from the rational young
reformer who would make a genuinely better world. Both wish to "grasp
this sorry scheme of things entire ... shatter it to bits — and then remold
it nearer to the heart's desire." But when the mathematician Omar
Khayyám says this, one senses an improved, rational and intricate substitute model in his mind, whereas when Shelley worships the west wind
as "uncontrollable" and identifies it with himself as "tameless, and
swift, and proud," one sees roots for youthful revolt that are unfortunately more common in the "reform" complex. (Evidence in Shelley's
case is in the delusive character of his social reform, and some origins
of the sorrows of Mary Godwin.)

An argument from history that if Beyondism—instead of becoming evangelical—is to remain only the viewpoint of the more sober intellectuals—of the "scholar and the gentleman"—then it cannot grow is not necessarily true. (One thinks initially, of course, of the ineffectiveness of Stoicism and Epicureanism, compared to the more emotional and dramatic appeal of Christianity.) For throughout history until this century, the citizen with ten or more years of education was nearer nine percent than ninety percent of the population, as now. And, even among the cultured, the attempt at finding a rational basis had to lean on the broken reed of philosophical speculation, rather than to utilize a firm, progressive body of science. Our problems of evangelism and action are today substantially different. In Western democracies, at least, the movement can take place in ninety percent of the population, in the ordinary course of reading and secular education. In other respects the present day has also some new difficulties. They reside in the greater economic and technical complexity of our society, which offers especial difficulties to the harnessing of the energy of the young in the interests of progress. The social reforms which took place in the last thousand years were— with all due respect to the reformers—obvious ones. (For example, not until the sciences of ecology, and more recent economics became subjects of discussion could reformers of the Henry Wallace and Roosevelt type begin to realize the strange ways in which their obvious reforms could boomerang.) Progress is everywhere becoming a matter for expert technical inquiry and decision, rather than simple political action.

What Beyondism needs to do today is to persuade youthful reformers to a willingness to gain education in social science that will make their efforts sophisticated and potent. At the same time, it needs to evangelize, to an altruistic Beyondist standpoint, the older and technically more proficient members of society, whose mature capacities will otherwise merely serve the uninspired *status quo*. By what strategies of publicity and education this might be done we may next discuss.

9.5   WHAT ARE THE ROLES OF AUTHORITY AND OF TOLERATION OF DEVIATION?

If the Beyondist goal is in principle (page 278) the maximum possible human progress it becomes appropriate to ask "What are the technical limiting conditions in regard to the *rate* of evolution?" First, there will be the efficiency of the research institutes which constitute the brain of society. Secondly, there will be a number of general parameters of the population itself; its intelligence; its ideational rigidity versus plasticity (as a psychological and neurological dimension independent of intelligence); its average length of life, its length of education and so on. Thirdly, there will be features of the structural organization of society — its institutions and mores. Chief among these last is a set of values in the area of freedom of thought and speech, mutual toleration, the agreed role of technical authority, and the function of censorship. They have been touched on in this book: in the preceding section, in the section on mass media of communication and in Chapter 6 on Hedonism. But, as far as such matters can be treated briefly without waiting for a glossary of precise definitions, it is time to ask how they would be handled in a Beyondist community.

In this area of "authority" more than in most the problem is one of clearing away rank growths of stereotypes, considered to be "accepted thinking." This is necessary before constructive ideas can be put forward. The immediate, mechanical Pavlovian reflex of the unthinking type of liberal or self-styled type of "intellectual" is that freedom and toleration are never to be questioned and that control, authority or censorship are bad from the word "go." But maturity realizes that there is an appropriate time and place for almost any degree of control. Toleration, for example, is a vital trait in those who would follow the Beyondist ideal of distinct experiments in culture and race. But in the pursuit of scientific truth, for example, a major virtue is complete intolerance for inconsistent concepts, factual untruths, sloppy experiments, or crude thought. And, as for authority, the authority of reason and fact must always take precedence over the wishes of the individual.

Tolerance in the field of morals can be reasonably justified partly by those real differences of goals which groups may have; but largely by uncertainties as to what kinds of behavior actually serve progressive ends[9]. However, as morality is brought increasingly — however slowly — into the area of exact science, so does tolerance in the moral field become increasingly pointless and undesirable. Of two societies, that which has worked out more precisely the inferences of survival in terms of desirable individual moral standards is likely to be the more progressive and the more endowed to survive. In goodness, as in truth, if the right answer is known with greater certainty, there is good reason to apply it with greater rigor. There is no virtue in tolerating known evil. And failure to apply moral laws with a degree of authority proportional to their certainty is as debilitating to the social organism as would be failure of the body to obey the psysiological laws of its functioning.

To admit that our knowledge in moral matters is of a low degree of certainty and that laws must be administered with regard to the psychological limitations of individuals for adjustment is a very different matter from questioning the right to *authority* of moral laws when accurately known. Questioning the authority itself is what revealed religions have rightly regarded as the supreme blasphemy—the sin against the Holy Spirit. Compared with this denial (the vaunted prerogative of the Marquis de Sade and clever young writers of every generation) the evasiveness of claiming that the uncertainty of inference is greater than it really is, can be viewed as a mere amusing defense mechanism.

Among the drearily reiterated confusions of this area, by which knaves utilize fools, is that between conformity as required for morality, and mere conformity to a statistical average of custom. In the former area conformity *is* goodness. But, if the moral values have some genuine free play of doubt, then, as we have seen, those who respond to the uncertainty by criticism of authority are of two very different species: 10% of progressive reformers and 90% of sloths, neurotics and criminals.

Running this source of confusion—of statistical with moral conformity—closely for first place is the logically and psychologically erroneous argument we have studied already: that since creativity is deviation, deviation is creativity. The error is a very old one. No modern man, however, reveres oddity in and of itself quite as much as does the primitive, who fears his shamans in proportion to their eccentricity or deformity. Even in deviation quite unconnected with moral standards, and concerned—as in convention—only with public convenience or inconvenience, the same excess of the pointless is likely to be found in most societies—deviations through vanity decidedly exceeding deviations from creativity.

With these observations in mind one has to recognize that the conditions required for maximum progress are actually very different in the field of license-vs-conformity from those which the authority-allergic lesser academic, or the spokesman for hidden criminal or subversive organizations vociferously demands. Even authority applied by sheer power, which is different from the authority of truth or of social morality, has this much virtue over sheer anarchy: that those who cannot overthrow it lack the truths of collective living possessed by those who can make a firm society workable. Caesar had more truth than Pompey.

No one who reads these words attentively will conclude that there is any support here for loss of vital freedoms, but he will find arguments for legitimate new powers for moral authority over individuals. And, as in the last chapter's example of the mass media, he will find need for legitimate censorship and community control over the "freedom of the press" to misinform without penalty, to sensationalize at the expense of decent and necessary values, or to grow rich on pornography or similar exploitations of the young and defenseless. The area of public communication[10], and that concerning the varieties of freedom and control in relation to human progress are the most morally and conceptually confused of any now needing social psychological research. Hypnotized by the phrase "freedom of the press" the public seems not to have asked why industries should be fined for mere physical pollution, and pharmaceutical researchers censored if healing drugs have possible ill effects, while the mass media are allowed to pollute the minds of the defenseless young, misrepresent without risk of public reprimand, and scream whenever an objective look at censorship is called for.

Numerous thoughtful observers, from Freud to Kant, have noted that among human sentiments conscience has a peculiarly uncompromising, categorical, even harsh character. At least this is evident where it is strongly developed, as in John Brown over the slavery question, or in Luther, Savonarola, and countless others who may occur to the reader. The ancient Hebrews projected this outward, consistently, into the image of an uncompromising, affectionate but angry Jehovah. From a biological standpoint, since the evidence points to a genetic basis for acquiring this learnt set of superego values, one is moved to speculate that natural selection has developed proneness to a sense of guilt as a necessary quality for group survival. Since the blandishments of temptation, if they are to be resisted at all, must be met uncompromisingly (for to permit an inch is commonly to permit a yard) this psycho-biological quality of fierce imperativeness in the superego may be a desirable functional adaptation. Contrary to the belief of the rationalist, that gentle reasoning can handle all problems of conscience, conscience should, for ought we yet know, be properly instilled culturally, as it appears to be genetically, with an implacable and incorruptible "bigotted" character. If this is so we are given a clue that authority in society in moral matters should be an uncontradictable authority. This implies that the individual is not free to flout moral values because he disagrees with them, since, as every tired psychiatrist knows, the force of the id and the ingenuity of the ego and its defenses are capable of multiplying convincing arguments for anything.

To ward off one last confusion, let it be pointed out that the argument here is obviously not supportive of a Fascist or Communist view on authority. For these treat ultimate authority as that of a civil government,

and psychologically it fits the primitive horde pattern of emotional development. But in some respects the above argument for respect for an informed authority *is* quite close to that of Catholicism. The fine yet important
line of separation that can be drawn within the range of freedom to question dogma mooted here is actually rather close to that between Protestantism and Catholicism. It concerns the relative trust to be assigned to
the councils of specialists and to the individual conscience. Let us note
the parallels to this distinction as they appear in the field of social organization. Here one may ask "Is not the World Social Science Research
Institute proposed above only a College of Cardinals with science degrees
added?" In answering let us recognize that the science degrees are,
after all, important, and so also is the recognition of the rights of the
dissenting conscience to a hearing. Through psychological research,
moreover, we can probably hope for new certainties in regard to applying
what we defined as the psycho-dialectic tests of genuineness to the dissenting conscience. Clearly, however, on the individual conscience one
cannot swallow the position of Rousseau that man's nature is entirely
good, or the position of the modern sentimental "liberal" that in clashes
of the thinking and sincere individual with authority the institutions are
always wrong, tyrannical and willfully oppressive.

In the last resort, if the dissenting conscience is culturally sanctioned,
considered, and indeed encouraged, the tragedy of martyrdom by a well-
meaning but blind institution, or, alternatively of the overthrow of an
institution actually wiser than the individual, will always remain an
inescapable possibility. But before we succumb to the final resignation of
William Blake's "To be an error, and to be cast out, is part of God's
design," let us see what advances in handling individual conscience
versus authority a scientific, Beyondist society could organize. Primarily
it could arrange through delicately monitored, deliberately-organized
experimental splinter groups to see if the extreme variant possesses the
primary requisite for any group – the touchstone of viability *as* a group.
Such a small experimental group may quickly show whether the current
rebellion against the parent group is actually merely a revolt against the
necessary characteristics of *any* successful group. In the case of A. S.
Neil's experiments with delinquents, Christian's group of mutineers
against Captain Bligh, and other historical experiments (e.g., the American
far west in 1800) it has turned out that the free groups of "escaping"
delinquents have proved unable to form a viable group. In the case of the
Pilgrim Fathers, on the other hand, the questioning of authority was in
the name of stricter values than those they rejected, and the same may

be said of the Mormons, so that social science might have predicted with high probability an expanding group future for these dissidents.

The answer of Beyondism to the problem of authority and deviation lies partly in this benevolent, but impartial and precise splinter group experimentation. Largely, however, it is the setting up of research and analysis organizations, national and international, which constitute the brain respectively of the national group and the federated group of societies. *The greater fraction of the possible diseases of authority is eliminated when authority itself is built for constant research and movement.* It has passed with little notice that in our generation the community of scientists has repeatedly moved rapidly, smoothly, and efficiently to handle such public discontents as those occasioned by epidemic disease, shortage of food and power needs, etc., and has leapt ahead of public opinion in developing ecological foresight about pollution, space exploration, etc. There is very little room indeed for rebellious discontent *of a meaningful kind* from below when authority is itself a piece of social machinery built for sensitive concern with progress. One must stress that all that has been said about "revolution from above" applies only to revolution of a meaningful and progressive kind. Explosive tensions will continue to occur both above and below in society from the surliness of the disturbed Hedonist and the emotional incapacity of the neurotic to adjust. But to the extent that the new design ensures that leadership is itself sincerely and effectively progressive, the tensions from these sources can be truly regarded as problems explicitly for delinquency control and for psychotherapy. It should not be beyond the advances possible in diagnosis in social psychology to discern the character of these discontents — pointless from a progressive viewpoint — before they reach disruptive proportions, and to treat them wisely.

An initial effect of greater education and freedom to reason, as envisaged here, is often some degree of social disturbance. For example, the impact of the translation of the Bible from Latin into the languages of the common man — the Vulgate editions — produced social disorder. It has been asked by thoughtful men whether the same might not be expected in this age from the popularization of science and social science. In the light of the ultimately good effects historically of "the Vulgate" the outcome of a wider biological and general scientific education is surely likely to be good too, and by the peculiar nature of science we can expect the stimulation of merely rhetorically, emotionally, stimulated discontent to be less. It is the *nature* of the enlightenment that matters. The mistake of the last three hundred years which Beyondism now needs to rectify in the field of freedom-vs-authority is that of allowing freedom of the mass media to drown out, with the sheer "noise" of gossip and bad taste, the freedom of thoughtful men and trained leadership to help direct opinion. On noisy public platforms freedom of "mathematical calculation" is completely lost through the claims of freedom for "rhetoric." An effective society with real freedoms has drastically to curb the mass media view that "freedom" is freedom to shout.

9.6   THE MUTUAL SERVICES OF BEYONDISM AND THE ARTS

Our discussion above has turned on two major Beyondist requirements: a vital social science *research activity* to discover and confirm moral values, and an educational-evangelical *communication activity* to propagate understanding of these values. Any student of the history of the arts — of music, art, drama and literature — observing their relations to the great religions and social value movements, realizes the tremendous and beneficial role which the arts have played in the acceptance of a religious world view. They may reasonably be expected to play a similar interpretive role in the understanding of Beyondism. It is partly through this avenue that we may hope to avoid what Royce (1964) has called "the encapsulated man" at any rate in the sense of an individual who, though "adjusted" in his society, has lost touch with the living pulse of its growing values.

However, the lover of the arts may be in revolt against the conclusion we already reached (Chapter 1, Section 1.6; Chapter 2, Section 2.7; Chapter 8, Section 8.7) that in matters of moral values the era in which the arts could be accepted as *sources* of creative truths is over. To this issue we need not return: the arts and the revealed religions alike depend on the unchecked inspirational, intuitive thought processes. Consequently, if those processes are superseded as sources of dependable truth, by controlled experiment and rational, explicit theory, in the domain of religion so must they be in that of truth through art. On the other hand, the importance of the arts grows rather than diminishes in teaching *emotional* understanding of the positions reached by moral science. Moreover, they retain in any case the legitimate objectives of pure art — of art for art's sake, in abstraction from any social questions — which, as we have seen, is also a contribution to emotional adjustment.

Thus although literature and the arts have functioned in the past — though not nearly to the extent that religious revelation has — as discoverers of new moral values, the contention here is that with the development of a more reliable, objective science of ethics, their inherent degree of unreliability can no longer be tolerated. To those accustomed to believe that they are absorbing new moral values — as distinct from gaining insight into values already established — when they read Tolstoy, Shaw, Goethe or Tagore this statement may be upsetting. And it will be ill-received by that wide and amorphous body called "writers" most of whom are incorrigibly unconsciously propagandist in intention. Yet it merely continues consistently the epistemological position taken in Chapter 1, Section 1.5, to state here that there is no validity in the position "I feel this profoundly: therefore it is true."

Here we return to understanding the sharp reversal between the Beyondist and the intuitive religions (and certainly the Humanistic) acceptance of human feelings in relation to the realities of Nature already stated. The revealed religions say "These are my deepest inner feelings. Therefore the world must have meaning in these terms." The Beyondist says "This is the nature of the universe and my position in it. If my feelings do not fit this, it is because my emotional structure has not yet evolved harmoniously with my capacity to understand. Therefore I need to evolve and learn emotionally." (This rejection of "instinct" *has* been recognized by some religions in the notion that primal man needs to be "born again.") Among persons of sensitive intellect, especially, this vital emotional relearning is an experience in which the arts — as practiced by those great artists from Michelangelo to Beethoven, who nevertheless do not assert that art is above morals — can play an enormous role. Even so, this does not commit us to the Shelleyan view that "poets are the unacknowledged legislators of the world." It does, however, recognize that the full emotional insight and adjustment to ideas which have been generated by more abstract, scientifically-analytical methods is likely to be through exposure to an ethically-integrated art.

Emotional education in the direction of the moral values which society pursues is, of course, not the *only* role of literature and the arts and, indeed, the social psychologist has to recognize a high complexity of functions, including (as discussed earlier, Chapter 1, Section 1.6): (1) Catharsis, as the classical philosophers recognized, in which the arts simply discharge, in acceptable form, anger, fear, sex and other primitive emotions too much thwarted by a complex society. (One is reminded here of the description of music as "a moony light for eyes tired by the harsh glare of this world's sun.") (2) Consolation, again as providing a form of emotional equilibrium, as in (1), but now as with a child's toys. ("And six or seven shells, a bottle of bluebells, and two French copper coins, ranged there with careful art, to comfort his sad heart.") (3) What we have called "condenser action," sustaining a complexity of repertoires of emotional action appropriate to an intellectually and socially complex society, and (4) education of the emotions to the moral and other realities of a philosophically complex view of life.

Art for art's sake belongs to the first three of these—in which the experience is seemingly its own reward, though in fact it is performing complex adjustive functions. But the deliberate interpretation of Beyondist or other rationally-reached values in terms of the life experience and intimate emotions of the individual is a different function of the arts and one which a sane society would encourage to the utmost[11]. Thus we see a division of labor in which the complex evaluation of values, and the calculations of where the stresses fall, will be matters for advanced specialists in the moral research institutions, but, the interpretation to the man in the street, and the personal emotional education could properly become the task of literature, of intelligent journalism, and of the arts generally.

A major educational problem exists in this generation in that neither the writers nor, at least, their more youthful readers, are likely to be aware that the above emergence of two types of specialists—"creators" in research, and "explainers" in the arts and journalism—is inevitable in the domain of moral values and reform. Furthermore, there can be little doubt that a general decline in ethical pointedness and internal morale and sense of purpose has occurred in the last fifty years in the arts—at least in literature and drama. This arises partly from the general scepticism of society about its revealed values, partly from the loss through recruitment of brains from the arts into the currently more vital and adventurous sciences, and partly from the lack of liberal scientific education in the arts' student. As to the last, there is little doubt that the continuing essentially classical, Humanistic education of the artist, and the university arts (and journalism) student leaves art, drama and journalism blind to the exciting vistas and the spiritual adventures seen by the scientist. (Contrast, for example, the fine popular writings of Eddington, (1929) and Huxley a generation ago). They remain uninterested in what is *there*, outside the domestic experience of man and his puny arts. They have not shared the scientists experience of cosmic wonder and awe.

Long ago Plato realized that the arts can corrupt, and suggested deleting even parts of Homer (but "the near-worship of Homer remained"—De Selincourt, 1962). One can imagine his reaction to our recent "kitchen sink" and beatnik writings, preoccupied with self-induced neuroses, with tedious problems in perverted sex, drugs, and violence that would not interest well-occupied people; and the countless tinsel trivialities of fashion. Instead of enriching the emotional interpretation and enjoyment of life for the healthy, intelligent man, fighting his way through the problems of our time, the arts become a pointlessly esoteric game, elaborating bizarre deviations which, even when not intended to abrade morale are at least trivial. Thus modern "writers," with a few splendid exceptions, have failed to do what classical, Medieval, and Renaissance art did for the spirit of those eras. Obviously this is not to say that the *best* in the arts today is not as splendid and as relevant as the best in any age, but only that what is read by the average man from the average best seller bookstall is woefully devoid of a positive message. When Haldane, a scientist of great breadth of cultural interest and imagination, said, "Not until our poets are once more drawn from the educated classes (I speak as a scientist) will they appeal to the average man by showing him the beauty in his own life" (Clark, 1969), he expressed the feelings of many of those in the vanguard of modern advance, who in their starved but precious leisure are offered the immature products of modern novel writing and drama. Among leading scientists the personal impression gained by the present writer is that the majority, like Haldane above, find only a vacuum (save for the classics) in the place where they would hope to find their recreational and inspirational reading[12].

Our generation is not the first in history in which art and literature have fallen out of touch with the true spirit and the moral values of their time. A socio-historical study would, indeed, be profitable on the conditions and origins of art being "in" and "out" of harmony. In Elizabethan times art and literature responded harmoniously to the Renaissance of knowledge and the great age of geographical discovery. Both Protestant Christianity, as with the Puritans and Cromwell, and Mohammedanism, however, found themselves at odds with some forms of art and drama as being either pointless or having a demoralizing influence; and even Plato specified that some art and music should be banned as unfavorable to the spirit.

The secret of why harmony appears in some times and cultures and not in others is, however, not what we primarily need to investigate here. The point that *is* important is to demonstrate a proposition, quite heretical to the "intelligentsia," that culture, both in the broader sense and in the narrower sense of literature and art, *can* become very ill-adapted to the survival of the group without its devotees being in the least aware of the irrelevance or malignancy of their intellectual elaborations. By contrast, scientific or economic (agricultural, industrial, etc.) culture is kept to some realism of adjustment by the more immediate evidences of society's resultant success or failure. The fact must never be overlooked that what a culture likes to think of as its pride of art, literature, and polite manners can unwittingly become, in biological terms, highly artificial, unreal or burdensome. It can be absurdly narcistic in its self-valuation, and narrow in emotional interest, as is more evident when one culture looks at another, e.g., when Western culture looked at the Chinese Mandarin culture, or at the intichuma ceremonies of the Australoids. Artistic elaborations can become virtually as unadaptive as the private fantasy world of a schizophrenic.

A cultural harmony and functionality exists when art and literature adjust themselves to the moral purpose—the survival values—of the group. This showed itself in finest form in the life of ancient Egypt, in classical Greece, and when architects, musicians, writers, and artists gave of their best for the Christian religion, as in Michelangelo, Dante, Milton, Bach, and the anonymous designers who gave their lives, generation after generation, to the mystery of the cathedrals. It expressed itself very explicitly in Milton's dedicating his poetry "to justify(ing) the ways of God to man."

The intellectual acceptance by thinking men of a Beyondist interpretation must surely lead, in due course, to the end of empty posturing in the arts. If the human evolutionary purpose informs our whole society the arts must surely gain a new momentum and a new vitality from their felicitous absorption in the great task of interpreting the new moral values in daily emotional life. By offering the individual vicarious experience of emotional integration after conflict, and the emotional adjustment that accompanies insight into basic values, they will perform an immense service. What the nature of that emotional integration may be, we can glimpse in the next section, but only truly great writers, artists, and musicians are likely to achieve it in a superb fashion, and in their hands it must be left[13].

9.7 THE EMOTIONAL MEANING OF BEYONDISM TO THE INDIVIDUAL

To the poet, the dramatist and the artist falls the spiritual adventure of richly expressing the emotional meaning of Beyondism. Nevertheless, as far as half a dozen pages will allow, let us try to define in more prosaic terms, the emotional nature of the individual satisfaction which the Beyondist possesses, letting it stand out in contrast—for sharper definition—against those consolations which traditional moral systems have offered.

The foremost satisfactions which a scientific morality peculiarly offers are (1) a continuously advancing, intellectually sophisticated, perspective on man and his universe, in which a scientific integrity of thought is preserved across all domains, and about which a sense of wonder forever grows. (2) A love of fellow man which has not degenerated into mutual narcism, or sentimentality, but which binds by a common understanding the men of high endeavor, companions or rivals alike, in the human evolutionary strategy, and (3) an orientation away from the sensuous and trivial satisfactions of the present toward the serenity of future achievement. This achievement is not individual but expresses itself in a rich variety of distinctive collective adventures, heading beyond the horizons of our present vision. Let us focus each of these emotional meanings in turn, bringing out their more detailed orchestration.

As science was born from a sense of wonder and a spirit of enquiry, so also is its offspring—an ethics begotten of a scientific perspective—a subject for continual intellectual provocation. Science is now centuries old; but only in the last hundred years has it knitted together sufficiently to create perspective and to permit us to read the longer sentences which Nature writes. Thus from the remotest depths of history until the present, man has lived, like all other living things, in an unquestioned, generally sufficient, but nevertheless cruel and indifferent Garden of Eden. Just as the bee, drunk with the aroma of honey from the fairest bloom in the garden, feels that all life rotates about himself, so man has viewed the garden, and even the sun and stars, as a well-arranged backdrop for the scene of his life. If the known time scale of our universe were condensed to a morning, then it is only in the last fraction of a second, by geological time, that man has lifted a startled face to the dawning truth that he is a trifle—albeit perhaps a tremendous trifle—in an ongoing evolutionary process of immense extent, and of which he is by no means the center.

So, in wonder, we ask "What is the purpose of it all?" and the Beyondist, while he begins to sense the answer, reaches the constructive conclusion that man needs to change and grow before he can hope more fully to grasp the wonder of what is happening. At least he perceives that in living matter and its "awareness" there is the hope of reaching understanding of the universe and himself. Furthermore, with a quickening surmise he recognizes that the privilege of gaining this understanding and control has a good chance of falling to his own variety of life.

Proud and petty men may reject this confession that the brain of man must *itself* evolve before it can understand *why* it should evolve. Yet this assent to the need for biological advance is wrung from us by the fact that the finest imaginations alive today boggle at the great and basic questions. If God is omnipotent and a god of love, why is there ghastly suffering in the world? If a particular mathematical selection among possible properties for basic matter, followed through by a causal, deterministic process pre-ordains what type of living matter, intelligence, and cultural-spiritual development must eventually inevitably follow, why is it necessary actually to stage this long evolutionary development, instead of creating that necessary pre-ordained product by an immediate construction? The fertile minds of scientists can, of course, already provide many intriguing speculations. Perhaps a particular dice-throw of properties of primary matter is thrown down by a great experimenter to see what it will generate in the form of mind and society, because he himself does not yet know. And, the result still being imperfect, the emergent mind itself re-sets the combination to a new value before blowing up the universe again. The search by trial and error would then go on, but where are the records kept of such a series of experiments unless in the mind of some entity we may call God?

In regard to these ultimates the man of science has to confess that he scarcely knows what questions to ask, and still less what tentative answer to offer. But he is at least justifiably certain that traditional religions over the last four thousand years have given man nothing but a childish collection of allegorical tales. Further, he recognizes that in their preoccupation with desires and frustrations they have been remarkably little concerned with the outer world, beneath or beyond this superficial spherical crust which is our accepted footing. Their main concern, indeed, has been to get half-evolved, ape-like men to live together with some modicum of amicability or "decency" and to conjure up at least enough man-to-man morality to make some degree of organized community life possible. For only on some thin crust of community order could the beginning of spiritual culture emerge, and only with spirituality can the quest of science begin.

What is new in the emotional meaning offered by Beyondism is thus in the first place the substitution for these crude static anthropocentric legends of a tested, realistic view of the universe, along with an organization of development and control. With this much self-consciousness, awareness of our world, and readiness to organize ourselves the possibility at long last arises of a common adventure in evolutionary diversity and progress, and of emotional participation for all mankind in a collective adventure. In contrast to this new demand the emphasis in the emotional life of the traditional religions was largely on palliative consolations—attempts to quiet the bawling infant preoccupied with the discomforts of his internal conflicts, and the squabbles of mankind. At last these squabbles—and all the rivalry of man with man—falls into meaningful perspective, as a diversification to be pursued with dignity but also with inexorable realism.

Nevertheless, the man in the street, in a mood of blunt scepticism, may still ask how this austere perspective and plan of action in Beyondism can hope to "sell itself" in competition with the traditional religious and moral systems. These offer individual immortality, the daily attentions of a benevolent deity, rewards in the hereafter, and an appeal to all the sentiment for mysticism in mankind. Furthermore, it may be asked whether these long term perspectives, and concern for remote cultures or men unborn can hope to have any grip on the mind of the average pleasure-seeking person. Social psychologists tell us that perhaps four-fifths of the leisure of this average person is spent in watching T.V. or sociable chatter, on trivial matters and local gossip, with his fellows, and that most not only do not desire, but positively avoid, any reflection on more remote matters and ultimates. Perhaps a Beyondist adjustment is something only for an educated minority, and its satisfactions will have power only with intellectuals in the true sense. Perhaps for a substantial fraction of mankind the transcendental conscience must remain feeble, and the main sanctions for moral behavior must spring from social vigilance, the need for good personal reputation, and such new social reward systems as psychologists can devise. One can only *hope* that education will prove capable of bringing ever larger numbers to a moral conviction truly based on this insightful perspective.

Meanwhile, for those who can embrace this vision, and act together, an enthralling common adventure begins of intelligent, planned variation, competition, objective evaluation and breakthrough. And as part of this vista the Beyondist reads, unappalled, the physical clock of the universe, which tells us we are already perhaps half-way to the close of day, when our solar system will sink into darkness. He believes there is time enough for the mind to expand its capacity to control matter and make the universe a home for mind. But while he looks forward in sober faith to an indefinitely expanding future for mankind, he fully realizes, as we have seen in Chapter 6, that the greatest danger to this future lies in man hedonistically betraying himself.

Indeed, before the great Odyssey of collective human adventure can begin, the new crew has to make the ship of society more ship-shape for such a voyage by more widespread planning in human affairs than has hitherto been contemplated. The sciences up till now have sharpened their methods and modes of thought upon the physical world. They are about to take hold, in a way that our imaginations cannot hope at this moment fully to grasp, of the understanding of mental and social life. As scientists we are about to witness a complex, multivariate mathematical understanding of the life and growth and inner laws of societies as they unfold before us. At last we shall monitor the heartbeats of Hobbes' Great Leviathan, and understand its physiology and the working of its brain. But as social science becomes more potent, an entirely new class of problems will appear, and an entirely new class of values has to be built up in the political sense of the common man. How is the democratic will of society to be effected when the direction of social affairs can be understood only by experts, thinking in terms of complexities of a greater order than those which a doctor has to consider in planning the health of the body? A tremendous adaptation, of the type discussed above under politics, in which scientific expertise has to be invoked, but freedom of the citizen somehow retained is going to be demanded in the next few generations.

From the standpoint of the emotional life of the individual in Beyond-ism, which we are now considering, this question of the increasing complexity of social direction raises a further issue of adjustment. To the Beyondist the splendid vision and the elegant scientific discipline connected with the evolutionary adventure into the unknown future strike the main chords of his emotional life. The emotional values of love for and solidarity with his fellow citizens — the main emotional theme of the teaching of the great religions — he will want to take for granted as the necessary foundation of the new realms of value — the luminous new avenues of knowledge and group adventure — in which he lives. That is to say he will want to take for granted that societies are orderly, that crime and violence are essentially phased out, that no one starves or lacks a sense that society cares for him. But he will want to be assured that this control and compassion is not achieved at the cost of making the whole of society a hospital, or a producer of dependent adults as immature as children, or of a fat bureaucracy of "welfare" converting substantial fractions of society into stall-fed, domesticated animals. Knowing that "love," as pity, can err like any other emotion, and even create what it needs to feed upon, he will depart from religions of the past and the habits of these Asian communities which make beggars an established feature of society. To eliminate these flaws in the basic spirit and order of society he will be determined to start more radical cultural and genetic steps than societies have yet contemplated.

Order, fair-play, honesty and the effective use of time and energy are indispensable elements in the platform on which the new superstructure of the human pursuit of vaster knowledge and extended control of the universe, as a group adventure, can alone grow. The "adventure" of the crime novel and the picaresque drama does not belong here, but should decay with the regressive elements which the mature mind leaves behind. The change and evolutionary adventure of independent and mature minds in collective undertakings is something very different indeed. And the magnitudes of the changes that will be brought about are not to be thought of in terms of violence or revolution. For as Francis Bacon said "It were good that men in their innovations would follow the example of time itself, which indeed innovateth greatly, but quietly" (Patrick, 1948).

The role of love and compassion is not to drag mankind into that morass of mutual hedonistic services which we have characterized as the Hedonic Pact, but to reach out to those who fall in the battles of mankind with nature, and to the sympathizer with the loneliness of leadership in little-understood ideals — a loneliness which is to some degree the burden of every highly developed being. To this Beyondist ethic: that our profoundest sympathy and love goes out to those striving to the limits of their powers for the future of man — as the greater love of the soldier, in Owen's verse is "knit in the welding of the rifle thong" — most traditional, conservative religions present a discordant contrast. We are asked to give all our consideration to "the publicans and sinners," the lost sheep, the prodigal sons and reprobates which Christianity so debatably cherishes. What would a rational sociology and psychology say of these? Today's newspaper tells us with piously approved optimism that "poor and rich, patriotic and alienated, criminals and good citizens; we all need one another." To which a society with any sense of direction whatever must reasonably add the amendment "Some more than others!" If this problem of the non-contributing were analyzed by social scientific principles and Beyondist values it would, of course, be recognized that a movement of culture in any direction necessarily creates a straggling "tail" who might indeed be likened to the rearguard of an army. It will also be recognized that both sociologically and genetically the more adventurous the variation, the greater the likelihood of failures. But there is a difference on the one hand, between the organized rearguard of an army, which falls back, and is perhaps annihilated, for inevitable functional reasons, and, on the other hand, the swarm of camp followers who could not care less about the forward striving of society, and who in fact, as criminals, hippies, drug addicts and professional welfare dependents, have often quite explicitly rejected society. The question of what society owes to these individuals needs finer analysis than space, or existing research on cause and effect, permit us here and now. The protocols of many young criminals and drug takers today suggest that sheer hubris — arrogance and contempt for the oft-repeated warnings by society — plays a much larger role than sentimentalists about the "disadvantaged" are willing to admit. To such types — as to Hitler — such appeasement is a futile weakness.

To the Beyondist what is presently clear is (a) that men are knit in a higher love — an agape of boundless mutual loyalty, love and respect — when they are pledged together to great tasks; (b) that society has been excessively propagandized by conflicting, irrational and unsupported arguments in traditional religions as to exactly what its moral obligations are toward supporting reprobate sub-groups that reject its values but remain parasitic upon it [14].

As regards the emotional life of Beyondism, therefore, what has been the central cultivation of traditional religions — love of fellow man as he is in an existing society — still remains the foundation, but it has to be shaped in a more positive way. To find what the heart of the Beyondist emotional outlook itself is, one has to ask: "What is the emotional life of a community of capable men and women engrossed in the enterprise of understanding their world, and advancing the very nature of the human type?" As yet we have little experience of the levels and kinds of emotional awareness that may be reached by every man when Beyondism, as a religion of constructive action and exploration, supersedes the finally revealed and, therefore, stagnant religions of consolation. Only the small communities constituted by scientists and doctors (Osler, 1958), e.g., the astronauts, some artists, educators, city planners, mass communications writers, and medical research laboratories etc., at present offer us some glimpse of what the values and mode of life is likely to be. In these and kindred groups we see today the eager sharing of an intellectually adventurous mode of life in a brotherhood of dedicated co-workers. While human nature does not change fundamentally in these settings the moral values become very different from those with which traditional religions have hitherto suffused society. Moreover, from among the half-alienated and frequently unimaginative journalists of our current newspapers, a more refined species of mass communicator has come forward who has shown that even in the mass of dimly educated

citizens it is possible to engender some sense of collective adventure —
e.g., regarding the astronauts, geophysics, nuclear energy, or the physiologists now attacking the cancer problem.

One is encouraged to believe that with a scientifically educated and
imaginative new species of journalist even the more complicated findings
from a massive *social* scientific attack would become of interest to the
public. One may begin to see discussion on issues such as intelligence and
birth rate, the advances in average general knowledge through new
teaching devices, the effect of sexual mores upon the cultural creativity
of a community, the effects of selective immigration, the setting up of
monitored cultural sub-experiments to determine the effect of this and
that economic incentive, taxation style, and so on. Healthy though
interest in sports and fashion may be, there is no intrinsic reason why the
dynamics of a living society could not be made just as absorbing. In times
of war trivialities wane and a nation follows the drama of its life or death
with passionate interest. Properly understood, the events of peace are
just as interesting and just as vital to the survival of the nation. An
educated and more idealistic mass communications service should be
able to bring them into an equally dramatic perspective.

Finally in the equipment of emotional understanding which the Beyondist needs we must not overlook the appropriateness of a deep loyalty —
call it patriotism if one will — to the particular culturo-racial experiment
with which he has cast his lot. As we have seen, the progress of scientific-
empirical research requires that living experiments be faithfully carried
out, recorded and evaluated. The selfish and stupid cosmopolitanism
which sneers at such steadfast allegiance of individuals to their "local"
cultures has much to learn. And a devotion to world citizenship in the
true but complex sense we have discussed, respects and endorses a
loyalty which begins at home.

With all his fascination with and engrossment in the ongoing, planful
adventure of mankind the Beyondist is bound to have wistful moments,
in, as it were, a lull in the battle, concerning those warmer, nostalgic,
simpler emotional answers, including personal immortality, which traditional religions offer, and which in the austerity of his reasoning, he cannot
accept as true. The "compensatory" Christian, Mohammedan (and even
the shadowy Greek) views of immortality are gone. The only immortality
we know of is in our children, and in that unfinished story of the acts of
lives, which, forever expanding, like waves from a pebble in the lake,
have their immortality in the acts of future generations. The personal
experience of consciousness, and of communication with one's loved

ones, are vouchsafed to us as part of a strangely brief pilgrimage from oblivion to oblivion. Against this strange brevity the human heart has mourned in the language of a thousand mystified poets. There could well be scientific reasons, utterly beyond our present imaginations, for this brevity being an illusion. Meanwhile, we are compelled to puzzle over it as an insoluble mystery, incomprehensible except in so far as we can at least see that evolution and simple individual continuity are incompatible. But at least each soul is given his hour of contact with immortality. The individual may enjoy fully his hour of vision and his moment of illumination from the great minds which shine down the corridors of history. And perhaps there is some eternal consequence of his experience of the miraculous fire of life in his own being.

In any case brevity is not meaninglessness; and indeed, perhaps all we need for complete peace of mind as we are borne on this flood of life is to learn to correct the powerful "optical illusion" of the self-contained and separate ego — an illusion necessitated in maintaining its very existence in an indifferent world. For plainly each is part of something far greater than himself. To each is given the opportunity to contribute, however humbly, his personal thoughts and acts to an eternal river of increasing human understanding. It is a river that visibly gathers breadth and depth by the tributaries of our lives, though we do not know yet to what ocean it is bound. But while life is with us, the supreme delight is possible which Plato knew (and which he attributed as fit life for the Deity) of watching in rapture the perfect working of the laws of an amazing creation.

9.8   SUMMARY

(1) Psychologists recognize that the conscience or superego has a dynamic unity; and a strength in individuals partly contributed to by innate components and partly by the moral strength and affection of upbringing, especially in infancy. But in the complex modern world it needs to develop various specialized emotional attitudes, and will encounter different value conflicts, in some three main areas: man to man within the culture; society to society; and man to his cosmic environment. Such behaviors as personal affection, hostility, conformity, frankness, compassion, self-assertion, will receive quite different values in different settings. In this respect the education of conscience is today relatively primitive. Actually, when the full roster of role relations for the individual in connection with the above three areas is worked out there are over two dozen different "relations-of-obligation."

(2) Corresponding to the internalized values in the branches of conscience there are in the external world group institutions and their contractual requirements and legal forms. The latter include relations to government, to fellow citizen in the same culture, to fellow citizen under federated world government, and to a united nations organization. Finally there is what we have called the transcendental conscience which is to the individual conception of cosmic purpose as it exists outside any human group. Both social organization and education still have to sort out these obligations.

(3) Since the value system in each is in need of continual development a necessity arises, not yet recognized in any formal establishments, for research centers devoted to interpreting evolutionary goals in terms of the specific mechanisms of interaction discovered by social science in each.

(4) This would indicate the need to inaugurate (i) research institutes for which the individual national or other community is a trustee and supporter, and which concern themselves with the most desirable moral behavior of individuals within a group (a) with respect to any group whatever, and (b) with respect to the particular experimental goal values of the given unique group. (ii) Research institutes attached to the world federation of united nations, and concerned with discovering the best rules (for evolutionary purposes) governing the relation of group to group. These would also coordinate comparatively the data from the numerous specific group research institutes and offer advice and direction on the specific interactions in what we have called the Grand Experiment. (iii) Independent research institutes, depending on private capital, free to work out theoretical positions in relation to the transcendental conscience.

(5) Since such institutes cannot exist in an anarchic world, this description of research implies a corresponding political power structure. The central federated world power structure will facilitate data gathering (since both within and between countries official support will be needed to obtain data otherwise reluctantly given) and the implementation of recommendations for international adjustments. The United Nations, and its auxiliaries, e.g., U.N.E.S.C.O., do not at present extend to the conceptions here proposed. Their conceptual span is limited both by taking a legalistic aim of preservation of the status quo rather than a dynamic view of adjusting to growth and in not having reached the conception of a very great emphasis on research here proposed.

The research institutes in morality — the king of sciences in its complexity and breadth of required knowledge — must not be viewed as routine, civil service, bureaucratic organizations, but as recruiting individuals of genius to pursue pioneer trails in research in social psychology, economics, sociology, genetics, social medicine, etc. For progress in social thought still depends on individuals, though they are not able as in previous times to achieve their creations alone in a garret, but need teams of helpers, computers, and laboratory facilities. It is vitally important that so far as research on the values of the transcendental conscience is concerned, such equipment, and the finances to maintain it, shall not prohibit the complete freedom of the single investigator possessed of genius.

(6) By any historical standard of what constitutes a cultural revolution, Beyondism is revolutionary. However, both in its inauguration and in its running it introduces mechanisms which make revolutionary degrees of change possible by peaceful, though vigorous evolutionary steps, which begin in scientific, intellectual progress at the top. These mechanisms are, first, that a research institute in the socio-moral area, needs to be linked to the legislative-executive political government of each society, thus assuring that change starts from above. Secondly, that secular education needs to deal more extensively with social science, providing a means for educating public opinion to a continuously rationally changing verdict of science.

Nevertheless, if reaction in religion (which now holds the endowments) should prove too obstructive, an evangelical force for a Beyondist viewpoint may become necessary. The energy of youth here needs to be enlisted, though realistically one must recognize that its revolutionary idealism is always dangerously mixed with reaction against normal, socially-inevitable restraints. In the complex world of today evangelical youth has to drop those whose persistence extends only to "instant satisfaction" and to cultivate that gifted fraction which carries its activism into a mature, scientific, technical examination of issues. This youthful group is indeed the Vanguard that will carry Beyondism into effect.

(7) The problem of authority is confused by popular stereotypes, e.g., that since creativity is deviation, deviation must be creativity. Toleration of deviation is partly justified by the need to experiment partly by our ignorance of precisely true moral values. As far as the latter is concerned, the only logical justification for "liberal" permissiveness is ignorance. If morality becomes a branch of science it has the authority of truth; and then should be enforced in practice as "tightly" as the degree of approximation at that point of scientific advance of the subject permits. The objections frequently raised to authority are actually to a dogmatic, non-

explanatory and unprogressive authority, and these vanish if authority itself has a built-in machinery for research movement and is more closely in touch with scientific advance than is the general public. Authority connotes the possibility of censorship or restraint which, though resisted by mass media in the name of "freedom of the press" constitutes the as yet missing "frontal lobes" in the institution we call the press and T.V. Either an internal or an external censorship has as important and legitimate a role here notably in relation to misrepresentation of fact, biased choice of what is news, and pollution of young minds, as it does already in respect to other professions and other goods supplied to the public. The press has no more right to be free of democratic "quality" controls than education, medicine, the legal profession or business. In the end an objective evaluation of a questionable deviation in manner of life consists in asking it to prove its capacity to maintain a viable, non-parasitic, experimental splinter culture composed solely of persons with those beliefs.

(8) The arts, being based like revealed religion upon intuition, cannot offer the reliably checked inferences and inventions in moral values that a scientifically-based Beyondism can do. In the process of adjustment — including moral adjustment — feeling must adjust to reality thinking not thinking to feeling. The role of the arts toward Beyondism remains what it has been to other moral systems accepted with authority, namely, that of subtly and sympathetically educating the emotions of the common man to the new and better adjustments demanded. (Not that this is the *sole* function of the arts, the roles of which cover catharsis, consolation, and condenser action also, these being present, for example, definitely in the functions of "art for art's sake.")

Unfortunately, though drama, music, art, architecture and literature have had their historical golden ages of harmony, powerfully interpreting great spiritual value movements, their activity in this role has become enfeebled in the last century as the credibility of traditional religion has declined. Although they have been given no chance as yet to discover how they would interpret Beyondism, since its creed has not hitherto been expressly developed, their relation to science, which is the parent of Beyondism, has so far been inauspicious. Glancing at science without any real education therein, they have been able to see it only as "mechanical knowledge," whereas in fact science has brought into the world a spirit previously absent — an austere spirit, commanding a basic integrity of thought; a patient spirit demanding a loving respect for factual realities; an undaunted, adventurous spirit conferring faith in the future of mind, and a generous spirit responsible for most of the improvement in man's lot in the last five hundred years. The arts of the present century have hitherto failed to tune themselves to this new message; but the work of a few pioneers now gives hope that they will soon create the emotional education and insight needed to enrich a Beyondist adjustment.

(9) The emotional life which Beyondism offers is very different from that of the traditional religions, which are concerned more with palliating failure and supplying consolations for frustrated instinct, by what are often intellectual illusions, e.g., of personal immortality and a personally loving deity. By contrast, Beyondism, while offering consolation to misfortune, gives emotional support to aspiration for human advance, supplies the means for an engrossing group adventure, and presents us with the deep satisfaction of an integrated vista of the universe, and our true part therein.

The two main contrasts: (a) Beyondism's argument that a gross love of man is not the whole of religion, and (b) its occupation of this extensive new world of evolutionary endeavor — more than compensate for the loss of an illusory type of immortality and the dangerously misleading concept of a Providence in the universe benevolent to man. As to the first, some revealed religions, it is true, have had an inspired perception that "love of God" is as important as love of man; but they have not been able to find operations, as in science and evolutionary social experiment, whereby this concept can be given service, other than by love of man. Although the individual needs to move from self love to love of mankind, the latter has to be intelligently interpreted in a Beyondist framework, else it is mere mutual narcism and subject to many hedonistic perversions. Indeed, there is almost as much need for research, education and control regarding *agape* as of *eros* or even of such perennially destructive emotions as pugnacity and envy. Yet more refined research is needed to find what the Beyondist position should be on extending love and succorance (as Christianity feels it should) to the deliberate, planful parasite and criminal.

(10) The passing of the individual human life, as well as that of noble races and great cultures, though not a meaningless tragedy, is yet an ever-grieving loss to the human heart. The abstract splendor of man climbs toward the empyrean, but the rose that scented a June night, and the face that meant for us the depths of human understanding sink back into the dusk into which all precious particular human memories crowd and fade. All that our present understanding permits us to see is that each individual has his rendezvous with life, in which he succeeds, according to his aspirations, in linking with the immortals of the past, participating in the drama of his hour and contributing by the immortality of his acts to the ever expanding racial and cultural stream. Meaning and emotional

warmth are given to his life as it is lived by what he can share with equally dedicated companions in the adventure of Beyondism, passing the present horizons of our world.

9.9 NOTES TO CHAPTER 9

[1] Although the general reader is unlikely to want to pursue this question of complexity and diversity of the roles of conscience into technicalities that the social scientist will ultimately have to face, the appended table will supply a brief systematic view of the problem.

Briefly, an individual may be said to have as many roles as there are groups to which he belongs — except that there are as many sub-roles, i.e., sets of values determining attitudinal behavior, as there are *relations* among groups. So far we have spoken of three forms of conscience corresponding to the individual's own cultural group, the united, federated world group, and our universe as it extends beyond these. In more detail this actually gives eight "groups" to interact and therefore, in mathematic potential, sixty-four relationships. However, several are identical or null, so that in fact these seem to be thirty-two, as follows:

Role	Roles, numbered as at side						
	1	2	3	4	5	6	7
1. Free Individual	1	22	22	23	24	24	25
2. Citizen of Society *A*	2	3	4	26	27	28	29
3. Citizen of Society *B*	2	4	3	26	28	27	29
4. Citizen of World Society	5	6	6	7	30	30	31
5. Government of Society *A*	8	9	10	11	—	12	32
6. Government of Society *B*	8	9	9	11	12	—	32
7. World Government	13	14	14	15	16	16	—
8. The Abstract Universe	17	18	18	19	20	20	21

In each column an entry is an attitude (duty, aspect of conscience) of the individual (or individual in government) indicated at the top to the individual (or individual in government) defined by the row. There is no column 8, because this is not a person. Government *A* to *B* is the same *type* as government *B* to *A*, and so on. (But government *A* to citizen of *A* is not the same as government *A* to citizen of *B*.) Whenever attitudes (obligations) are of the same type (class) they are given the same number. However, government to citizen, though the same *combination* as citizen to government is not the same *permutation* and not psychologically the same, and is thus given a different group number. Each number entered in the cells therefore represents a specific set of attitudes and intelligently specialized ethical values to incorporate in conscience. It will be evident that an educated conscience is indeed complex in a progressive world and that much has to be done by schooling to give goodwill its maximum effect in the behavior of the individual.

[2] In his verse "The Philanthropist," based on that of Leigh Hunt, Chesterton transforms the original, thus (Beauchamp, 1924):

> Abou ben Adhem . . .
> Mellow with learning lightly took the word
> That marked him not with them that love the Lord
> And told the angel of the book and pen
> "Write me as one that loves his fellow-men:

> For them alone I labour; to . . .
> Plot out the desert into streets and squares;
> And count it a more fruitful work than theirs
> Who lift a vain and visionary love
> To your vague Allah in the skies above."

> Gently replied the angel of the pen:
> "Labour in peace and love your fellow-men:
> And love not God, since men alone are dear,
> Only fear God; for you have cause to fear."

[3] A practical political problem, requiring realism and ingenuity, arises in connection with ensuring the proper trustee mechanisms and appropriate freedoms for these three types of institutions. The scale on which the national and the world federation research institutes need to operate, as well as their need for official and easy access to all manner of data-collecting departments, leaves no alternative to their reporting to a political financial organization. However, even though they are accountable to political power one would hope that tradition would grant them the freedoms which the culture insists upon in, say, the government of state-run universities.

The question of what is implied in what are designated below the Free Enquiry Institutes, reporting to a *private* endowment (Cattell, 1972c), deserves more study. The alternatives seem to be the charity of private capital or the "subscription" of a religious congregation. Freedom in any practicable sense is financial freedom, and countries which deny the capitalist right to private means (one of Locke's freedoms) hardly need to resort to political prisons or exile in order to stifle opposition. The majority of social innovations have been made by people with the capital to support themselves in the creative period of their lives. Buddha had his family fortune (given up when he reached the "teaching" stage), Mohammed his rich widow, Marx his Engels, Christ a culture obsessed with the Hebrew tradition of supporting wandering religious men, Bernard Shaw had a hard-working mother supporting him until he was thirty-five, and the English liberals, such as Mill and Bentham, possessed prudent and successful middle class business parents who placed them in undemanding occupations.

As the roots of intellectual and moral leadership shift from what can be achieved by individual philosophical insight to that which requires extensive technical research, the possibility should be considered that one source of the necessary more substantial endowment is support by an idealistic religious movement. As to the practicable magnitude of research support from this source one notes that the wealth now tied up in the churches of the world is enormous, and if diverted from training in dogmatic theology to research on moral values would be more than adequate. The hope of such a development is greater in some of the more flexible Protestant sects and the Unitarians, who might decide that research and teaching in the realm of values are appropriately financed together. Now as the need for national loyalties is attenuated, by measures of collective, federated security, the religious or ethical value groups are likely to become more prominent and more organic in nature. Indeed, we must recognize, as both Darlington (1969) and Fisher (1930) have done, that both religious sects and social classes are also to some degree genetically and culturally differentiated inbreeding groups. (*Vide* the distinctiveness of the Jewish people, the Parsees and perhaps even the very recent Mormons after persisting as a community largely of a religious unity.) The seeking of independence for the support of the transcendental conscience research in some sort of congregation might then fail, for the research institutes designed as Free Enquiry units would find themselves as much tied to religious community values and specific group service functions as the national cultural research centers.

[4] We have hopefully seen the end of attempts to impose a single world wide set of particular cultural values by the brute force of a single military or sectarian force; but some danger, though less, still remains with a federation. (The reactions of U.N.O. to such diverse cultures as China and South Africa illustrate this.) The first sectarian attempt was probably by Israel, with its conviction that nations should have no other gods but Jehovah. The fate of Israel, blasted apart over the earth, suggests that there is some inherent force in societies inimical to monopolistic domination. The ultimate fate of the attempt by Mohammed along the same lines adds reassurance. For reasons partly already discussed, sheer military imposition of a single culture as by Alexander, Ghengis Khan, Napoleon or Hitler has also failed. Nevertheless, these suggest the improbability rather than the impossibility of a cultural monopoly. What our present degree of sophistication calls for is an explicit and world wide assertion of the ideal that there shall be no imperial monopoly, but only a federation. This expresses the position reached by Beyondism that a variety of cultures needs to be maintained in a federated political power structure which establishes only a minimal "order for experiment" and concentrates most on organizing research and information.

Some observers feel forced to conclude today that, alas, what has so far transpired with U.N.O. and U.N.E.S.C.O. does not augur well either for order with minimal interference on the one hand or active research and information services on the other. The principal real contributions of U.N.E.S.C.O. have been education for world health, while those of U.N.O. have been gestures and ineffectual pressures toward world peace. The latter, however, has to its credit agreement on the peaceful uses of outer space and the setting up of the nuclear test ban. On the other hand, it has been powerless to prevent political solutions by force of arms in Nigeria, Manchuria, Vietnam, and Czechoslovakia. One should not be critical if an idea fails to work itself out — especially in a lukewarm climate of world opinion — in one generation, or even half a dozen.

Unfortunately, one has to point out that time is not the whole problem, and that systematic defects of conception and ideals exist in these international organizations. These have been brought here in the light of Beyondist ideals, e.g., in the one vote-one-country of U.N.O. and the anti-biological bias of U.N.E.S.C.O. *Biomedical News* in 1971 notes "The 16th General Conference of U.N.E.S.C.O. (October–November, 1970) may have destroyed an important and mutually beneficial relationship between U.N.E.S.C.O. and International, Nongovernmental Organizations [in countries with internal racial policies which U.N.E.S.C.O. does not agree with]. History indicates that these [latter] organizations have frequently been the key forces in leading public opinion to more dynamic concepts of peace, freedom and human welfare." Here U.N.E.S.C.O. has set itself up as a complete dictator of a monopolistic denial of any form of dissent and experimental variation in values in specific countries. The output of research from U.N.E.S.C.O., in any basic sense, is utterly trivial when it is considered that it has 45,000 on its payroll and 828 million dollars (1969) as an annual expenditure. Certainly its research on the psychology of racio-cultural variation should have been far greater than that of the few individuals outside U.N.E.S.C.O. who have worked alone on these problems in the meantime.

The main role proposed here for the World Research Institute has been to analyze in dynamic terms the data from the historically given mosaic of culturo-genetic experiments. But, as we have indicated in the Grand Experiment, it might also want to *manipulate* this presently given fortuitous collection, as yet imperfectly adapted to a factorial experiment, by *suggesting* certain experimental culturo-genetic combinations. If circumstances should make it wise for the international center thus to move from passively recording and analyzing to advising and creatively designing, the issue of its political power and the origin of that power would be sharper. This is not overlooking that a great deal of real control can be exercised by advice. We have cited (in discussing war, Chapter 5, Section 5.5) the powerful effect in reducing war that could follow from predictions of military outcome that have become respected as close to infallible. France's 1870 intention of "revanche" on Germany; the American South's bitter loss after 1860, and probably the recent Egyptian "six day war" with Israel, could probably have been avoided by reading the extreme odds which a socio-psycho-economic analysis would have predicted. But, indeed, in its role of discoverer of what forms of international behavior are morally desirable the world inter-group behavioral research center is bound to get connected sooner or later to the judiciary in a world police system. Here society must make sure that the price of freedom from anarchy is never allowed to become monopoly. The freedom of the nation, as of the individual, to deviate, when it is solemnly prepared to pay the heavy cost of doing so, must be respected.

[5] As the net of data gathering necessary for more sophisticated scientific inference has to be cast ever more widely the problems of (a) suspicious or churlish non-cooperativeness and (b) failure to contribute a share to the necessarily large expenditures, will become sharper.

In the within-group realm, and as between individuals, we can expect—despite the presently given fortuitous collection, as yet imperfectly adapted to a factorial experiment, zens of the value of relinquishing certain narcistic "privacies" that all but the usual delinquent minority will gladly help social scientific research. Indeed, it is probable that certain voluntary-membership societies within the larger society, e.g., a Beyondist congregation, will lead the way by showing the considerable personal advantages from participating in a far more intensive medical, psychological and social periodical check-up service, designed both for personal help and guidance and social research.

In the between-group domain the problem is more difficult. Only those who, like the present writer, have been engaged in international, cross cultural research, perhaps have any idea how far down the scale some countries go in the matter of playing even a minimally responsible role in the world task of contributing to the common wealth of knowledge, health and order. The British Commonwealth countries, the Scandinavian countries, South Africa, Germany, Japan, and especially the U.S.A. and U.S.S.R., have deliberately and systematically contributed in the region of one to six percent of the total tax income to scientific research. Other countries—and they are by no means only the "underdeveloped" or the "deteriorated"—seem to experience no obligation to contribute. For instance, the old Mediterranean countries, e.g., Spain, Greece, and the new Mediterranean cultures in South America, contribute trivially to world science relative to the prosperity and even luxury of their cultures.

It has been argued above, in the name of a Beyondist analysis of inter-group morality, that the statesman is right in maintaining that the "liberal intellectual" is foolish in criticizing the maintenance of differential rewards *from* science in favor of those peoples and cultures which contribute more *to* science. In this connection, it is interesting to watch the stay-at-home (but cosmopolitan-thinking) "liberal intellectual"—a believer in all cultures being treated as equal contributors—when he is forced to the discomforts and dangers of travelling in some of them. Rich cities with poor public libraries, artistic elegance with typhoid in the water supply, awaken him painfully to the truth that concern for public expenditure on knowledge and the advance of knowledge is peculiar to a minority of cultures, to which he would do well to express more gratitude and loyalty. New habits will spread to others, as any psychologist familiar with Skinnerian learning research realizes, not by continual charity, but by making the rewards of science increasingly available to each country as precisely as possible in proportion to the increases in the fraction of its national income that it is prepared to contribute to basic science. At present (say in the last three centuries) the counts of such as Knapp (1963) show a quite disproportionate contribution to science, over the whole world, from Protestant and Jewish cultures, and perhaps more recently Communist Russia and Buddhist Japan would be added. From an evolutionary point of view this sacrifice to general scientific knowledge needs to accrue rewards in something more than intangible "prestige" — in fact in greater real growth of land and population for the cultures concerned.

[6] The growth of new institutions corresponding to the internalized national, international and transcendental consciences requires a new and more developed education of the emotions. The educational process whereby societies can achieve the delicate but reliable balance of loyalties and superego values in the individual has so far not been studied at all by personality and learning psychology. One can be reasonably sure from present measurements, however, that a normal emotional development toward supra-group conscience is one in which early, concrete, and local loyalties grow in strength and extend to more abstract loyalties. Thus a healthy emotional development today must aim at a balance between a later dedication to mankind and an earlier growth of special attachment to the values and people of the parent culture, from which the former can grow. In instances where a loyalty to a relative abstraction gives the appearance of having matured with no previous local loyalty closer inspection usually shows the more concrete loyalty preceding it. Charity begins at home, and the person defective in early home loyalties seldom grows strong loyalties to wider communities later. Napoleon's empire was knit by a collection of princes showing an extension of Corsican family loyalties; Lenin's devotion to the proletariat began when the Tsar executed his brother. Dag Hammarskjöld's services to the United Nations seemed built on an abstract love of man, but as Snow's biography shows, it began in an unusually well-knit family life. And the poet (E. Hilton Young) has given us a poet's insight that the dying vision of Christ dwelt not on all mankind but on the companions of his childhood in the village street of Nazareth. Even those perversions of loyalty, as in Coriolanus, Benedict Arnold and some more recent instances, seem to reveal to the psychologist the presence of the primary group loyalty by the strength of their "reaction formation" against it.

Much remains to be learnt about the dynamics and growth of human sentiments, but it is at least clear that one does not begin to establish an international loyalty in the adult by destroying patriotism in the child. The men of all nations who are recruited to the international research and police forces must thus be the recipients of a superbly developed education of emotional loyalties and conscience, based on a social psychological understanding of the dependencies required in a federation of nations.

[7] Throughout the proposals for development of politically trusted research institutes in this final chapter there are implied contrasts with U.N.O. and its agencies, the Economic and Social Council, the International Court of Justice, U.N.E.S.C.O., etc., which, at the present stage where we draw inferences for practice, should be frankly brought into the open.

With all its obvious shortcomings, U.N.O. today nevertheless fits the Beyondist conception of a *federated* power better than the more "efficient" totalitarian solution for which some would be ready to abolish U.N.O.: namely, a single world dominion by a victorious Communistic or Democratic military power (despite Russell's argument (1968) for the latter). Actually such a monolithic pyramid of power by conquest happens easily enough, as in the Nile Delta around 2000 B.C., and in China and Rome — in each case covering the "known" world. For some reason — perhaps one may say "fortunately" — they never last; and the reason probably is the psychological one that absence of an external threat interferes with the role which fear has of cementing otherwise disruptive inner variations.

The inadequacies of U.N.O. and U.N.E.S.C.O. discussed here arise partly from its being begotten, as all things must be, partly in expediencies, and partly in its failure explicitly to incorporate Beyondist evolutionary values. Both defects, having results that are remediable, are a small price to pay for this organization appearing perhaps a century earlier than it otherwise might have done. Nevertheless let us as graciously as possible raise some unpleasant criticisms.

The assignment of one vote to one country, in the General Assembly (which controls funds) is "corrected" to reality by erecting an inner Security Council of eleven, in which, however, the veto power of the Big Five is all important. One would prefer that instead of this crude *ad hoc* corrective structure the basic question had been faced from the beginning, of evaluating, according to cultural and power considerations, the specific voting power of different countries in the General Assembly. For (as suggested on page 142 above) this would bring to the fore the need for more basic moral principles than have yet been thought through. These principles are also needed in the maintenance of peace by something less inherently false and foredoomed to failure than the keeping of the *status quo*.

In the agencies of U.N.O. there already exist the germs of the institutions, notably the research institutions, here advocated. It requires only that U.N.O. press for basing its judgments and actions more on substantial research evidence. Unfortunately, in one of these, U.N.E.S.C.O., which is nearest to a research institute though primarily a propagator of values, the bias given at its birth, from historical accidents and expediences, has been particularly grave. That bias arose principally from its being over-run at the time of its formation by a humanistically-oriented, methodologically weak, and superficially-thinking species of scientist instead of men of basic creativity and force as scientists, possessed of a crusty integrity of opinion. For example, no behavioral geneticist of standing today would have signed the "manifesto" denying human behavior genetic differences. While maintaining a firm agnostic position that we do not know what "goodness" to associate with these differences, he would consider it absurd to deny that they could be highly significant. Similarly in the field of economic science its prescription of absolute "economic rights" seems to have little relation to realities. The fact is that seriously committed scientists do not drop their tools and run when an organization such as this is formed, and even after the fashionable rush, the selection of politically interested scientists is one in favor of those less basically interested in scientific integrities as such.

[8] A proper appraisal and unbiased appreciation of the values that can be contributed by activist student youth, by revolt or by true drives for reform is difficult at this time of writing when the general public is thoroughly tired of a complexly determined, confused and violent behavior by the young. Horn and Knott's (1970) deep and sympathetic psychological analysis is one of the most recommendable, but nevertheless does not take sufficient account of the "energy economics" of the current scene.

Before asking what the latter may mean let us note that in general, since fluid intelligence is mature at fifteen to sixteen years, sheer *reasoning power*, but not trained, disciplined reasoning, is ample in the adolescent. Since the whole thrust of our observations in this book, however, is that man at any age fails not through lack of intelligent logic, but through lack of facts, experience, and the capacity to think in unbiased and unemotional fashion, the conclusion must unhappily be drawn that the age from sixteen to twenty-four is not an ideal one for producing ideas to solve the world's problems.

A typical modern example of mistaking the revolutionary enthusiasm of youth for capacity for reform and progress is seen in the catastrophe of the Proletarian Cultural Revolution in China. Here Mao thought to bring about, through the hordes of his young "Red Guards," the progressive changes for which no one else had much enthusiasm. For a moment it seemed to work, but soon the rationalizations wore out, the naked regressiveness of defiance of *any* kind of order broke through, and the episode ended by restoration of order by the military. The final result was, thus, a reactionary swing toward military control of the nation as a whole.

Nevertheless, although the wholesale rejection of the establishment by the young is likely to be misguided, basic questioning of authority is essential if the individual is later to contribute to solving social problems, and it is educationally important that he exercise his independence and his powers at this time. (Provided his elders demonstrate tactfully that this is a learning period, and that he is bound to make mistakes.) Actually, it is well documented by the work of Lehman (1936), Dennis (1956), Roe (1953), and others, that although creative work of the highest rank can be done by young men in such fields as mathematics, which require sheer intelligence and little storage of content; the top performances in the sciences, particularly those where much has to be mastered, absorbed, and digested, as in the biological and social sciences, are typically not reached until forty and later.

But even when one is aiming at evolution rather than revolution, some shaking of the house is at times necessary and appropriate. To get re-structuring that requires shake-up it is functional to harness the revolutionary energies of the young to the insight of the middle aged, as occurred with the Marxian and other revolutions. For youth has three sources of drive for change. Though it is not particularly intellectually original (compared to older persons), it does have the excess energy which experimentation needs. Secondly, being frustrated more by even the *necessary* restraints of reality, because they are new to them, they are rebellious enough to take the trouble to search out any possible *legitimate* defense for change. And thirdly, since in any society motivated in a way that will work, rewards of property and prestige must necessarily *follow* demonstrated service, youth finds itself possessed of no "vested interests" to lose by whatever change seems indicated. Decidedly more social psychological study is needed on the standard mechanics of social change (Hirsh, 1964).

To these factors, present throughout history, has been added in the last generation a circumstance justifying comment, since it is not merely local and temporary but could recur. This is what we have called the "cultural lushness" or sheltered affluence phenomenon. We have discussed the meaning of this already in connection with the oscillations of cultural-environmental pressure defined in Chapter 6, and as a factor in the growth of Humanism in Chapter 7. In an accelerating period of affluence, due to new gifts from science (which, however, Bernard Darwin assures us will be but a brief moment of expansiveness and ease in the tribulations of the next two million years) generations grow up with an unprecedented freedom from suffering, pain (for example, few have lost a tooth without an anesthetic, though this was common with their grandparents), and economic insecurity, and with an excess of leisure compared to the sixty hour work week which sobered their grandparents. A good deal of the sad failure to get the real meaning of current revolt springs from overlooking this substantial "economic" contribution to behavior (as much as the failure to understand the increase in delinquency springs from overlooking the basic decay of moral authority).

As observant historians have noted (e.g., Crane Brinton in *The Anatomy of a Revolution*, 1938), the stereotyped belief that revolutions occur with dire poverty and oppression is wrong. They occur when the man in the street is frustrated yet sufficiently above the grind of poverty to "feel his oats." Experience of a sufficiently "lush" youth, bringing about those personality changes toward surgency, dominance and premsia which psychology records

(Cattell and Barton, *In preparation*) as following from those conditions, can bring about a revolution much more quickly than oppressiveness. Between the Madison Avenue-culture of salesmen calling for the "good life" of unnecessary wants fulfilled, and this trend in personality to "instant satisfaction," intolerance of deeper thought, and irreverence for what cannot at once be rationally explained, a mass psychological state has been created in which the restraints adequate for earlier generations fail.

Only slowly is it being realized by both the genuine reformer among the young, and the permissive (open-minded or ethically lost) educator who should have been leading, that what happened among the street marchers of the 'sixties was the merest hoax on progressive intentions. One of the nation's leading educators, President Henry of the University of Illinois quickly recognized in these parades "the demand for quick remedies which do not take into account the complexities and realities involved," and in the stupid violence of these youths a common outlook with "violence reflected in the increase in crime, in mob action on the streets of our cities, in the use of terror as a political weapon, and in the defiance and degradation of our courts." Another eminent educator prepared, more light-heartedly, to change to the language of his student listeners (with some alliteration added) describes the alleged activist "reformers" as a "collection of cacophonous citizens seeking simplified, speedy solutions satisfying multitudinous masses of miscellaneous malcontents each of which poses as the righteous, wronged representative of the really responsible residue of a shattered society." (!) Nevertheless, despite all this exploitation by the extro-vert showmen of sheer-rebelliousness a core of genuine, idealistic, thoughtful student reformers exists, to be matured some day into vital progressive leaders.

More to be deplored in the long run than the violence which sometimes accompanies forthrightness is the habit of intellectual evasion and double-talk which in the end develops into a schizoid rejection of logic and scientific fact. A popular modern "philosopher," Djilas, tells his willing listeners (1962) that "The world is satiated with dogmas [read logic], but people are hungry for life" and notes with approval that young European politicians are emerging "ready to act more out of instinct and conscience than knowledge and experience." The "hunger for life" is an excellent thing—the first essential for evolution—so long as it is not a mere phrase, as is evidently intended here, properly meaning the refusal (under the pleasure principle) to come to grips also with the facts of life, frustration and death—all essential in evolution. The rhetoric of the big lie, whether used as here by a rebel behind the Iron Curtain, or by Hitler, or in Rousseau's assertion that "Man is everywhere in chains" is everywhere the same. Generally, the revolt against science and learning is not direct but by a devious pseudo-intellectualism, and often a strange use of standard words. Sometimes, it is simply naive, as in the sad case of the young student martyr in Czecho-slovakia who demanded of the authorities "An instant abolition of censorship and a ban on the distribution of *Zpravy*" (this last being the Russian language newspaper in Czecho-slovakia).

With this glance at some of the mechanisms and sources one may well ask whether Beyondism wants to harness to itself all forms of revolutionary discontent. Many, indeed most social and political revolutionary movements have not hesitated to employ discontent from any source, and this has been particularly so with Communism. As Churchill observed in 1930, "The anatomy of discontent and revolution has been studied [by the Bolsheviks] in every phase and aspect."

[9] Toleration of differences over *ends* i.e., ultimate values from which others are de-rived, even when the differences are rooted in honest differences of conscience and values, should have a limit. Many a complete stalemate in social development could be avoided by

some "sorting" (migration of each individual to a society which better expresses his ideals) and by splinter group formation. The requirement of some draft tribunals that conscientious objectors to war demonstrate that they belong to a religious or other sub-group known to maintain systematically that self-defense is evil, is a sound one. For as we have seen, a splinter group should be able to show by practical living that it has the properties of a viable group. Actually a conscientious objector group, by its very nature, is viable only until the last patriot in the larger society which shelters it is killed by the enemy.

Apart from such sociological considerations there are psychological considerations which justify considering the present popular permissiveness to indefinitely extended deviation a form of unrealism and indeed superstition. In a recent article the organizer of a group of conscientious objectors ruefully reports his discovery that they commonly turn out to be "objectors to any form of community authority." That they are not "much able to understand others or take care of matters of administrative requirement." After continuing a fairly faithful clinical description of the "authority-allergic personality" he takes the familiar Pollyanna rationalization that "outstanding individuals tend to be unbalanced" and that "creative imagination comes from inner inconsistency." Psychological study of creativity does not support this in the least. The Newtons, Darwins, Pasteurs and Edisons are unusually stable, and capable of great patience and organized application to tasks. Deviation should be examined on its own merits, with a clear realization that society can suffer badly from too much of it — and without dragging in obsolete superstitions about a bond of madness and genius.

[10] There is no sub-group, profession or institution within society that can justifiably claim to be free of community control in the way that a substantial number of young journalists naively claim to be when they use the phrase "freedom of the press." As usual the alternatives are external control by law and censorship or internal control, by the profession. Many higher professions — priests, doctors, lawyers, psychologists — have elected to control themselves by their own ethical "courts." Though this should properly still leave them liable to courts of the state, or international law courts, it has the advantage of introducing a technical appropriateness of ethics as seen by a keen peer group.

An internal regulation is, however, only as good as the morale of the group concerned. In the English speaking world it has frequently been noted that the American press, in which one has no alternative but to include its "yellow press", has shown altogether lower standards of restraint and regard for the public good (when tempted by a scoop or a sensation) than has characteristically been exercised in the English press. Consequently, there is more ground for demanding a national censoring authority outside the press itself in such a group. At the moment of writing the courts have decided that the *New York Times* and other newspapers may publish documents illegally obtained (one *might* say stolen) from government files. Regardless of the issues of cost to the country in lives and treasure which exposure of war secrets may bring, in this case, the noteworthy point is that every newspaper to which the man in the street may turn endorses, with "trade union" solidarity, the right to publish data so obtained. Only in the mature citizen does one hear the question that a democracy should ask: "Who elected the *New York Times*?"

[11] The psychologist may be interested to perceive a parallelism here between the future guidance of morality in society as a whole and what one may venture to foresee as the future in individual psychotherapy. As structural personality analysis advances, and the psychometric precision of assessing individual abilities, temperament indices and dynamic adjustments, the complex interactions of emotional learning, body chemistry and adjustment will undoubtedly be handled in the individual case by complex equations indicating what meas-

ures are necessary at each step in therapy. The psychological specialists who handle this, perhaps at some central clinical consulting and computing institute are likely to be different from those who actually see the patient as psychiatrists or social workers, and whose task it will be by artistic skills to aid the patient's insight, and convey the specific persuasions and rewards or "conditionings." Similarly in the case of the common adjustments to ethical standards of society as a whole there will be on the one hand, the researchers on morality and on the other, the artists who assist and bring out in emotional terms the meaning of the adjustment.

[12] If space permitted it would be intriguing to roam through literature exploring on the one hand, the fumbling and futile attempts of the novel and the stage to create values, and, on the other, their degree of success in interpreting the emotional impact of dynamic scientific perspectives. In the former there has been no lack of imaginative inventiveness or persuasiveness—as in the writings of Tolstoy, Dickens, Goethe, Shaw, Tagore, Dostoyevsky, Musset, Aldous Huxley, Orwell and others. What vitiates them as moral leadership, as argued earlier, is their lack of any marshalling of scientific evidence, such as would differentiate them from the conclusions of revealed religion. And when Dickens presumes to draw conclusions as an economist, Orwell as a political scientist, Mailer as a psychologist or Shaw as an immunologist (*The Doctor's Dilemma*) we tend to get an obviously emotionally biased, nonsensical fabrication.

Most references in the literature or drama of the last thirty years to what science might mean in terms of values are amazingly stereotyped and obtuse. Science is there seen on the one hand as "mechanical"—either merely cold and abstract or dangerously inhuman and even hostile—or on the other, as a utilitarian concern—either crass or contributory to luxuries. That it brings to man a new spiritual discipline in perception; a new integrity of intellectual life, and a vast panorama enriching our emotional life and sense of purpose, is lost on the typical arts writer. And the further perception developed in this book—that science is the foundation for a research derivation of moral values, in the opening up of a science of human society—is *completely* lacking. The sad consequence is a spiral of futility in popular music, art and literature, where misunderstanding begets alienation from the vastly growing world of science and increased alienation begets more bemused lack of understanding. And the scientists—except the few poseurs for the dernier cri—go back to Renaissance or Greek art, to Shakespeare or Goethe, and to Beethoven for something firmly in tune with their sense of the drama of life. From decade to decade in the era of retreat from firm revealed values, one sees art run off into schizoid and tortuous expressions from which no busy scientist or indeed any other well occupied, constructively minded citizen can receive the emotional recreation that the robust and participating—as distinct from the neurotic and alienated—rightly need.

Of course, one is aware that, beginning a generation or more ago with Verne and Wells a literature began which took ample note of the meaning of *physical* science and its constructive impetus to society. Unfortunately, this frittered itself away in a firework display of largely emotionally uncreative and trivial science fiction. Meanwhile the writers who sense the deeper emotional meaning—Jeans, de Kruif, Clarke, Asimov, and some imaginative T.V. dramatists such as the Star Trek group—are scarce indeed. In any case what we are asking of literature is something in addition to an exercise in emotional adjustment to the way in which physical science is transforming our world, namely, a sense of the spiritual values in scientific exploration itself. The emotional creativity of the arts is now needed in revealing to the man in the street also what *social* scientific research is going to mean in his life. It needs to bring home to him the consolations, insights and emotional convictions appropriate to founding moral values on the broad developments of biological and social science.

[13] The intuition that art belongs with science and religion – that goodness, truth and beauty share a common excellence to the point of being the same entity in different contexts – has persisted since at least ancient Greek times. The basic argument, in these pages has, one hopes, successfully brought the first two together, in that scientific truth is made the basis for moral truth. However, beauty has so far been left in our discussions in the domain of the intuitive, except for what some readers may find a recondite argument, namely, that for each species the sense of beauty accompanies behavior that is functionally true, i.e., adaptive, for that species. The present writer would make a strong argument for this position, though it would be hard to prove that a male crab finds nothing so beautiful as a female crab, or a kestrel nothing more harmonious than his vertical dive through the air. Indeed, in fairness one must quote an eminent philosophical psychologist who found it impossible to conceive the perception and creation of beauty as related, directly or indirectly to true, evolutionarily valuable biological function, saying "it is unconceivable that musical . . . genius . . . can be the mere by-product of a naturalistic evolution – of a struggle for life" (McDougall, 1934). Nevertheless, granted that the definition of beauty refers, as it must, to an experience peculiar to each species, the old philosophical intuition, in scientists such as Haeckel, and countless artists, that "beauty is truth; truth beauty" and that a philosophical trinity, of goodness, truth and beauty exists, can be meaningfully argued.

[14] The need expressed above searchingly to question the widely accepted traditional religious value that succorance and affection in all circumstances is a desirable ideal is based on recognition of gross human differences in goodwill and good intention, not differences of competence or temperament. Furthermore, it would be a mistake to confuse the right to membership in a society – in the sense of deserving its fullest care and succor – with mere prominence or "success" in that society. The janitor in a famous laboratory, and the girl who washes the bottles, in so far as they share in and aid the high morale of the group, *belong* in as full a sense as the leading scientists. The flower must have roots and the blade a strong hilt. The issue above is not whether men "belong" more or less to a community by virtue of more and less important roles, but of their claim to succor when they have deliberately chosen roles – as the parasite and the criminal have – *against* society. Except in the hope of repentance and reform, as Christ looked at sinners, what would an objective Beyondist ethic say about giving affection and support to anti-social types simply on the basis that "they are human?" At a first glance, an objective, functional evaluation certainly suggests that a Beyondist would pervert his principles by feeling that affection is owed to individuals of this kind.

Undoubtedly one of the most urgent tasks of a science of ethics at the present juncture is to analyze and separate the role of true compassion from that of mutual indulgence, Hedonism, and the rejection of any painfully exacting standards. The majority of writers, including many in science, e.g., Dobzhansky, page 99 above, seem to have accepted without fundamental examination the thesis that love of fellow man (and sometimes of all creatures, including the insect and the cobra) is a primary requirement of all ethics. One has only to think of the life of various sects in India, for example, to see that this essentially "sentimental" attitude – especially extended to all living matter – produces contradictions which even in theory – and still more obviously in practice – deny the utility of the premise. Love has to be discriminating and wise. Psychologically "affection" still needs to be examined in a meaningful "psychological taxonomy of feelings," which the dynamic calculus will doubtless some day do.

An understanding mutual love is the most important commodity in any society that is to achieve both stability and creativity. Most unfortunately, however, the advocacy of "love" as a panacea is the stock in trade of every hypocrite, and the selling point of every charlatan of moral values. Moreover, it seems to change its meaning and connoted values with every historical pressure. Nothing can be so hideously corrupt as pseudo-compassion. When men were a few fragile groups precariously existing in a vast hostile animal world, human life, just as human life, naturally acquired a scarcity value. In the condition to which we are heading, in which there will be physical glut of humanity over the whole earth, there may naturally arise, as the crowded Orient shows us on a smaller scale, a pressure toward a lesser intrinsic valuation of human life just *as* life. And as medicine and economics lead to those in active life having to support an increasing number of truly senile aged, will not euthanasia become more seriously debated? Under these conditions the real meaning of love will be put to the test.

A special aspect of the compassion issue is that by the biological origins of the drive it tends to be directed more to the obviously defective concrete case than to the less obvious stresses where it could more reasonably bring succor, e.g., to the hungry cat rather than the overworked mother. In particular the traditional ethics fail to direct it to what we have argued elsewhere to be a vital need — the growth of sympathy rather than envy for the eminent. (Lloyd even argues that Christ was forced into a manic episode by the insecurities created by envy and jealousy of his superiority of character, 1971.) A higher morality requires in numerous ways a toleration of just inequalities. Differences and inequalities are an inescapable requirement of a differentiated, complex, and progressively changing society, dedicated to the fullest average self-expression across society. For example, some individuals, such as doctors, priests and researchers, will work a seventy hour week, others may be required only to do an eight hour day and a four day week; men will be called to risk their lives in war while women will not, some will earn more, because few can produce what they produce, and so on. The countries which tolerated aristocracy longest — Sweden, Denmark, Britain, Holland, for example — rank well in their contribution to civilized life and settled government — compared, say, to Colombia, Mexico, Venezuela and many others whose intolerance of eminence has been endemic. An interesting contrast between Communism and the early Christian Church — which justifies calling Communism a heresy from the Christian concept of love and which reciprocally caused Marxians to call Christianity the opium of the masses — is the explicit emphasis which the latter placed on a cheerful acceptance of one's fair — or even "unfair" — status in civil life. Equality of opportunity and regard for the worth of every citizen as a *person* are fundamental to a good society. But love of fellow man in a common endeavor actually expresses itself by acceptance of functional inequalities as a vital aid to the common purpose. This is only one instance, but perhaps an important one, of the need to define the word "love," in the ethical domain, by more precisely what its exercise is meant to achieve.

References

Adelson, M. P-technique analysis: A P-technique study of social change in the U.S., 1845–1942. Master's Thesis, University of Illinois, Urbana, 1950.

Adorno, T. W., Frenkel-Brunswik, E., Levinson, D. J., and Sanford, R. N. *The authoritarian personality*. New York: Harper, 1950.

Alker, H. and Russett, B. *World politics in the general assembly*. New Haven: Yale University Press, 1965.

Allee, W. C. *The social life of animals*. New York: Norton, 1938.

Andreski, S. *Military organization and society*. London: Routledge & Kegan Paul, 1954.

Andreski, S. Origins of war. In J. D. Carthy and F. J. Ebling, Jr. (Eds.), *The natural history of aggression*. New York: Academic Press, 1964.

Archer, J. C. *Faiths men live by*. New York: Ronald Press, 1958.

Ardrey, R. A. *African genesis*. New York: Dell, 1961.

Ardrey, R. A. *Territorial imperative*. New York: Dell, 1966.

Ardrey, R. A. *The social contract*. New York: Athenaeum Press, 1970.

Arehart, J. L. Pre-natal diagnosis: How fast, how far? *Scientific News*, 1971, **100**, 44–45. (a).

Arehart, J. L. Genetic engineering: Myth or reality. *Science News*, 1971, **100**, 152–153. (b).

Aristotle. *Politics*. New York: Everyman's Library, 1943.

Aronfreed, J. *Conduct and conscience*. New York: Academic Press, 1968.

Baber, H. H. *Images of good and evil*. Translated by Michael Bulloch. London: Routledge & Kegan Paul, 1952.

Bacon, F. *The new Atlantis*. New York: Routledge and Sons, 1893.

Baetke, W. Die aufnahme des Christentums durch die Germanen; ein Beitrag zur Frage der Germanisierung des Christentums. Sonderaugabe. Darmstadt, Wissenschaftliche Buchgesellschaft, 1962.

Bagehot, W. *Physics and politics or thoughts on the application of principles of natural selection and inheritance to politics. Selected Essays*. London: Nelson, 1873 (1927).

Bales, R . F. *Interaction process analysis*. Cambridge: Adelson-Wesley, 1950.

Barker, E. *National character and the factors in its formation*. London: Methuen, 1948.

Barth, K. *The word of God and the word of man*. Translated by Douglas Horton. New York: Harper, 1957.

Bavelay, R. *Apology for true Christian divinity.* London, 1701.

Beadle, G. W. *Physical and chemical basis of inheritance.* Eugene, Oregon: Oregon State System of Higher Education, 1957.

Beadle, G. W. *Genetics and modern biology.* Philadelphia: American Philosophical Society, 1963.

Beauchamp, Joan. *Poems of revolt.* London: Noel Douglas, 1924.

Becker, C. L. *Heavenly city of the 18th century philosophers.* New Haven: Yale University Press, 1932.

Benedict, R. *Patterns of culture.* Boston: Houghton Mifflin, 1934.

Bentham, J. *Deontology or the science of morality.* London: Longman, Rees, Orme, Browne, Green, and Longman, 1834.

Berlin, I. *Karl Marx, his life and environment.* New York: Time, 1963.

Bogart, E. L. *Direct and indirect costs of the great world war.* New York: Oxford University Press, 1919.

Bolitho, W. *Twelve against the gods.* New York: Simon and Schuster, 1929.

Borgatta, E. F., Cottrell, L. S., Jr., and Meyer, H. J. On the dimensions of group behavior. *Sociometry,* 1956, **19,** 223–240.

Borgatta, E. F., and Meyer, H. J. *Sociological theory.* New York: Knopf, 1956.

Boulding, K. E. *The organizational revolution.* New York: Harper, 1953.

Boulding, K. E. *The meaning of the twentieth century; the great transition.* New York: Harper & Row, 1964

Bowra, C. M. Classical Greece. *Time,* 1966, 102.

Brinton, C. *The anatomy of a revolution.* New York: W. W. Norton & Co., Inc., 1938.

Bronfenbrenner, U., and Devereux, E. C. Interdisciplinary planning for team research on constructive community behavior: the Springdale project. *Human Relations,* 1952, **5,** 187–203.

Brooke, R. *Collected poems.* New York: Dodd & Mead, 1943.

Browning, R. *Complete poetical works.* Boston: Houghton Mifflin, 1895.

Burt, C. L. *Three reports on distribution and relations of educational abilities.* London: P. S. King, 1917.

Burt, C. L. *The young delinquent.* London: University of London Press, 1925.

Burt, C. L. *Factors of the mind.* London: University of London Press, 1940.

Bury, J. B. *The idea of progress.* London: Macmillan, 1920.

Butler, S. *Erewhon.* London: Trübner, 1872.

Cancro, J. (Ed.) *Intelligence: genetic and environmental influences.* New York: Grune & Stratton, 1971.

Carritt, E. F. *The theory of morals.* London: Oxford University Press, 1928.

Carritt, E. F. *Morals and politics.* London: Oxford University Press, 1952.

Carr-Saunders, A. M. *World population: Past growth and present trends.* London: Oxford University Press, 1936.

Carthy, J. D., and Ebling, F. J. *The natural history of aggression.* New York: Academic Press, 1964.

Cartwright, D. S., and Cartwright, Carol F. *Psychological adjustment: Behavior in the inner world.* Chicago: Rand McNally, 1971.

Cattell, R. B. *Psychology and social progress.* London: Daniel, 1933. (a).

Cattell, R. B. Education: The conquest of obstruction. In R. B. Cattell (Ed.), *Psychology and social progress.* London: Daniel, 1933. (b).

Cattell, R. B. Is national intelligence declining? *Eugenics Review,* 1936, **27,** 181–203.

Cattell, R. B. *The fight for our national intelligence.* London: King, 1937. (a).

Cattell, R. B. Some further relations between intelligence, fertility and socio-economic factors. *Eugenics Review*, 1937, **29**, 171–179. (b).

Cattell, R. B. *Psychology and the religious quest.* New York: Nelson, 1938.

Cattell, R. B. The concept of social status. *Journal of Social Psychology*, 1942, **15**, 293–308.

Cattell, R. B. The place of religion and ethics in a civilization based on science. In R. Wulsin (Ed.), *A revaluation of our civilization.* Albany: Argus Press, 1944. Chapter 2.

Cattell, R. B. The cultural functions of social stratification. I. Regarding the genetic basis of society. *Journal of Social Psychology*, 1945, **21**, 3–23. (a).

Cattell, R. B. The cultural functions of social stratification. II. Regarding individual and group dynamics. *Journal of Social Psychology*, 1945, **21**, 22–55. (b).

Cattell, R. B. *The description and measurement of personality.* New York: World Book Co., 1946.

Cattell, R. B. Concepts and methods in the measurement of group syntality. *Psychological Review*, 1948, **55**, 48–63.

Cattell, R. B. The dimensions of culture patterns by factorization of national characters. *Journal of Abnormal and Social Psychology*, 1949, **44**, 443–469.

Cattell, R. B. The principle culture patterns discoverable in the syntal dimensions of existing nations. *Journal of Social Psychology*, 1950, **32**, 215–253. (a).

Cattell, R. B. The scientific ethics of "Beyond." *Journal of Social Issues*, 1950, **6**, 21–27. (b).

Cattell, R. B. *Personality, a systematic theoretical and factual study.* New York: McGraw-Hill, 1950. (c).

Cattell, R. B. A quantitative analysis of the changes in the cultural pattern of Great Britain, 1837–1937, by P-technique. *Acta Psychologica*, 1953, **9**, 99–121.

Cattell, R. B. *Personality and motivation structure and measurement.* New York: World, 1957.

Cattell, R. B. The dynamic calculus: concepts and crucial experiments. In M. R. Jones (Ed.), *The Nebraska symposium on motivation.* Lincoln: University of Nebraska Press, 1959. Pp. 84–134.

Cattell, R. B. Group theory, personality and role: a model for experimental researches. In F. Geldard (Ed.), *Defence psychology.* Oxford: Pergamon Press, 1961. Pp. 209–258.

Cattell, R. B. *The scientific analysis of personality.* London: Penguin Books, 1965.

Cattell, R. B. *Handbook of multivariate experimental psychology.* Chicago: Rand McNally, 1966.

Cattell, R. B. *Abilities: Their structure, growth and action.* Boston: Houghton Mifflin, 1971.

Cattell, R. B. The Organization of advanced research institutes symbiotic with universities. *Higher Education*, 1972. In press.

Cattell, R. B., and Adelson, M. The dimensions of social change in the U.S.A. as determined by P-technique. *Social Forces*, 1951, **30**, 190–201.

Cattell, R. B., and Barton, K. Developmental psychology. Monograph, in preparation.

Cattell, R. B., Blewett, D. B., and Beloff, J. R. The inheritance of personality: A multiple variance analysis determination of approximate nature-nurture ratios for primary personality factors in Q-data. *American Journal of Human Genetics*, 1955, **7**, 122–146.

Cattell, R. B., Bolz, C., and Korth, B. Behavioral types in pure bred dogs objectively determined by Taxonome. *Human Behavioral Genetics*, 2, 1972, **2**, 10–19.

Cattell, R. B., Brace, C. L., and Korth, B. The isolation of temperament dimensions in dogs. In press.

Cattell, R. B., Breul, H., and Hartmann, H. P. An attempt at more refined definition of the cultural dimensions of syntality in modern nations. *American Sociological Review*, 1952, **17**, 408–421.

Cattell, R. B., and Butcher, J. *The prediction of achievement and creativity*. Indianapolis: Bobbs-Merrill, 1968.

Cattell, R. B., De Young, G., and Burdsal, C. Theory of motivation analysis: An experimental approach. *Multivariate Behavioral Research Monographs*, In press.

Cattell, R. B., Eber, H. W., and Tatsuoka, M. Handbook for the Sixteen Personality Factor Questionnaire. 1970 Edition. Champaign: Institute for Personality and Ability Testing, 1970.

Cattell, R. B., and Gorsuch, R. The definition and measurement of national morale and morality. *Journal of Social Psychology*, 1965, **67**, 77–96.

Cattell, R. B., and Gorsuch, R. Personality and socio-ethical values: the structure of self and superego. In R. B. Cattell and R. M. Dreger (Eds.), *Handbook of Modern Personality Theory*. New York: Teachers' College Press, 1973. Chapter 30.

Cattell, R. B., Kawash, G. F., and De Young, G. E. Validation of objective measures of ergic tension: Response of the sex erg to visual stimulation. *Journal of Experimental Research in Personality*, 1972.

Cattell, R. B., and Nesselroade, J. R. Likeness and completeness theories examined by 16 P.F. measures on stably and unstably married couples. *Journal of Personality and Social Psychology*, 1967, **7**, 351–361.

Cattell, R. B., and Nesselroade, J. R. *Methodology of human behavior genetics*. In preparation, 1972.

Cattell, R. B., and Radcliffe, J. Factors in objective motivation measures with children. A preliminary study. *Australian Journal of Psychology*, 1961, **13**, 65–76.

Cattell, R. B., Radcliffe, J., and Sweney, A. B. The nature and measurement of components of motivation. *Genetic and Psychological Monographs*, 1963, **68**, 49–211.

Cattell, R. B., and Scheier, I. H. *The meaning and measurement of neuroticism and anxiety*. New York: Ronald Press, 1961.

Cattell, R. B., and Stice, G. F. The dimensions of groups and their relations to the behavior of members. Ann Arbor, Michigan: University Microfilms, 1969.

Cattell, R. B., Stice, G. F., and Kristy, N. F. A first approximation to nature-nurture ratios for eleven primary personality factors in objective tests. *Journal of Abnormal and Social Psychology*, 1957, **54**, 143–159.

Cattell, R. B., and Tatro, D. F. The personality factors objectively measured which distinguish psychotics from normals. *Behavioral Research Therapy*, 1966, **4**, 39–51.

Cattell, R. B., Tatro, D. F., and Komlos, E. The diagnosis and inferred structure of paranoid and non-paranoid schizophrenia from the 16 P.F. profile. *Indian Psychological Review*, 1965, **1**, 108–115.

Chamberlin, E. H. *The theory of monopolistic competition*. Cambridge: Harvard University Press, 1948.

Chassell, C. F. *Relation between morality and intellect*. Columbia, New York: Teachers College, 1935.

Chicago Tribune. Editorial, June 10, 1970.

Childe, G. *What happened in history*. Harmondsworth, Middlesex, England: Penguin, 1950.

Chomsky, N. *Syntactic structures*. s-Gravenhage: Mouton & Co., 1957.

Clark, C. *The conditions of economic progress*. London: Macmillan, 1957.

Clark, R. W. *J. B. S.: The life and work of J. B. S. Haldane*. New York: Coward-McCann, 1969.

Clausewitz, K. von. *On war.* New York: The Modern Library, 1943.

Clemenceau, G. *In the evening of my thought.* Boston: Houghton Mifflin, 1929.

Comte, A. *Cours de philosophie positive,* 1829 (*The Positive Philosophy*). London: Routledge, 1905.

Cooley, C. H. *Human nature and the social order.* New York: Scribner, 1922.

Coon, C. S. *The origin of races.* New York: Knopf, 1962. (a).

Coon, C. S. *The story of man.* New York: Knopf, 1962. (b).

Cottrell, L. *The anvil of civilization.* New York: New American Library, 1957.

Cranston, M. *What are human rights?* New York: Basic Books, 1963.

Croce, B. *Politics and morals.* New York: Philosophical Library, 1945.

Curry, W. B. *The school and a changing civilization.* London: John Lane, 1937.

Darlington, C. D. *The conflict of science and society.* New York: Basic Books, 1947.

Darlington, C. D. *Genetics and man.* London: Allen & Unwin, 1966.

Darlington, C. D. *The evolution of man and society.* New York: Simon & Schuster, 1969.

Darwin, B. *The Next Million Years.* Garden City, New York: Doubleday, 1952.

Darwin, C. *The descent of man.* New York: Modern Library, 1871.

Darwin, C. *The origin of species.* London: Murray, 1859 (1917).

Darwin, L. *The need for eugenic reform.* London: J. Murray, 1926.

Davis, K. *Human society.* New York: Macmillan, 1949.

Delhees, K. Conflict measurement by the dynamic calculus model, and its applicability in clinical practice. *Multivariate Behavioral Research,* 1968, 3, 73–96.

Delhees, K., and Nesselroade, J. Methods and findings in experimentally based personality theory. In R. B. Cattell (Ed.), *Handbook of Multivariate Experimental Psychology.* Chicago: Rand McNally, 1966. Chapter 19.

Dennis, W. Age of productivity among scientists. *Science,* 1956, 123, 724–725.

Deutsch, Martin. *Social class, race and psychological development.* New York: Holt, Rinehart and Winston, 1968.

Deutsch, Morton. *Theories in social psychology.* New York: Basic Books, 1965.

Dickson, L. *H. G. Wells, his turbulent life and times.* New York: Atheneum, 1969.

Diderot, D. *Oeuvres choisies.* Vol. II. *De l'interpretation de la nature.* Paris: Editions sociales, 1953–64 (1775).

Dielman, T. E., and Krug, S. Trait description and measurement in motivation and dynamic structure. In R. B. Cattell and R. M. Dreger (Eds.), *Handbook of Modern Personality Theory.* New York: Teachers' College Press, 1973. Chapter 5.

Digman, J., and Tuttle, D. W. An interpretation of an election by means of transposed factors. *Journal of Social Psychology,* 1961, 53, 183–194.

Djilas, M. *Conversations with Stalin.* New York: Harcourt, 1962.

Dobzhansky, T. H. *The biological basis of human freedom.* New York: Columbia University Press, 1960.

Dobzhansky, T. H. *Mankind evolving.* New Jersey: Yale University Press. 1962.

Dodds, E. R. *The Greeks and the irrational.* Berkeley: University of California Press, 1951.

Draper, J. W. *History of the conflict between science and religion.* New York: D. Appleton & Co., 1898.

Drawbridge, C. P. *The religion of scientists.* London: Rationalist Press, 1932.

Dreikurs, R. Equality: The challenge of our time. Mimeograph, 1960.

Dunn, L. C., and Dobzhansky, T. *Heredity and the nature of man.* New York: Harcourt, Brace & World, 1964.

Durkheim, E. *The elementary forms of the religious life.* Glencoe, Illinois: The Free Press, 1915.

Ebling, F. J. *Biology and ethics.* New York: Academic Press, 1969.

Eddington, A. S. *Science and the unseen world.* New York: Swarthmore Lecture, 1929.

Ehrlich, P. *The population bomb.* New York: Ballantine Books, 1968.

Eibl-Eibesfeldt, I. *Ethology, the biology of behavior.* New York: Rinehart and Winston, 1970.

Eisenhower, D. Farewell Address of the President. Proceedings of Congress, January 17, 1961.

Ellis, A. Viewpoints in modern psychology, *Mensa,* 1970, 138.

Emmet, Dorothy. *Rules, roles and relations.* London: Macmillan, 1966.

Epstein, S. Anxiety and achievement. In C. D. Spielberger (Ed.). *The psychology of anxiety.* New York: Academic Press, 1972.

Eysenck, H. J. General social attitudes. *Journal of Social Psychology,* 1944, **19**, 207–277.

Eysenck, H. J. *The scientific study of personality.* London: Routledge & Kegan Paul, 1952.

Eysenck, H. J. *The psychology of politics.* London: Routledge & Kegan Paul, 1954.

Eysenck, H. J. *The I.Q. argument.* New York: Library Press, 1971.

Festinger, L., Schacter, S., and Back, K. *Social pressures in informal groups.* New York: Harper & Bros., 1950.

Fiedler, F. E. Leadership — a new model. *Discovery,* April 1965.

Fisher, R. A. *Genetical theory of natural selection.* Oxford: Clarendon Press, 1930.

Fosdick, H. E. *As I see religion.* New York: Harper & Bros., 1932.

France, A. *The garden of Epicurus.* London: Lane, 1920.

Frazer, J. G. *The golden bough.* London: Macmillan, 1890.

Frenkel-Brunswik, Else. Meaning of psychoanalytic concepts and confirmation of psycho-analytic theories. *Scientific Monthly,* 1954, **79**, 293–300.

Freud, S. *Totem and taboo.* Leipzig: H. Helle, 1913.

Freud, S. *Future of an illusion.* London: Hogarth Press, 1928.

Freud, S. *Civilization and its discontents.* London: L. and Virginia Woolf at the Hogarth Press, 1930.

Freud, S. *Moses, an Egyptian.* Translated from the German by Katherine Jones. New York: Vintage Books, 1939 (1961).

Friedman, M. *Dollars and deficits.* New York: Prentice Hall, 1968.

Fulbright, J. W. *The arrogance of power.* New York: Random House, 1967.

Fuller, J. L., and Thompson, W. R. *Behavior genetics.* New York: Wiley, 1960.

Galton, Sir F. *Inquiries into human faculty and its development.* London: Dent, 1883 (1951).

Gardner, J. *No easy victories.* New York: Harper & Row, 1968.

Gibb, C. A. Changes in the culture pattern of Australia, 1906–1946 as determined by P-technique. *Journal of Social Psychology,* 1956, **43**, 225–238.

Gibb, C. A. *Leadership.* London: Penguin Books, 1969.

Gibbon, E. *The decline and fall of the Roman empire.* London: Dent & Sons, 1910.

Gissing, G. *By the Ionian Sea.* London: Richards Press, 1956.

Gordon, M. *American people's encyclopedia.* Chicago: Spencer Press, 1953.

Gorsuch, R. L. The clarification of some super ego factors. Unpublished doctoral dissertation, University of Illinois, Urbana, Illinois, 1965.

Gorsuch, R. L., and Cattell, R. B. Personality and socio-ethical values: the structure of self and superego. In R. B. Cattell and R. M. Dreger (Eds.), *Handbook of Modern Personality Theory.* Teachers' College Press, 1972. Chapter 30.

Gould, J. B. *World News,* June 3, 1970.

Gouldner, A., and Peterson, R. A. *Notes on technology and the moral order.* Indianapolis: Bobbs-Merrill, 1962.

Graham, R. K. *The future of man.* North Quincy: Christopher, 1970.

Graves, R. *King Jesus.* New York: Creative Age Press, 1946.

Gregg, P. M., and Banks, A. S. Dimensions of political systems: Factor analyses of a cross-polity survey. *American Political Science Review*, 1965, **59**, 602–614.

Gregor, A. Proceedings of London symposium on individual differences. Institute of Psychiatry, August 19–21, 1970, In press.

Grinker, R. R., and Spiegel, J. P. *Men under stress.* Philadelphia: Blakiston, 1945.

Gubser, C. J. *Congressional Record*, No. 117, July 15, 1969.

Guilford, J. P. *Personality.* New York: McGraw-Hill, 1959.

Hadden, J. K., and Borgatta, E. *American cities: Their social characteristics.* Chicago: Rand McNally, 1965.

Haeckel, E. *The riddle of the universe.* London: Watts, 1929.

Haldane, G. B. S. *Daedalus or science and the future.* London: Kegan Paul, 1925.

Haldane, G. B. S. *Possible worlds.* London: Harper & Bros., 1928.

Hall, C. S., and Lindzey, G. *Theories of personality.* New York: Wiley, 1970.

Hardin, G. J. *Population, evolution, and birth control; a collage of controversial readings.* New York: W. H. Freeman, 1964.

Hardy, T. *The dynasts.* New York: Macmillan, 1904.

Harlow, H. F. The formation of learning sets. *Psychological Review*, 1949, **56**, 51–65.

Harrison, G. A. *Genetical variation in human populations.* New York: Pergamon Press, 1961.

Haskins, C. P. Advances and challenges in science in 1969. *American Scientist*, July–August, 1970, **58**, 365–377.

Hayek, F. A. von. *The road to serfdom.* Chicago: University of Chicago Press, 1945.

Heape, W. *Emigration, migration and nomadism.* Cambridge: W. Heffer & Sons, 1931.

Heidegger, M. *Existence and being.* Chicago: H. Regnery Co., 1949.

Hendricks, B. The sensitivity of the dynamic calculus to short term change and interest structure. Unpublished Master's thesis, University of Illinois, Urbana, Illinois, 1971.

Henry, D. D. Annual report of the President of the University of Illinois, 1970.

Hess, E. H. The relationship between imprinting and motivation. In M. R. Jones (Ed.), *Nebraska symposium on motivation.* Lincoln: University of Nebraska Press, 1959. Pp. 47–77.

Higgins, J. V., Reed, E. W., and Reed, S. C. Intelligence and family size: a paradox resolved. *Eugenics Quarterly*, 1962, **9**, 84–90.

Hirsch, N. D. M. A study of natio-racial mental differences. *Genetic Psychology Monographs*, 1926, **1**, 231–406.

Hirsch, N. D. M. *Genius and creative intelligence.* Cambridge, Massachusetts: Science-Art Publishers, 1931.

Hirsh, W. *Explorations in social change.* Boston: Houghton Mifflin, 1964.

Hoagland, H. The human adrenal cortex in relation to stressful activities. *Journal of Aviation Medicine*, 1947, **18**, 5.

Hoagland, H., and Burloe, R. W. (Eds.) *Evolution and man's progress.* New York: Columbia University Press, 1962.

Hobbes, T. *The great leviathan.* New York: Liberal Arts Press, 1958.

Hofstadter, R. *Anti-intellectualism in American life.* New York: Knopf, 1963.

Hogben, L. *Science for the citizen.* New York: W. W. Norton, 1951.

Hooton, E. A. *The American criminal.* Cambridge: Harvard University Press, 1939.

Hooton, E. A. *Man's poor relations.* New York: Doubleday, 1942.

Hooton, E. A. *Up from the ape.* New York: Macmillan, 1946.

Horn, J. L. Fluid and crystallized intelligence: a factor analytic study of the structure among primary mental abilities. Unpublished doctoral dissertation, University of Illinois, Urbana, Illinois, 1965.

Horn, J. L. Motivation and dynamic calculus concepts from multivariate experiment. In R. B. Cattell (Ed.), *Handbook of multivariate experimental psychology.* Chicago: Rand McNally, 1966. Chapter 20.

Horn, J. L., and Knott, P. D. Activist youth of the 1960's: Summary and progress. *Science,* 1970, **171**, 977–985.

Hornbein, T. *Everest, the west ridge.* San Francisco: Sierra Club, 1965.

Hume, D. *Treatise on human nature.* Cambridge: The University Press, 1938.

Hundleby, J., Pawlik, K., and Cattell, R. B. *Personality factors in objective test devices.* San Diego: R. R. Knapp & Co., 1965.

Huntington, E. *Mainsprings of civilization.* New York: New American Library, 1945.

Huxley, A. *Jesting Pilate.* London: Chatto and Windus, 1926.

Huxley, A. *Brave new world.* New York: Garden City Publishing Co., 1932.

Huxley, J. *Evolution in action.* London: Chatto and Windus, 1953.

Huxley, J. *Knowledge, morality and destiny.* New York: New American Library, 1957.

Huxley, J., and Huxley, T. *Touchstone for ethics, 1893–1943.* New York: Harper, 1947.

Huxley, J., and Kettlewell, H. B. D. "Naturalness" of separate, segregated development. In *Charles Darwin and his world.* New York: Viking, 1965.

Huxley, T. *Evolution and ethics.* London: Macmillan, 1893.

Huxley, T. *Science and education.* New York: Appleton, 1901.

Inge, W. R. *Science and ultimate truth.* London: Longmans, Green & Co., 1926.

Inge, W. R. *Assessments and anticipations.* London: Cassell, 1929.

James, W. The moral equivalent of war, 1910. In *Essays on faith and morals.* New York: Meridian, 1962.

James, W. *The varieties of religious experience.* New Hyde Park, New York: University Books, 1963.

Jefferson, T. *Crusade against ignorance; Thomas Jefferson on education.* New York: Bureau of Publications, Teachers' College, Columbia University, 1961.

Jinks, J. L., and Fulker, D. W. Comparison of the biometrical genetical, MAVA and classical approaches to the analyses of human behavior. *Psychological Bulletin,* 1970, **73**, 311–349.

Joad, C. E. M. *The present and the future of religion.* London: E. Benn, 1930.

Johnson, D. M. Reasoning and logic. In *International Encyclopedia of Social Scientists.* New York: Macmillan, 1968. Pp. 344–350.

Johnson, R. C., Dokecki, P. R., and Mowrer, O. H. *Conscience, contract and social reality.* New York: Holt, Rhinehart and Winston, 1972.

Jonassen, C. T. Functional unities in 88 community systems. *American Sociological Review,* 1961, **26**, 399–407.

Julian, J. W., Bishop, D. W., and Fiedler, F. E. Quasi-therapeutic effects of inter-group competition. *Journal of Personality and Social Psychology,* 1966, **3**, 321–327.

Kahn, H. *On thermonuclear war.* Princeton, New Jersey: Princeton University Press, 1960.

Keith, Sir J. A. *Essays on human evolution.* London: Watts, 1946.

Keith, Sir J. A. *A new theory of human evolution.* New York: Philosophical Library, 1949.

Kierkegaard, S. A. *Sickness unto death.* London: Oxford University Press, 1941.

Khrushchev, N. S. *Khrushchev remembers*. Edited and translated by C. Talbott. *Life*, November 27, 1970, **69**, 32–9.

King, R. C. *Genetics*. New York: Oxford University Press, 1965.

Kipling, R. *Inclusive verses, 1885–1932*. New York: Sun Dial Press, 1940.

Knapp, R. R. Demographic cultural and personality attributes of scientists. In C. W. Taylor and F. Barron (Eds.), *Scientific creativity: It's recognition and development*. New York: Wiley, 1963.

Kretschmer, E. *The psychology of men of genius*. London: Kegan Paul, 1931.

Kroeber, A. L. *Anthropology*. New York: Harcourt, 1958.

Kropotkin, P. *Mutual aid: a factor in evolution*. New York: McClure, Phillips and Co., 1902.

Krug, S. An examination of experimentally induced changes in ergic tension levels. Unpublished doctoral dissertation, University of Illinois, Urbana, Illinois, 1971.

Kuhn, T. S. *The structure of scientific revolutions*. Chicago: University of Chicago Press, 1962.

Ladd, J. *The structure of a moral code*. Cambridge: Harvard University Press, 1957.

Lanier, L. New trends in university education. Honors Day Address, University of Illinois, April 30, 1971, mimeograph.

Lasswell, T. E. *Class and stratum: An introduction to concepts and research*. Boston: Houghton-Mifflin, 1965.

Leary, T. Politics and ethics of ecstasy. *Cavalier*, July, 1966.

Lederberg, J. Genetic engineering controlling man's building blocks. *Today's Health*, November 1969, **47**, 24–27.

Lehman, H. G. The creative years in science and literature. *Scientific Monthly*, 1936, **43**, 151–162.

Lentz, T. F. The relation of I.Q. to size of family. *Journal of Educational Psychology*, 1927, **18**, 486–496.

Lepley, R. *Verifiability of value*. New York: Columbia University Press, 1944.

Lerner, D., and Lasswell, H. S. *The policy sciences: recent developments in scope and method*. Stanford: Hoover Institute, 1951.

Lerner, I. M. *Heredity, evolution and society*. New York: Freeman, 1968.

Levine, R. A. *The poor, ye need not have with you*. Boston: MIT Press, 1970.

Lewin, K. *Resolving social conflicts*. New York: Harper, 1948.

Light, R. J., and Smith, P. V. Social allocation models of intelligence: A methodological inquiry. *Harvard Educational Review*, 1969, **39**, 484–510.

Lindbergh, C. A. *The wartime journals of Charles A. Lindbergh*. New York: Harcourt Brace Jovanovich, 1970.

Lindsey, B. B., and Evans, W. *The revolt of modern youth*. New York: Boni & Liveright, 1925.

Linton, R. *The study of man*. New York: Appleton-Century, 1936.

Lloyd, R. Cross and psychosis. *Faith and Freedom*, 1971, **24**, 1–40.

Lombroso, C. *Criminal man*. New York: G . P. Putnam's Sons, 1911.

Lorenz, K. *On aggression*. New York: Harcourt, Brace and World, 1966.

Luria, A. R. L. S. Vygotsky and the problem of localization of functions. *Neuropsychologia*, 1965, **3**, 387–392.

Lynn, R. *Personality and national character*. New York: Pergamon Press, 1971.

MacArthur, D. *Reminiscences*. New York: McGraw-Hill, 1964.

Macaulay, T. B. *The works of Lord Macaulay complete*. Edited by Lady Trevelyon. New York: Longmans, Green & Co., 1897.

Malinowski, B. Anthropology as basis of social science. In R. B. Cattell, J. Cohen and R. W. M. Travis (Eds.), *Human affairs*. London: Macmillan, 1937. Pp. 199–252.

Malraux, A. *The psychology of art*. New York: Pantheon Books, 1949.

Mangasarian, M. M. *The bible unveiled*. Chicago: Independent Religious Society, 1960.

Mannheim, K. Present trends in the building of society. In R. B. Cattell, *et al.* (Eds.), *Human affairs*. London: Macmillan, 1937.

Marx, K. *Das Kapital*. Hamburg: O Meissner, 1890–94.

Maslow, A. H. *Toward a psychology of being*. New York: Van Nostrand, 1960.

Mather, K. The genetical structure of populations. *Symposium of the Society for Experimental Biology*, 1953, 7, 63.

Mathews, Dean W. R. *God in Christian thought and experience*. Digswell Place, England: J. Nisbet, 1963.

Mazzini, G. *The duties of man*. Everyman's, New York: Dutton, 1907.

McDougall, W. *National welfare and national decay*. London: Methuen, 1921.

McDougall, W. *Ethics and some modern world problems*. London: Methuen, 1924.

McDougall, W. *Janus, the conquest of war*. London: Today and Tomorrow Series, 1925.

McDougall, W. *Religion and the sciences of life*. London: Methuen, 1934.

Mead, G. H. The ideal of social integration. In C. W. Morris (Ed.), *Mind, self and society*. Chicago: University of Chicago Press, 1934.

Mead, Margaret. Some relationships between social anthropology and psychiatry. In T. Alexander and H. Ross (Eds.), *Dynamic psychiatry*. Chicago: University of Chicago Press, 1952.

Mead, Margaret. *Culture patterns and technological change*. New York: New American Library, 1955.

Meadows, D. H., *et al*. *The limits to growth: A report of the Club of Rome Project on the predicament of mankind*. New York: Universe Press, 1972.

Meeland, T., Egbert, R., *et al*. Task Fighter; Description of Tests. OCAFF, Ford Ord, California, Human Research Unit No. 2, April, 1954.

Merritt, R. L. *Systematic approaches to comparative politics*. Chicago: Rand McNally, 1970.

Merton, R. K. *Social theory and social structure*. Glencoe, Illinois: Free Press, 1957.

Michael, D. *The unprepared society*. New York: Harper, 1968.

Mill, J. S. *Utilitarianism*. London: Parker & Bourn, 1863.

Miller, J. G. Living systems, basic concepts. *Behavioral Science*, 1965, **10**, 193–237.

Moltmann, J. *Theology of hope; on the ground and the implications of Christian eschatology*. New York: Harper & Row, 1967.

Monod, J. *Chance and necessity*. New York: Knopf, 1971.

Montesquieu, de C. L. *Spirit of the laws*. London: F. Wingrave, 1793.

Morgan, C. L. *Emergent evolution*. New York: H. Holt, 1923.

Morgan, L. H. *Ancient society; or researches in the lines of human progress from savagery, through barbarism to civilization*. New York: H. Holt & Co., 1878.

Morris, C. *Varieties of human value*, Chicago: University of Chicago Press, 1956.

Morris, D. *The naked ape*. London: Cape, 1967.

Mowrer, O. H. Some constructive features of the concept of sin. *Journal of Counselling Psychology*, 1960, 7, 185–188.

Mowrer, O. H. *The new group psychotherapy*. New York: Van Nostrand, 1964.

Mowrer, O. H. *Morality and mental health*. Chicago: Rand McNally, 1967.

Muller, H. J. Our load of mutations. *American Journal of Human Genetics*, 1950, **2**, 60–70.

Muller, H. J. *Out of the night.* New York: Vanguard, 1953.

Muller, H. J. What genetic course will man steer? Proceedings of 3rd International Congress on Human Genetics, Chicago, September, 1966.

Murray, J. C. *Freedom and man.* New York: P. J. Kennedy, 1965. (a).

Murray, J. C. *The problem of religious freedom.* Westminister, Maryland: Newman Press, 1965. (b).

Myrdal, G. *Challenge to affluence.* New York: Vintage Books, 1965.

Myrdal, G. *Asian drama; an inquiry into the poverty of nations.* New York: Pantheon Press, 1968.

Nadel, S. F. *The theory of social structure.* Glencoe, Illinois: Free Press, 1957.

National Geographic Editorial, January 1969. p. 10.

Neary, J. A scientist's variations on a disturbing racial theme. *Life,* June 12, 1970, 68, 58b.

Needham, J. *The sceptical biologist.* London, 1929. (No publisher listed.)

Newbolt, H. *Collected poems, 1897–1907.* London: Nelson & Sons, 1908.

Nietzsche, F. *Also sprach Zarathustra.* Leipzig: Kroner, 1930.

Noüy, L. de.*Human destiny.* New York: Longmans Green, 1947.

Oppenheimer, J. R. *The open mind.* New York: Simon & Schuster, 1955.

Osler, Sir W. *A way of life.* New York: Dover, 1958.

Paddock, W. *Famine, 1975!* Boston: Little, Brown & Co., 1967.

Paley, W. *Natural theology, as evidences of the existence and attributes of the deity, collected from the appearances of nature.* London: Faulder, 1802.

Pannenberg, W. *What is man? Contemporary anthropology in theological perspective.* Philadelphia: Fortress Press, 1970.

Parkinson, C. N. *Parkinson's law, and other studies in administration.* Boston: Houghton Mifflin, 1957.

Patrick, J. M. (Ed.) *Selected essays of Francis Bacon.* New York: Appleton-Century, 1948.

Patton, M. *War as I knew it.* Boston: Houghton Mifflin, 1947.

Pearson, K. *The groundwork of eugenics.* London: Dulace & Co., 1909.

Pendell, E. *Population on the loose.* New York: Funk, 1951.

Penrose, L. S. The supposed threat of declining intelligence. *American Journal of Mental Deficiency,* 1948, **53,** 114–118.

Petrie, W. M. F. *The revolutions of civilization.* London: Harper, 1911.

Plato. *The Republic.* Cleveland: World, 1946.

Polyani, M. *Personal knowledge.* London: Routledge & Kegan Paul, 1958.

Popper, K. R. *The open society and its enemies.* London: Routledge & Kegan Paul, 1957.

Price, D. de Solla. *Science since Babylon.* New Haven: Yale University Press, 1961.

Ramsey, P. *Fabricated man; the ethics of genetic control.* New Haven: Yale University Press, 1970.

Reed, S. C., and Reed, Elizabeth W. *Mental retardation, a family study.* Philadelphia: Saunders, 1965.

Ricardo, D. *The principles of political economy and taxation.* London: Dent, 1817 (1911).

Richardson, L. F. The distribution of wars in time. *Journal of the Royal Statistical Association,* 1946, **107,** 242–250.

Richardson, L. F. Statistics of deadly quarrels. In Q. Wright and C. C. Lieman (Eds.) Pittsburgh: Boxwood Press, 1960.

Riddle, O. *The unleashing of evolutionary thought.* New York: Vantage, 1948.

Roberts, M. *The behavior of nations.* London: J. M. Dent & Sons, 1941.

Roe, A. *The making of a scientist.* New York: Dodd, 1953.

Rosenfeld, A. *The second geneses.* Englewood Cliffs: Prentice-Hall, 1969.

Rosenthal, D. *Genetic theory and abnormal behavior.* New York: McGraw-Hill, 1970.

Royce, J. R. *Man and his nature.* New York: McGraw-Hill, 1961.

Royce, J. R. *Psychology and the symbol.* New York: Random House, 1965.

Royce, J. R. *The encapsulated man.* Princeton: Van Nostrand, 1964.

Royce, J. R. Metaphoric knowledge and humanistic psychology. In J. F. T. Bugental (Ed.), *Challenges of humanistic psychology.* New York: McGraw-Hill, 1967. Pp. 20–28.

Rousseau, J. J. *The social contract.* London: J. M. Dent & Sons, 1913.

Rummel, R. J. Dimensions of conflict behavior within and between nations. *General systems: Yearbook of the Society for General Systems Research*, 1963, **8**, 1–50.

Rummel, R. J. *Dimensions of nations*, Beverly Hills: Sage, 1972.

Rummel, R. J. *Applied factor analysis.* Evanston: Northwestern University Press, 1970.

Rundquist, E. E. Inheritance of spontaneous activity in rats. *Journal of Comparative Psychology*, 1933, **16**, 1–23.

Russell, B. *Human society in ethics and politics.* New York: Simon & Schuster, 1955.

Russell, B. *Marriage and morals.* New York: H. Liveright, 1957.

Russell, B. *Autobiography.* Boston: Little, Brown & Co., 1968.

Sarason, S. B., Mandler, G., and Craighill, P. G. The effect of differential instructions on anxiety and learning. *Journal of Abnormal and Social Psychology*, 1952, **47**, 561–565.

Sartre, J. P. *Existentialism and humanism.* London: Methuen & Co., 1948.

Scheier, I. H. A nationwide testing system for the objective study of delinquency. In press.

Schoeps, H. J. *An intelligent person's guide to the religions of mankind.* London: Gallancy (Victor) Ltd., 1967.

Schröder, C. M. *Die religionen der Menschheit.* Stuttgart W. Kohlhammen, 1960–68.

Schweitzer, A. *Christianity and the religions of the world.* New York: H. Holt, 1939.

Sélincourt, A. de. *The World of Herodotus,* Boston: Little Brown, 1962.

Sells, S. *Stimulus determinants of behavior.* New York: Ronald, 1962.

Shaw, G. B. *Everybody's political what's what.* New York: Dodd Mead, 1944.

Shaw, G. B. *Back to Methusaleh.* London: Constable Co., 1949.

Shaw, G. B. *The complete prefaces of Bernard Shaw.* London: Hamlyn, 1965.

Shirer, W. L. *The collapse of the third republic.* New York: Simon & Schuster, 1969.

Shockley, W. B. Human quality problems, research taboos and eugenics. Convocation Lecture, University of Bridgeport, Conn., December 10, 1969.

Short, J. F., and Strodbeck, F. L. *Group process and gang delinquency.* Chicago: University of Chicago Press, 1965.

Shwayder, D. S. *The stratification of behavior.* London: Routledge & Kegan Paul, 1965.

Sidgwick, H. *The methods of ethics.* London: Macmillan, 1893.

Simpson, G. G. *The meaning of evolution.* New Jersey: Yale University Press, 1949.

Simpson, J. *Landmarks in the struggle between science and religion.* New York, 1926. (No publisher given.)

Singer, J. D. (Ed.) *Human behavior and international politics.* Chicago: Rand McNally, 1965.

Snow, C. P. *The 2 cultures and the scientific revolution.* Cambridge: University Press, 1959.

Snow, C. P. *Science and government.* Cambridge: Harvard University Press, 1961.

Sorokin, P. A. *Social mobility.* New York: Harper, 1927.

Sorokin, P. A. *Social and cultural dynamics.* Vol. 3, New York: American Book Co., 1937.

Spearman, C. *Creative mind*. London: Nisbet, 1930.

Spencer, H. *Principles of ethics*. London: Williams & Norgate, 1892.

Spengler, O. *Decline of the West*. New York: Knopf, 1928.

Spielberger, C. D. (Ed.) *Anxiety and behavior*. New York: Academic Press, 1966.

Spuhler, J. N. (Ed.) *Genetic diversity and human behavior*. New York: Viking Fund Publication, 1967.

Stephen, J. F. *Liberty, equality, fraternity*. New York: Holt, 1873.

Stevens, S. S. The NAS-NRC and psychology. *American Psychologist*, April 1952, **7**, 119–124.

Sturtevant, A. H., and Beadle, G. W. *Introduction to genetics*. Dover: Constable, 1964.

Sweney, A. B., and Cattell, R. B. Relationships between integrated and unintegrated motivation structure examined by objective tests. *Journal of Social Psychology*, 1962, **57**, 217–226.

Taylor, C. W., and Barron, F. (Eds.) *Scientific creativity: It s recognition and development*. New York: Wiley, 1963.

Tennyson, A. L. *Poetical works*. London: Macmillan, 1908.

Terman, L. M. *Genetic studies of genius. I. A thousand gifted children*; II. (With Catherine M. Cox, *et al.*) *The early mental traits of three hundred geniuses*. London: Harrap, 1926.

Tharp, R. G. Dimensions of marriage roles. *Marriage and Family Living*, 1963, **11**, 389–404.

Thomson, G. M. *The Twelve Days: 24 July to 4 August, 1914*. New York: Putnam, 1964.

Thorndike, E. L. *Human nature and social order*. New York: Macmillan, 1939.

Thurstone, L. L. *A factorial study of perception*. Chicago: University of Chicago Press, 1944.

Tillich, P. *Love, power, and justice*. New York: Oxford University Press, 1954.

Tillich, P. *Biblical religion and the search for ultimate reality*. Chicago: University of Chicago Press, 1955.

Tillich, P. *Christianity and the encounter of the world religions*. New York: Columbia University Press, 1963.

Tinbergen, N. *The study of instinct*. Oxford: Clarendon Press, 1951.

Tinbergen, N. Behavior, systematics and natural selection. *Ibis*, 1959, **101**, 119.

Tocqueville, de. *Democracy in America*. New York: Oxford University Press, 1947.

Tolstoy, L. *The kingdom of God is within you*. Translated by A. Delano. London: W. Scott, 1894.

Toynbee, A. J. *A study of history*. New York: Oxford University Press, 1947.

Treitschke, H. G. *Politics*. New York: 1916. (No publisher given.)

Tryon, R. C. Genetic differences in maze learning ability in rats. In *Yearbook of National Society for the Study of Education*. Vol. 39, 1940. Pp. 111–119.

Truman, H. S. Congressional Record, March 12, 1947.

Tuchman, Barbara. Address to National Conference on Higher Education, Chicago, November 7, 1967.

Udry, J. R. *The social context of marriage*. Philadelphia: Lippincott, 1966.

Unamuno, M. de. *The tragic sense of life*. Paris: Gaurmard, 1926.

United States Office of Education. United States Office of Education Report for 1970. Washington, D.C.: Government Printing Office, 1971.

Unwin, J. D. *Sex and culture*. London: Oxford University Press, 1934.

Vandenberg, S. G. *Methods and goals in human behavior genetics*. New York: Academic Press, 1965.

Veblen, V. *The theory of the leisure class.* New York: Macmillan, 1899.

Vidal, F. *Problem solving: méthodologie général de la creativité.* Paris: Dunod, 1971.

Voltaire, F. *Candide.* New York: Appleton-Century-Crofts, Inc., 1946.

Waddington, C. H. Genetic assimilation of an acquired character. *Evolution*, 1953, 7, 118–126.

Waddington, C. H. *The nature of life.* New York: Atheneum, 1962.

Wallas, G. *The great society.* London: Macmillan, 1914.

Ward, M. *Robert Browning.* New York: Holt, 1967.

Watson, J. B. *Behavior: An introduction to comparative psychology.* New York: Holt, Rinehart & Winston, 1914.

Weber, M. *The protestant ethic.* New York: Scribner & Sons, 1904 (1956).

Weber, M. Objectivity in social science. In E. A. Shils and H. A. Finch (Eds.), *The methodology of the social sciences.* New York: Free Press, 1949 (1968).

Wells, H. G. *Mankind in the making.* London: Chapman & Hall, 1903.

Wells, H. G. *A modern utopia.* London: Chapman & Hall, 1905.

Wells, H. G. *The outline of history.* London: Cassell, 1920.

Wells, H. G. *The open conspiracy.* London: Hogarth, 1930.

Wells, H. G. *What are we to do with our lives?* London: W. Heineman, 1931.

Wells, H. G. *The rights of man; or, what are we fighting for?* New York: Penguin Books, 1940.

Westermarck, E. A. *Ethical relativity.* New York: Harcourt, Brace & Co., 1932.

White, A. D. *A history of the warfare of science with theology in Christendom.* London: Constable & Co., 1896.

White, M. and L. *The intellectual vs. the city.* New York: Mentor, 1962.

Wiener, N. *The human use of human beings.* Boston: Houghton Mifflin, 1954.

Wilkie, W. L. *One world.* New York: Simon & Schuster, 1943.

Williams, R. J. *Biochemical individuality.* New York: Wiley, 1956.

Williams, R. J. Heredity, human understanding and civilization. *American Scientist*, 1969, 57, 237–243.

Wilson, W. Congressional Record, April 2, 1917.

Winborn, B. B., and Jansen, D. G. Personality characteristics of campus political action leaders. *Journal of Counselling Psychology*, 1967, 14, 509–518.

Wolfle, D. *The uses of talent.* Princeton, New Jersey: Princeton University Press, 1971.

Woodstock music and art fair. *Chicago Tribune* editorial, June 10, 1970.

Wrigley, C. A multivariate study of United Nations General Assembly voting records. East Lansing: Michigan State University, 1963. Mimeograph.

Name Index

Adams, J. C. 23
Adelson, M. 129, 130, 275, 366
Adorno, T. W. 46, 47, 66
Aeschylus 272
Agnew, S. 382
Ahknaton 406
Alexander The Great 103, 406, 445
Alker, H. 130, 131, 331, 343
Allee, W. C. 82
Anderson, V. E. 290
Andreski, S. 198, 201, 206
Aquinas, T. 52, 243
Archer, J. C. 13, 50
Archimedes 233
Ardrey, R. A. 83, 99, 129, 182, 202, 203, 206, 321
Arehart, J. L. 350, 390
Aristarchus 418
Aristophanes 175
Aristotle 23, 63, 73, 319, 374, 378, 389
Arnold, B. 447
Arnold, M. 36, 58, 375
Aronfreed, J. 403
Asimov, I. 452
Ataturk, K. 195
Augustine, St. 53, 63, 396

Baber, H. H. 292
Bach, J. S. 430
Bacon, F. 32, 53, 63, 263, 319, 378, 388, 435

Baetke, W. 12, 13, 292
Bagehot, W. 385
Bales, R. F. 129
Banks, A. F. 343
Barker, E. 177
Barron, F. 24
Barth, K. 292
Barton, K. 450
Baudelaire, P. C. 26, 210
Bavelay, R. 229
Beadle, G. W. 158, 368, 390
Beauchamp, Joan 443
Becker, C. L. 42, 52, 68
Beethoven, L. van 26, 427, 452
Bellamy, E. 53
Beloff, J. R. 112, 142, 241, 251
Benedict, R. 115, 130, 182
Bentham, J. 43, 56, 75, 76, 175, 269, 326, 444
Berlin, I. 57
Bishop, D. W. 293
Bismarck, O. von 339
Blake, W. 306, 424
Blewett, D. B. 112, 142, 241, 251
Boerma, A. H. 169
Bogart, E. L. 202
Bolitho, W. 309
Bolz, C. 144
Borgatta, E. F. 55, 129, 130, 183, 342
Boulding, K. E. 338
Bowra, C. M. 293

Subject Index